Springer Proceedings in Complexity publishes proceedings from scholarly meetings on all topics relating to the interdisciplinary studies of complex systems science. Springer welcomes book ideas from authors. The series is indexed in Scopus.

Proposals must include the following:

- name, place and date of the scientific meeting
- a link to the committees (local organization, international advisors etc.)
- scientific description of the meeting
- list of invited/plenary speakers
- an estimate of the planned proceedings book parameters (number of pages/articles, requested number of bulk copies, submission deadline)

Submit your proposals to: christoph.baumann@springer.com

More information about this series at http://www.springer.com/series/11637

Hugo Barbosa · Jesus Gomez-Gardenes ·
Bruno Gonçalves · Giuseppe Mangioni ·
Ronaldo Menezes · Marcos Oliveira
Editors

Complex Networks XI

Proceedings of the 11th Conference on Complex Networks CompleNet 2020

Springer

Editors

Hugo Barbosa
Department of Computer Science
University of Exeter
Exeter, UK

Jesus Gomez-Gardenes
Dept de Fisica de la Materia Condensada
Univ de Zaragoza, Fac de Ciencias
Zaragoza, Zaragoza, Spain

Bruno Gonçalves
Center for Data Science, Inc.
New York, NY, USA

Giuseppe Mangioni
University of Catania
Catania, Catania, Italy

Ronaldo Menezes
Department of Computer Science
University of Exeter
Exeter, UK

Marcos Oliveira
Leibniz Institute for the Social Sciences
Cologne, Germany

ISSN 2213-8684 ISSN 2213-8692 (electronic)
Springer Proceedings in Complexity
ISBN 978-3-030-40942-5 ISBN 978-3-030-40943-2 (eBook)
https://doi.org/10.1007/978-3-030-40943-2

This Springer imprint is published by the registered company Springer Nature Switzerland AG
The registered company address is: Gewerbestrasse 11, 6330 Cham, Switzerland

Preface

The International Workshop on Complex Networks—CompleNet (www.complenet.org) was initially proposed in 2008, and the first workshop took place in 2009 in Catania. The initiative was the result of efforts from researchers from the (i) BioComplex Laboratory in the Department of Computer Sciences at Florida Institute of Technology, USA, and the (ii) Dipartimento di Ingegneria Informatica e delle Telecomunicazioni, Università di Catania, Italy. CompleNet aims at bringing together researchers and practitioners working on complex networks or related areas. In the past two decades, we have indeed witnessed an exponential increase of the number of publications in this field. From Biology to Computer Science, from Economics to Social Systems, Complex Networks are becoming pervasive in many fields of science. It is this interdisciplinary nature of complex networks that CompleNet aims at addressing. CompleNet 2020 was the eleventh event in the series and was hosted at University of Exeter, from March 31–April 3, 2020.

This book includes the peer-reviewed list of works presented at CompleNet 2020. We received 241 submissions from 48 countries. Each submission was reviewed by at least three members of the Program Committee. Acceptance was judged based on the relevance to the symposium themes, clarity of presentation, originality and accuracy of results and proposed solutions. After the review process, 29 full papers and four short papers were selected to be included in this book. The 33 contributions in this book address many topics related to complex networks and have been organized in eight major groups: (1) Theory, (2) Processes, (3) Biomedical Applications, (4) Social Media Analysis, (5) Mobility Networks, (6) Economical Networks, (7) Social Problems and (8) Science of Science. We would like to thank to the Program Committee members for their work in promoting the event and refereeing submissions. We are grateful to our speakers: Michael Batty, Laura Alessandretti, Robin Dunbar, Andrea Migliano, Tiago de Paula Peixoto, Yamir Moreno, Hiroki Sayama, Dirk Brockman, Adilson Motter,

Clara Granell, Ciro Cattuto and Heather Harrington; their presentation is one of the reasons CompleNet 2020 was such a success.

March 2020

Hugo Barbosa
Jesus Gomez-Gardenes
Bruno Gonçalves
Giuseppe Mangioni
Ronaldo Menezes
Marcos Oliveira

Contents

List of Contributors

André F. de Angelis School of Technology, University of Campinas (UNICAMP), Limeira, São Paulo, Brazil

Hiroki Arihara Department of Computer Science, School of Computing, Tokyo Institute of Technology, Meguro, Tokyo, Japan

Firas Aswad Department Computer Engineering and Sciences, Florida Tech, Melbourne, USA

A. P. F. Atman Programa de Pós-Graduação em Modelagem Matemática e Computacional, Centro Federal de Educação Tecnológica de Minas Gerais, Belo Horizonte, Brazil;
Instituto Nacional de Ciência e Tecnologia de Sistemas Complexos, Rio de Janeiro, Brazil

Hugo Barbosa BioComplex Lab, Department of Computer Science, University of Exeter, Exeter, UK

Sharmodeep Bhattacharyya Oregon State University, Corvallis, OR, USA

Bethany Boettner The Ohio State University, Columbus, OH, USA

Fabian Braesemann Saïd Business School & Oxford Internet Institute, University of Oxford, Oxford, UK

Markus Brede School of Electronics and Computer Science, University of Southampton, Southampton, UK

Christopher Browning The Ohio State University, Columbus, OH, USA

Martin Cadeiras Department of Internal Medicine, University of California, Davis, USA

Alessio Cardillo Department of Engineering Mathematics, University of Bristol, Bristol, UK;
Department of Computer Science and Mathematics, University Rovira i Virgili, Tarragona, Spain;
GOTHAM Lab, Institute for Biocomputation and Physics of Complex Systems, University of Zaragoza, Zaragoza, Spain

Catherine Calder University of Texas at Austin, Austin, TX, USA

Chico Q. Camargo Oxford Internet Institute, University of Oxford, Oxford, UK

V. Carchiolo Dip. di Matematica e Informatica, Università degli Studi di Catania, Catania, Italy

Laura C. Carpi Programa de Pós-Graduação em Modelagem Matemática e Computacional, Centro Federal de Educação Tecnológica de Minas Gerais, Belo Horizonte, Brazil

Batyr Charyyev School of Systems and Enterprises, Stevens Institute of Technology, Hoboken, NJ, USA

Shirshendu Chatterjee City University of New York, New York, NY, USA

Alex Dawn City Science, Exeter, UK

Joyati Debnath Winona State University, Winona, MN, USA

Stefan Dernbach College of Information and Computer Sciences, University of Massachusetts Amherst, Amherst, USA

Bruce A. Desmarais The Pennsylvania State University, University Park, State College, PA, USA

Mohamed El Marraki LRIT, Rabat IT Center, Faculty of Sciences, Mohammed V University in Rabat, Rabat, Morocco

Bojan Evkoski Faculty of Computer Science and Engineering, Ss. Cyril and Methodius University, Skopje, North Macedonia

Josemar Faustino Florida Institute of Technology, Melbourne, FL, USA

Michael Frommelt IBM, AI Core, GER, New York, USA

Takayasu Fushimi School of Computer Science, Tokyo University of Technology, Hachioji, Japan

Ralucca Gera Department of Applied Mathematics, Naval Postgraduate School, Monterey, CA, USA

M. Grassia Dip. Ingegneria Elettrica Elettronica Informatica, Università degli Studi di Catania, Catania, Italy

Will Gray-Roncal Johns Hopkins University Applied Physics Laboratory, Laurel, MD, USA

Mehmet Hadi Gunes School of Systems and Enterprises, Stevens Institute of Technology, Hoboken, NJ, USA

Saket Gurukar The Ohio State University, Columbus, OH, USA

Harith Hamoodat Department Computer Engineering and Sciences, Florida Tech, Melbourne, USA

Chengyuan Han Institute for Energy and Climate Research (IEK-STE), Forschungszentrum Jülich, Jülich, Germany;
Institute for Theoretical Physics, University of Cologne, Köln, Germany

Yu Han Beijing University of Technology, Beijing, China

Ryan Hartman Department of Internal Medicine, University of California, Davis, USA;
Department of Computer Science, University of Exeter, Exeter, UK

John A. Hird University of Massachusetts Amherst, Amherst, MA, USA

Marisa J. Hughes Johns Hopkins University Applied Physics Laboratory, Laurel, MD, USA

Nandini Iyer University of Illinois at Urbana–Champaign, Champaign, IL, USA

Shahin Jalili University of Exeter, Exeter, UK

Erik C. Johnson Johns Hopkins University Applied Physics Laboratory, Laurel, MD, USA

Lidija Jovanovska Faculty of Computer Science and Engineering, Ss. Cyril and Methodius University, Skopje, North Macedonia;
Jožef Stefan International Postgraduate School, Ljubljana, Slovenia

Edward Keedwell University of Exeter, Exeter, UK

Moniba Keymanesh The Ohio State University, Columbus, OH, USA

Yara Khaluf IDLab - Department of Information Technology, Ghent University - imec, Ghent, Belgium

Wilhelm Kuckshinrichs Institute for Energy and Climate Research (IEK-STE), Forschungszentrum Jülich, Jülich, Germany

Vesa Kuikka Finnish Defence Research Agency, Riihimäki, Finland

A. Longheu Dip. Ingegneria Elettrica Elettronica Informatica, Università degli Studi di Catania, Catania, Italy

Laura Lotero Faculty of Industrial Engineering, Universidad Pontificia Bolivariana, Medellín, Colombia

Dounia Lotfi LRIT, Rabat IT Center, Faculty of Sciences, Mohammed V University in Rabat, Rabat, Morocco

Jorge Louça Information Sciences, Technologies and Architecture Research Center, ISCTE-IUL, Lisbon, Portugal

Ard A. Louis Rudolf Peierls Centre for Theoretical Physics, University of Oxford, Oxford, UK

Mariana Macedo BioComplex Lab, Department of Computer Science, University of Exeter, Exeter, UK

M. Malgeri Dip. Ingegneria Elettrica Elettronica Informatica, Università degli Studi di Catania, Catania, Italy

G. Mangioni Dip. Ingegneria Elettrica Elettronica Informatica, Università degli Studi di Catania, Catania, Italy

Edoardo Manino School of Electronics and Computer Science, University of Southampton, Southampton, UK

Miguel Martins CRACS & INESC-TEC DCC-FCUP, Universidade do Porto, Porto, Portugal

Paulo S. Martins School of Technology, University of Campinas (UNICAMP), Limeira, São Paulo, Brazil

Juan Mendonza Central Washington University, Ellensburg, WA, USA

Ronaldo Menezes BioComplex Lab, Department of Computer Science, University of Exeter, Exeter, UK

Natalia Meshcheryakova National Research University Higher School of Economics, Moscow, Russia;
V.A. Trapeznikov Institute of Control Sciences of Russian Academy of Science, Moscow, Russia

Miroslav Mirchev Faculty of Computer Science and Engineering, Ss. Cyril and Methodius University, Skopje, North Macedonia

Igor Mishkovski Faculty of Computer Science and Engineering, Ss. Cyril and Methodius University, Skopje, North Macedonia

Raihana Mokhlissi LRIT, Rabat IT Center, Faculty of Sciences, Mohammed V University in Rabat, Rabat, Morocco

Tsuyoshi Murata Department of Computer Science, School of Computing, Tokyo Institute of Technology, Meguro, Tokyo, Japan

Samadhi Nallaperuma University of Exeter, Exeter, UK

Jordi Nin ESADE, Universitat Ramon Llull, Barcelona, Catalonia, Spain

Laurence Oakes-Ash City Science, Exeter, UK

William R. Paiva School of Technology, University of Campinas (UNICAMP), Limeira, São Paulo, Brazil

Laurent Park Johns Hopkins University, Baltimore, MD, USA

Srinivasan Parthasarathy The Ohio State University, Columbus, OH, USA

Diego Pinheiro Department of Internal Medicine, University of California, Davis, USA

Oriol Pujol Department of Mathematics and Computer Science, Universitat de Barcelona, Barcelona, Catalonia, Spain

Devin Ramsden Johns Hopkins University, Baltimore, MD, USA

Ilja Rausch IDLab - Department of Information Technology, Ghent University - imec, Ghent, Belgium

Arash Ravandi Division of Orthopeadic Rheumatology, Friedrich-Alexander University Erlangen-Nuremberg, Waldkrankenhaus Erlangen, Erlangen, Germany

Babak Ravandi Network Science Institute, Center for Complex Network Research, Northeastern University, Boston, MA, USA

Elizabeth P. Reilly Johns Hopkins University Applied Physics Laboratory, Laurel, MD, USA

Eraldo Ribeiro Department Computer Engineering and Sciences, Florida Tech, Melbourne, USA

Pedro Ribeiro CRACS & INESC-TEC DCC-FCUP, Universidade do Porto, Porto, Portugal

Erick Romero Department of Internal Medicine, University of California, Davis, USA

Guillermo Romero Moreno School of Electronics and Computer Science, University of Southampton, Southampton, UK

Harvey Sanchez-Restrepo Faculty of Sciences, University of Lisbon, Lisbon, Portugal

Malte Schröder Chair for Network Dynamics, Center for Advancing Electronics Dresden (cfaed) and Institute of Theoretical Physics, Technical University of Dresden, Dresden, Germany

Sergey Shvydun National Research University Higher School of Economics, Moscow, Russia;
V.A. Trapeznikov Institute of Control Sciences of Russian Academy of Science, Moscow, Russia

Pieter Simoens IDLab - Department of Information Technology, Ghent University - imec, Ghent, Belgium

Mustafa Solmaz Computer Science and Engineering, University of Nevada Reno, Reno, NV, USA

Sucheta Soundarajan Department of Electrical Engineering and Computer Science, Syracuse University, Syracuse, USA

Niklas Stoehr IBM, AI Core, GER, New York, USA

Johannes Többen Gesellschaft für Wirtschaftliche Strukturforschung, Osnabrück, Germany;
Potsdam Institute for Climate Impact Research, Social Metabolism and Impacts, Potsdam, Germany

A. S. O. Toledo Programa de Pós-Graduação em Modelagem Matemática e Computacional, Centro Federal de Educação Tecnológica de Minas Gerais, Belo Horizonte, Brazil;
Instituto Brasileiro de Segurança Pública, Sao Paulo, Brazil

Don Towsley College of Information and Computer Sciences, University of Massachusetts Amherst, Amherst, USA

Long Tran-Thanh School of Electronics and Computer Science, University of Southampton, Southampton, UK

Irene Unceta BBVA Data & Analytics, Barcelona, Catalonia, Spain;
Department of Mathematics and Computer Science, Universitat de Barcelona, Barcelona, Catalonia, Spain

Brock Wester Johns Hopkins University Applied Physics Laboratory, Laurel, MD, USA

Pivithuru Wijegunawardana Department of Electrical Engineering and Computer Science, Syracuse University, Syracuse, USA

Dirk Witthaut Institute for Energy and Climate Research (IEK-STE), Forschungszentrum Jülich, Jülich, Germany;
Institute for Theoretical Physics, University of Cologne, Köln, Germany

Lizhi Xing Beijing University of Technology, Beijing, China;
Indiana University, Bloomington, IN, USA

Masaya Yazaki School of Computer Science, Tokyo University of Technology, Hachioji, Japan

Shi Zhou Department of Computer Science, University College London, London, UK

Theory

Condensed Graphs: A Generic Framework for Accelerating Subgraph Census Computation

Miguel Martins and Pedro Ribeiro(✉)

CRACS & INESC-TEC DCC-FCUP, Universidade do Porto, Porto, Portugal
`mlmartins@fc.up.pt`, `pribeiro@dcc.fc.up.pt`

Abstract. Determining subgraph frequencies is at the core of several graph mining methodologies such as discovering network motifs or computing graphlet degree distributions. Current state-of-the-art algorithms for this task either take advantage of common patterns emerging on the networks or target a set of specific subgraphs for which analytical calculations are feasible. Here, we propose a novel network generic framework revolving around a new data-structure, a *Condensed Graph*, that combines both the aforementioned approaches, but generalized to support any subgraph topology and size. Furthermore, our methodology can use as a baseline any enumeration based census algorithm, speeding up its computation. We target simple topologies that allow us to skip several redundant and heavy computational steps using combinatorics. We were are able to achieve substantial improvements, with evidence of exponential speedup for our best cases, where these patterns represent up to 97% of the network, from a broad set of real and synthetic networks.

Keywords: Subgraph frequency · Subgraph census · Condensed graph

1 Introduction

Many complex real world problems can be modelled with networks, from which we need to extract information. Several graph mining methodologies rely on understanding the importance of subgraphs as a very rich topological characterization. Two broadly known examples are network motifs [12] and graphlet degree distributions [15]. At the core of these approaches lies the subgraph census problem, that is, computing the frequencies of a set of subgraphs. However, this is a fundamentally hard computational task that is related so the *subgraph isomorphism problem*, which is NP-complete [3].

Current algorithms for counting subgraphs typically rely on one of two different conceptual approaches. Several algorithms, such as G-tries [17], QuateXelero [9] or FaSE [13], are based on a subgraph enumeration phase intertwined with isomorphic testing to discover the topological class of each enumerated subgraph occurrence. These algorithms are very general and take advantage of

H. Barbosa et al. (Eds.): *Complex Networks XI*, SPCOM, pp. 3–15, 2020.
https://doi.org/10.1007/978-3-030-40943-2_1

common topologies to speedup the isomorphism computation. Other methods such as ORCA [6] or Escape [14] do not depend on fully enumerating all subgraph occurrences, but at the same time are geared towards more specific and less general sets of subgraphs, taking advantage of some of their analytical properties.

In this paper we propose an hybrid methodology that draws inspiration from both the enumeration and analytical approaches, that we called the **C**ondensation-**D**econdensation **F**ramework (**CFW**). The core motivation is to take advantage of combinatorial effects that create substantial speedups while at the same time being able to keep the general applicability of enumeration based algorithms, not constraining the subgraphs or networks being analyzed. For this we also introduce a new generic data structure, *Condensed Graphs*, that compresses subgraphs in a lossless way, capturing multiple occurrences of the same subtopology. More specifically, and as a proof of concept, we condense star-like peripheral structures, that commonly emerge in real networks. Our framework encapsulates existing enumeration based methods, speeding up their computation by taking advantage of operations enabled through the use of Condensed Graphs. Here we show how it could be coupled with both ESU [19] and FaSE [13] algorithms, but in principle it could be applicable with other enumeration methodologies.

With all of this in place, we achieved very promising results on representative sets of real world and synthetic networks, showcasing the applicability of our methodology. In terms of compression, we depend on the existence of star-like peripheries. Here we show that these are very common on real world networks, and we are able to reach up to 97% of compression. Regarding speedup, our experiments show that the gains can be substantial and that when the networks exhibit this kind of topology we are able to achieve exponential gains. We also try to quantify the desired structural properties that make networks more amenable to our proposed approach.

2 Backgroud

2.1 Notation

A graph $G = (V, E)$ is comprised by a set of edges $E(G)$ connecting vertices $V(G)$. A k-graph is a graph with k vertices. In this paper we will only address simple undirected graphs, with at most a single edge connecting the same pair of vertices. Our data structure could however straightforwardly be extended to support directed graphs, multigraphs or even more complicated graph representations.

For two graphs G and S such that $V(S) \subseteq V(G)$, then it is said that S is a *subgraph* of G. Moreover, if $\forall u, v \in V(S) : (u,v) \in E(G) \iff (u,v) \in E(S)$, then S is an *induced subgraph* of G. Two graphs, G and G', are said to be *isomorphic* if and only if there is a bijection $f : V(G) \longrightarrow V(G')$, such that $(u,v) \in E(G) \iff (f(u), f(v)) \in E(G')$. An *automorphism* is an isomorphism from a graph on to itself, and the automorphisms of a graph G form a group called $Aut(G)$. Consider a vertex $u \in V(G)$, then the *automorphism orbit* of u is:

$$Orb(u) = \{v \in V(G) | v = g(u), g \in Aut(G)\} \quad (1)$$

Simply put, if u and v are in the same orbit, they are topologically equivalent, which means one could swap their labels without altering the graph topology.

2.2 Problem Definition

In this paper we tackle the following computational problem:

Definition 1 ***(Subgraph Census Problem).*** *Given some positive integer k and a graph G, count the exact number of distinct occurrences of each of all possible connected induced k-subgraphs of G. Two occurences are distinct if there is at least one vertex that they do not share.*

2.3 Related Work

Subgraph census computation has been studied for more then 15 years. In 2002 Milo et al. [12] coined the term *network motifs* as frequent overrepresented induced subgraph patterns and offered the `mfinder` subgraph enumeration algorithm as a first practical approach for computing subgraph frequencies. The first major breakthrough was introduced by Wernicke [19] with the `ESU` algorithm, which avoided graph symmetries and enumerated each subgraph only once. Isomorphism tests for each discovered subgraph occurrence are made trough the third party package `nauty` [11], a highly efficient isomorphism algorithm. In order to reduce the number of needed isomorphism tests, approaches such as `QuateXelero` [9] or `FaSE` [13] encapsulate the topology of the current subgraph match, grouping several occurrences as belonging to the same isomorphic class. If we know beforehand the set of subgraphs that we are interested on (which can possibly be smaller than the entire set of all possible k-subgraphs), the `g-tries` data structure [17] could be used, allowing for further improvements.

All the aforementioned approaches are general (i.e, are applicable to any subgraph size and also allow direction) and rely on doing a full subgraph enumeration. However, for more specific sets of subgraphs there has been an increasing number of more analytical algorithms that take into account the subgraphs topology and its combinatorial effects. For example, `ORCA` [6], which counts orbits and not directly subgraph occurrences, can tackle up to size 5 undirected subgraphs and relies on a derived set of linear equations that relate the orbit counts. This was also generalized for other small undirected orbits [7]. `PGD` [1] (up to size 4) and `Escape` [14] (size 5) are other examples of state-of-the-art analytical algorithms specialized on counting undirected subgraphs.

Our approach differs from these two conceptual approaches, as it tries to combine the general applicability of the enumeration algorithms with combinatorial improvements. However, instead on focusing on the topology of the subgraphs we are looking for, we focus on how to compress the network we are analyzing, targeting specific substructures than can be resumed as a combinatorial object.

All the aforementioned algorithms perform exact computations, but it should be said that there are also methodologies that can trade accuracy for speed, providing approximate results. Furthermore, some algorithms exploit parallelism. For the purposes of this paper we pursue exact sequential census computation, as to improve the baseline algorithm, but our approach could be further extended on the future towards other directions. For a more detailed survey of the state of the art on subgraph counting we refer the reader to [16].

3 A Novel Framework for Subgraph Census

3.1 Peripheral Stars

Our methodology revolves around peripheral areas of a network, which are topologically self contained. This allow us to perform combinatorial calculations to quickly identify a larger number of occurrences of the same subgraph topology, avoiding the need to explicitly pass trough each single occurrence. As an initial proof of concept, here we will focus on *star subgraphs* on undirected networks, but we envision many other potential extensions to more complex peripheries.

Definition 2 ***(Peripheral Star Subgraphs).*** *An induced subgraph S of a graph G is said to be a* ***peripheral star*** *of size m if it is comprised by m vertices of degree one, called the* ***peripheral vertices****, that are connected only to the same vertex s, called the* ***seed vertex****.*

The terms peripheral star and star will henceforth be used interchangeably. An induced star within a peripheral star will be called a *substar*. Furthermore, we will use $P(G)$ to denote the set of all peripheral vertices on a graph G.

A peripheral star only has two orbits: the seed vertex orbit, and the peripheral orbit, This simple topology lies at the core of our speedup. Suppose you have a star of size m. Then, for all $i \in [1, m]$, we know that the number of i-substars is precisely $C_i^m = \binom{m}{i}$, all of them with the exact same isomorphic class. A visual example is given in Fig. 1.

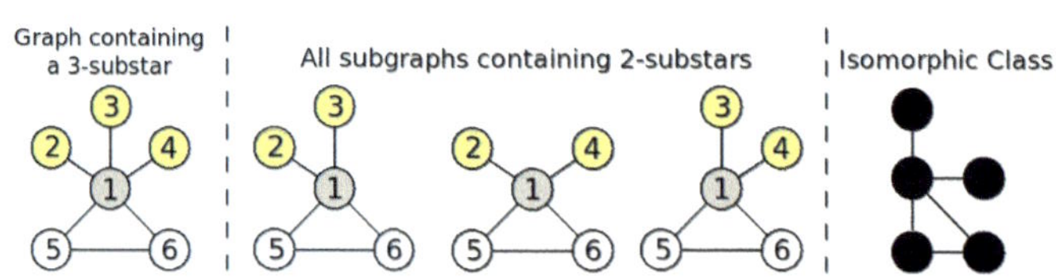

Fig. 1. (Left) A graph containing a peripheral star (seed vertex: gray, peripheral vertices yellow, regular vertices: white). (Middle) All possible 2-substars (Right) The corresponding isomorphic class (the same for all subgraphs in the middle)

As the size m of the star, and the size k of the subgraph increase, the number of combinations $\binom{m}{i}$ increases exponentially, a property that we will exploit.

3.2 Condensed Graphs

The first step on our methodology is to compress the original graph, such that all peripheral star subgraphs are discovered and reduced to identifying its size. This process is exemplified in Fig. 2. Let SQ_u be the number of peripheral vertices connected to a vertex u (which correspond to the numbers inside parenthesis in the figure). Condensing a graph can be thought as the process of eliminating all peripheral vertices and adding extra information to all other nodes in the form of SQ_u for all vertices u of the condensed graph.

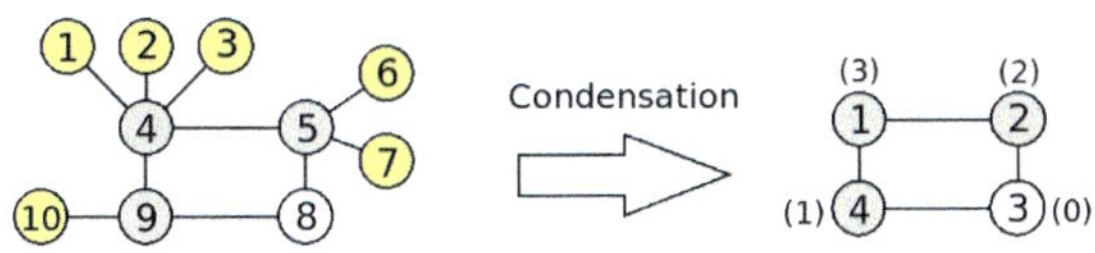

Fig. 2. (Left) A graph. (Right) the resulting Condensed Graph (seed vertices in gray).

We can trivially condense any graph in $\mathcal{O}(|V|+|E|)$ time by iterating through all vertices, reassigning labels to non-peripheral vertices based on the order they were visited. This is in fact a *lossless* compression scheme, since we still maintain all the original topological properties and we can easily decompress back to a graph isomorphic to the original one in time $\mathcal{O}(|V| + |E|)$, an operation we describe as *decondensation.*

3.3 Taking Advantage of Condensation

Classical enumeration algorithms try to explicitly traverse all subgraph occurrences, effectively increasing the frequency by one each time. The key point of our work is precisely to account for multiple occurrences at the same time, taking advantage of the combinatorial effects of self contained peripheries, avoiding the costly explicit traversal of all topologically equivalent substars.

Our framework is general and can be applied to any enumeration algorithm that builds subgraphs by adding one vertex at a time. Consider that we are performing a k-subgraph census and that we already have a partially enumerated vertex set $V_{subgraph}$ of size $d < k$, that we want to extend up to size k. When we add a seed vertex, we can consider all the possible substars that this new vertex may induce. Figure 3 exemplifies this concept. Condensed graphs' properties allow to proceed with the extension and simultaneously tracking multiple occurrences.

With all of these concepts in place, we are now ready to explain our ***C****ondensation-****D****econdensation* ***F****ramework* (***CFW***), that is able to improve an existing baseline subgraph census algorithm. An overview of our approach is given in Algorithm 1, which describes, in a k-census of graph G, how to extend any partially enumerated set of nodes $V_{subgraph}$, whose current frequency is given by $cur_{frequency}$.

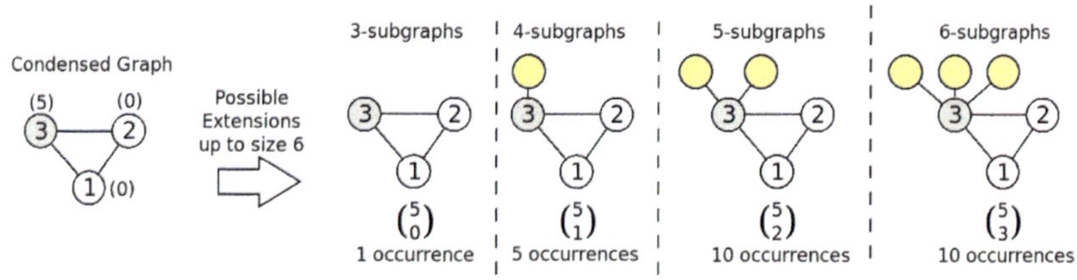

Fig. 3. Extending a condensed graph to subgraphs up to size 6.

Algorithm 1 Condensation-Decondensation Framework

```
1: procedure EXTEND_SUBGRAPH(G, k, V_subgraph, cur_frequency)
2:     if |V_subgraph| = k then
3:         Frequency[V_subgraph] += cur_frequency
4:     else
5:         for all vertex u extending V_subgraph do      ▷ Using baseline enumeration algorithm
6:             for all i ∈ {0 ... min(k − |V_subgraph| − 1, SQ_u)} do
7:                 V_extended ← V_subgraph ∪ {u} ∪ {i peripheral nodes attached to u}
8:                 extend_subgraph(G, k, V_extended, cur_frequency × (SQ_u choose i))
```

To start the process, we should start by calling $extend\_subgraph(G, k, u, 1)$ for all nodes $u \in V(G)$, that is, we try to create a subgraph starting from every node. Now, for each node we add (line 5), we take into account all possible substars that extend up to size k (lines 6 and 7), and we are able to directly identify how many isomorphic occurrences of that particular substar can be obtained (line 8), as previously explained. Notice how we multiply by the current frequency, which allows to consider subgraphs that incorporate multiple substars from different seed vertices. The process stops when we reach the desired subgraph size (line 2), when we can safely increment the frequency by a value reflecting how many multiple isomorphic occurrences we are considering (line 3), as opposed to simply incrementing by one in the baseline enumeration algorithm.

3.4 Baseline Enumeration Algorithms

For the purposes of this paper we will be adapting two well known subgraph counting algorithms that fit into our framework: they explicitly enumerate all occurrences and they work by extending subgraphs one vertex at a time. Given the space constraints of this paper, we will only give a very high level description of the two algorithms, and we refer the reader to the respective original papers for more in-depth detail.

The first of these algorithms is `ESU` [19], which uses carefully chosen restrictions on the way it extends subgraphs, to guarantee that each set of k connected nodes is only enumerated once. Each of these occurrences is then run trough `nauty` [11], a very efficient third-party isomorphism algorithm, so that we identify the topological class for which we need to increment the frequency.

The second algorithm we adapted was `FaSE` [13], which improves the previous approach by avoiding the need of doing one isomorphic test per occurrence. In order to do that, while still using the same baseline enumeration procedure as ESU, it uses the *G-Trie* data structure [17], which can be briefly described as a trie of graphs. In this way, node sets that induce the exact same adjacency matrix will give origin to the same path in the g-trie, allowing us to group many occurrences as belonging to the same topological class. Due to naturally occurring symmetries in the subgraphs, several different paths on the g-trie may still correspond to the same topology, and we still need to identify this. However, we only need one isomorphism test per group of occurrences (a path in the g-trie), which allows for a substantial speedup when compared to the classical `ESU` algorithm.

4 Experimental Results

Our main goal is to compare the baseline algorithms `ESU` and `FaSE`, to the adaptations using our framework, respectively called `Co-ESU` and `Co-FaSE`. All experiments were done on a machine with a 2.4 GHz Intel i5 CPU, 8 GB 1600 MHz DDR3 RAM running macOS High Sierra 10.13.6. All algorithms were implemented in C++ using clang-902 as the compiler.

4.1 Real World Networks

We will first test the algorithms on a broad set of real networks from different backgrounds, which are described in Table 1. For the purposes of this paper we ignored both weights and direction. Thus, we transformed `econpoli` (directed) to an undirected network and ignored the weights in `rtobama`.

Table 1. The set of used real networks, in decreasing order of compression ratio (CR).

Name	Type	Description	$\|VG)\|$	$\|E(G)\|$	$\|P(G)\|$	CR	μ_{star}	$\max_{star}$	Source
`facebook`	Social	Friendships	2888	2981	2790	97%	279.0	756	[18]
`rtobama`	Social	Retweets	9631	9772	9104	93%	69.2	7413	[18]
`reality`	Social	Phone calls	6809	7680	6284	92%	77.6	233	[18]
`mvcortex`	Brain	Fiber tracts	194	214	160	82%	14.6	23	[18]
`econpoli`	Economic	Transactions	15575	17468	12187	74%	10.7	490	[18]
`genefusion`	Biological	Gene Interact	291	279	203	56%	3.3	29	[10]
`gridworm`	Biological	Gene Interact	3518	6531	1887	45%	6.6	323	[18]

For each network, the previous table reports the number of nodes ($|V(G)|$) and edges ($|E(G)|$), as well as the number of peripheral nodes ($|P(G)|$). To indicate the potential for speedups using our framework, we give an idea of much we are compressing the graph in the form of a compression ratio

$CR = \frac{|P(G)\setminus P_1(G)|}{|V(G)|}$, where $P_1(G)$ are stars of size 1, which we disregard given that they are not combinatorially exploitable. Furthermore, we report the average size of stars larger than size 1 (μ_{star}) and the size of the largest star ($\max_{star}$).

Table 2 summarizes the experiments done with real networks, reporting the execution time (in seconds) of all algorithms, as well as the speedup of our adaptations when compared with the respective baseline algorithm. In each network, we show the results obtained for different subgraph sizes k.

Table 2. Speedup of our adaptations vs baseline algorithms on real networks.

Network	k	k-census execution time (s)		Speedup	k-census execution time (s)		Speedup
		ESU	Co-ESU		FaSE	Co-FaSE	
facebook	3	0.23	0.01	23.3x	0.04	0.01	3.8x
	4	44.91	0.07	641.6x	4.72	0.01	471.5x
	5	9720.98	1.08	9000.91x	1006.70	0.05	20133.9x
	6	>5 h	13.11	N/A	>5 h	0.31	N/A
rtobama	3	19.96	0.19	105.07x	2.058	0.1	20.6x
	4	>5 h	2.98	N/A	8439.36	0.23	36692.9x
	5	>5 h	254.79	N/A	>5 h	7.62	N/A
reality	3	0.14	0.09	1.54x	0.02	0.05	0.39x
	4	7.60	0.58	13.09x	1.00	0.09	11.11x
	5	451.61	10.08	44.80x	43.72	0.53	82.5x
	6	>5 h	244.19	N/A	2307.81	12.48	184.9x
mvcortex	7	1.69	0.1	16.94x	0.50	0.01	49.6x
	8	16.63	0.79	21.06x	6.96	0.05	139.2x
	9	120.36	3.67	32.79x	65.80	0.22	299.1x
	10	933.74	18.27	51.11x	662.87	0.83	798.6x
	11	6340.48	51.91	122.14x	4067.36	3.47	1172.2x
econpoli	3	0.07	0.47	0.16x	0.04	0.34	0.1x
	4	2.62	3.2	0.82x	2.23	0.53	4.2x
	5	165.15	66.58	2.48x	130.04	2.61	49.8x
genefusion	8	2.67	0.45	5.92x	1.19	0.04	29.7x
	9	21.45	1.93	11.11x	6.85	0.16	42.8x
	10	81.85	7.72	10.60x	38.60	0.67	57.6x
	11	209.43	29.18	7.18x	211.60	1.73	122.3x
gridworm	4	28.63	7.51	3.81x	1.25	0.38	3.3x
	5	5307.49	391.71	13.55x	125.64	16.94	7.4x
	6	>5 h	>5 h	N/A	>5 h	16726.42	N/A

The first major insight is that our adaptations are always quicker than their original counterparts for all non-trivial cases (>1 s), confirming we are indeed improving the baseline algorithms. Furthermore, our speedup tends to increase superlinearly with the size k in both algorithms, with more gains on the cases where the computation time is already higher. This is due to the fact that larger subgraphs will naturally correspond to an (exponentially) larger number of occurrences that we can combinatorially exploit. We also note that our speedup is typically higher with `FaSE`, which is already substantially faster than the `ESU` algorithm. We suspect this might be caused due to synergies between our condensation-decondensation operation and the way `FaSE` operates, which might result in smaller g-tries and less isomorphic tests needed.

For the two top performing networks `facebook` and `rtobama`, we focus on `FaSE` and `Co-FaSE`, since `ESU` did not perform fast enough in our time constraints to draw significant conclusions. Although `facebook` has higher CR, both show evidence of exponential speedup. However, in `rtobama` speedup seems to grow faster, reaching 4 orders of magnitude for $k = 2$ while `facebook` only matches this results for $k = 3$. Going into further detail, even the precise values of speedup favor `rtobama`, (36692.87x *versus* 20133.9x). Note that the $\max_{star}$ in `facebook` is 756, while in `rtobama` is 7413, accounting for ≈26% and ≈77% of the total networks' sizes respectively. Moreover, $\binom{n}{k}$ scales exponentially with n with regards to k. Although μ_{star} is higher for `facebook` the difference in size of $\max_{star}$ completely overshadows the impact of the former metric.

The next pair of networks analysed `reality` and `mvcortex`, that have a CR disparity of 10% between them. Regarding the former, in both comparisons, speedup seems to increase in a linear fashion, with 1 order of magnitude of speedup improvement measured for `Co-ESU` for $k = 5$, and 2 orders for `Co-FaSE` for $k = 4$. Addressing the latter network, speedup in both cases grows in a linear fashion with k, but in different orders of magnitude. In the case of `Co-ESU`, we measured up to 2 order of magnitude. In `Co-FaSE`, the results are more dramatic, reaching 3 orders of magnitude. We suspect that, `mvcortex` showed considerably better results than `reality`, even with less CR, because we were able to measure values of k that were closer to the optimum value of $\binom{\max_{star}}{k}$ and by extension, took full advantage for smaller stars.

Focusing on `econpoli` and `genefusion` they differ 18% in CR. For the former network, due to the size discrepancy among them and our hardware and time limitations, this led to a smaller number of observations. The consequences in speedup remain very similar for both adaptations, with a spike for $k = 7$, that we once attribute to the order of growth of maxima of $\binom{n}{k}$. Addressing `genefusion`, we were able to draw measure performance up to $k = 11$. Keep in mind that $\text{argmax}_k\binom{\max_{star}}{k} = 15$ and, as theory predicts concerning `Co-FaSE`, speedup grows linearly steady up to $k = 10$, but for $k = 11$ it almost doubles, since it is a point of ramp-up for the gradient of $\binom{n}{k}$. Surprisingly, this was not observed for `ESU`, and we do not yet have any credible theory regarding this phenomenon.

Finally, our last and worst performing network, `gridworm`, that has 11% less condensation ratio than `genefusion`. Its $\max_{star}$ accounts for ≈11% of the overall network. However, due to its size and complexity, we were only been able to measure speedup for a small range of k. Surprisingly, it is one of the few examples (along with `econpoli`, but much more drastic), that benefits `ESU` the most which is improved by one order of magnitude $k = 5$. For the same k, `Co-FaSE` follows behind closely measuring $7.4x$ speedup.

4.2 Synthetic Networks

To gain more insight into the specific properties that benefit our approach, we follow the same experimentation workflow, but for synthetic networks, generated using the NetworkX package [5]. We considered the following network models:

Barabási-Albert [2]**.** This model generates scale-free networks, who emerge in a plethora of phenomena in the real world, using *preferential attachment* as its connection mechanism. We will refer to it as BA(n, m), with n being the number of nodes and m the number of initial edges on each newly added vertex.

Holmes-Kim [8]**.** This model extends the BA model to produce networks with an higher clustering coefficient: after an edge is created between the newly added vertex v and another vertex w, a random neighbour of w is selected and an edge between it and v is created with probability p, thus forming a triangle between these three vertices. The alias for this model will be HK(n, m, p).

Random Power-Law Trees [4]**.** This model generates trees with a power law degree distribution. The model is too intricate to summarize, but essentially NetworkX's implementation takes three parameters, n the size of graph, γ the exponent of the power-law and *tries*, the number of tries necessary to ensure the degree sequence forms a tree. The alias for this model will be PLTrees$(n, \gamma, tries)$.

To make comparisons fair, we generated all networks with 1000 vertices. Table 3 gives an overview of the used synthetic networks, including the model parameters and the topological characteristics of the generated networks.

Table 3. The set of used synthetic networks generated using NetworkX package.

Model	$\|VG)\|$	$\|E(G)\|$	$\|P(G)\|$	CR	μ_{star}	$\max_{star}$
BA$(1000, 1)$	1000	999	686	53%	4.27	49
HK$(1000, 1, 0.9)$	1000	999	677	53%	4.05	44
PLTrees$(1000, 3, 100000)$	1000	1052	526	45%	3.17	54

We purposefully chose an high p parameter in HK, to see how well our framework would work on a scale-free network with high average clustering coefficient. Note that for both BA and HK the m parameter is 1, since its a necessary condition for emergence of peripheries. Regarding PLTrees, the γ is set to 3 by default to result in a scale-free network.

To avoid visual clutter, we will not include the model parameters in Table 4, and they will be referred simply by BF, HK and PLTrees

Concerning BA, FaSE clearly benefits from our framework, showing evidence of superlinear speedup, up to 2 orders of magnitude of improval. In the case of ESU, a slight increase from $k = 5$ up to $k = 7$ was measured. In $k = 8$ the trend shifts in the opposite direction. We suspect that, for larger values of k, a similar

Table 4. Speedup of our adaptations vs baseline algorithms on synthetic networks.

Network	k	k-census execution time (s)		Speedup	k-census execution time (s)		Speedup
		ESU	Co-ESU		FaSE	Co-FaSE	
BA	5	2.14	0.36	5.96x	0.17	0.03	5.6x
	6	53.68	5.98	8.98x	3.6	0.21	17.2x
	7	1084.62	120.31	9.02x	118.87	2.88	41.3x
	8	11756.72	1607.72	7.31x	4621.29	31.03	150.0x
HK	5	3.47	0.39	8.91x	0.28	0.02	14.2x
	6	44.86	7.76	5.78x	11.86	0.[illegible]	19[illegible].0x
	7	1027.72	232.21	4.43x	106.52	2.86	37.2x
	8	8413.58	2821.62	2.98x	2406.00	42.5	56.6x
PLTrees	6	3.25	0.02	162.70x	0.20	<0.01	N/A
	7	29.85	0.05	596.98x	2.40	0.01	240.4x
	8	264.26	0.16	1651.63x	19.90	0.02	995.2x
	9	1569.30	0.51	3077.07x	213.22	0.05	4264.5x
	10	11843.88	2.4	4934.95x	2666.02	0.25	10664.1x

pattern would occur, with an average of 1 order of magnitude of improvement with slight shifts in the trend of speedup. Note that has $k \rightarrow \left(\lceil\frac{\max_{star}}{2}\rceil = 500\right)$ the results can change drastically.

Addressing `HK`, the measurements are relatively similar to the ones observed in `BA`. Once again, `FaSE` benefits the most from our implementation. In this case, the trend does not appear to be strictly monotone. We suspect it will vary between 1 and 2 orders of magnitude of speedup as k grows, and then an upwards shift improvement when the gradient ramps up as k approaches 500. Unfortunately, we are limited once again by our hardware and time constraints to make an concrete comparison.

`PLTrees` display our best results on synthetic data. `ESU` shows evidence of a non-linear relationship regarding speedup, reaching 3 orders of magnitude of improvement. Although speedup seems to grow more slowly, between $k = 6$ and 8. Concerning `FaSe`, it is once again favoured, and shows evidence of super-linear speedup, reaching 4 orders of magnitude of performance improvement.

5 Conclusion

The goal of this paper was to build a generic framework adaptable to current subgraph census algorithms, from which we selected and effectively adapted `ESU` and `FaSE`. We have experimentally shown that our adaptations are significantly faster for a diverse set of networks extracted from different contexts. The framework enhanced significantly both algorithms on our experiment, up to 4 orders of magnitude speedup for both `Co-ESU` and `Co-Fase`, with indications of exponential speedup for our best cases. Note also that `Co-ESU` does not uses g-tries, and it still outperformed `FaSE` in all networks except `gridworm`, which further outlines the potential of our approach.

The condensation ratio of a network is highly correlated with performance, but does not fulfill a causal relationship. We refer back to the properties of the binomial the function that, coupled with the size of the largest star, affect speedup drastically. Note that max_{star} only gives a lower bound insight for potential speedup, since it does not account for the remaining smaller stars. From this, it is easy to see why networks with higher condensation ratio like `reality` are outperformed by others with less condensation like `mvcortex`, since we are able to explore values of k close to the maximum of for its max_{star}.

On our set of synthetic networks, we observed that our framework improves performance on scale-free networks, that are very recurrent on a plethora of real world phenomena, with the best case being the `PLTrees`.

The results are very promising and indicate this is a viable path for improving existing enumeration algorithms without losing generality. For the close future, we intend to tackle other types peripheries and to extend our approach to more complex networks, including aspects such as edge direction, temporal information and multiple layers of connectivity.

Acknowledgements. This work is financed by National Funds through the Portuguese funding agency, FCT - Fundação para a Ciência e a Tecnologia within project: UID/EEA/50014/2019.

References

1. Ahmed, N.K., Neville, J., Rossi, R.A., Duffield, N.: Efficient graphlet counting for large networks. In: International Conference on Data Mining, pp. 1–10. IEEE (2015)
2. Barabasi, A.L., Albert, R.: Emergence of scaling in random networks. Science **286**(5439), 509–512 (1999)
3. Cook, S.A.: The complexity of theorem-proving procedures. In: ACM Symposium on Theory of Computing, STOC 1971, pp. 151–158. ACM (1971)
4. Gao, Y.: The degree distribution of random k-trees. Theor. Comput. Sci. **410**, 688–695 (2009)
5. Hagberg, A., Schult, D., Swart, P., Conway, D., Séguin-Charbonneau, L., Ellison, C., Edwards, B., Torrents, J.: NetworkX. High productivity software for complex networks. Webová strá nka (2013). https://networkx.lanl.gov/wiki
6. Hočevar, T., Demšar, J.: A combinatorial approach to graphlet counting. Bioinformatics **30**(4), 559–565 (2014)
7. Hočevar, T., Demšar, J.: Combinatorial algorithm for counting small induced graphs and orbits. PloS One **12**(2), e0171428 (2017)
8. Holme, P., Kim, B.J.: Growing scale-free networks with tunable clustering. Phys. Rev. E **65**(2), 026107 (2002)
9. Khakabimamaghani, S., Sharafuddin, I., Dichter, N., Koch, I., Masoudi-Nejad, A.: Quatexelero: an accelerated exact network motif detection algorithm. PloS One **8**(7), e68073 (2013)
10. Kunegis, J.: Konect: the Koblenz network collection. In: Proceedings of the 22nd International Conference on World Wide Web, pp. 1343–1350. ACM (2013)
11. McKay, B.D.: Nauty user's guide (version 2.2). Technical report, TR-CS-9002, Australian National University (2003)

12. Milo, R., Shen-Orr, S., Itzkovitz, S., Kashtan, N., Chklovskii, D., Alon, U.: Network motifs: simple building blocks of complex networks. Science **298**(5594), 824–827 (2002)
13. Paredes, P., Ribeiro, P.: Towards a faster network-centric subgraph census. In: International Conference on Advances in Social Networks Analysis and Mining, pp. 264–271. IEEE (2013)
14. Pinar, A., Seshadhri, C., Vishal, V.: ESCAPE: efficiently counting all 5-vertex subgraphs. In: International Conference on World Wide Web, pp. 1431–1440. International World Wide Web Conferences Steering Committee (2017)
15. Pržulj, N.: Biological network comparison using graphlet degree distribution. Bioinformatics **23**, e177–e183 (2007)
16. Ribeiro, P., Paredes, P., Silva, M.E., Aparicio, D., Silva, F.: A survey on subgraph counting: concepts, algorithms and applications to network motifs and graphlets. arXiv preprint arXiv:1910.13011 (2019)
17. Ribeiro, P., Silva, F.: G-tries: a data structure for storing and finding subgraphs. Data Min. Knowl. Discov. **28**, 337–377 (2014)
18. Rossi, R.A., Ahmed, N.K.: The network data repository with interactive graph analytics and visualization. In: AAAI (2015). http://networkrepository.com
19. Wernicke, S.: Efficient detection of network motifs. IEEE/ACM Trans. Comput. Biol. Bioinform. **3**(4), 347–359 (2006)

Group Cohesion Assessment in Networks

V. Carchiolo[1], M. Grassia[2](✉), A. Longheu[2], M. Malgeri[2], and G. Mangioni[2]

[1] Dip. di Matematica e Informatica, Università degli Studi di Catania, Catania, Italy
[2] Dip. Ingegneria Elettrica Elettronica Informatica, Università degli Studi di Catania, Catania, Italy
marco.grassia@studium.unict.it

Abstract. Networks measurement is essential to catch and quantify their features, behaviour and/or emerging phenomena. The goal of *cohesiveness* metric introduced here is to establish the level of cohesion among network nodes. It comes from the Black-Hole metric introduced as a solution of the normalization problem that affects PageRank; in particular, here we present an extension that leverages a set of black hole nodes to assess intra- and inter-group cohesion in partitioned networks. We carried out the evaluation in several real-world networks, also considering temporal dynamics and group size.

1 Introduction

Since the beginning of the complex networks era [1,2], several measures and metrics were introduced to characterize networks, explore their features and understand their behaviour. Some of them, as the number of vertexes and the degree distribution, are basic and can be applied to any type of network; others are used to characterize and study specific networks, e.g. those based on Scale–Free, Erdos–Renyi or Watts–Strogatz models [3,4].

In this work, the *cohesiveness* of a network is investigated. This property aims at measuring the level of cohesion among groups of network nodes, which may arise naturally (i.e., communities) or be imposed externally (e.g., co-workers or people living in the same city). While the concept is rather intuitive, we propose a formal way to investigate which groups are more cohere internally and with other groups, that is an important problem in various domains, for instance in trust networks. Our proposal is based on the new metric presented in [5] as a solution of the normalization problem, which deals with the arc weights normalization effect of the PageRank by introducing a bogus node, called Black Hole, together with a weights transformation that keeps the network adjacency matrix stochastic.

Here we present an extension to that work, where several black holes nodes are adopted and network *cohesiveness* is defined using the PageRank value of such black holes; this allows to define cohesiveness as a measure of intra- and inter-group cohesion in partitioned networks.

H. Barbosa et al. (Eds.): *Complex Networks XI*, SPCOM, pp. 16–25, 2020.
https://doi.org/10.1007/978-3-030-40943-2_2

The progress discussed here also concerns the investigation on cohesiveness in real-world networks (monoplex as well as multilayer), its evolution in temporal networks and its meaning for a group, also in relation to group size. Results highlight cohesiveness role in capturing the global distribution of nodes relationships.

The paper is organized as follows: in Sect. 2 the Black Hole metric is recalled and in Sect. 3 we introduce the *cohesiveness*, whose assessment and related consideration is discussed in Sect. 4; Sect. 5 provide our conclusions and future works.

2 The Black Hole Metric

As discussed in Sect. 1, the Black Hole metric addresses the PageRank's normalization effect, which, in fact, has a twofold negative effect: (1) it is not able to preserve the absolute arc weights, modifying the weight distribution asymmetry and (2) it shadows the social implications of assigning low weights to all of a node's neighbours, that, for instance, would be a clear indication of bad social relationships. For a more in depth discussion of the Black Hole metric advantages, we refer to [5].

Shortly, the Black Hole metric consists in an arc weights scaling procedure and in the introduction of a new special node named *Black Hole*. The arc weights transformation only requires the knowledge of the maximum and the minimum value each weight can assume. This range may be global (each node has the same scale) or local (each node has its own weight scale). However, since these transformations (detailed below) do not preserve the out-strength s_i of each node i, the authors also introduce a new node, called Black Hole, that "absorbs" the missing out-strength value to reach 1 for each node. That is, every node is connected to the Black Hole with a strength of $1 - s_i$.

More formally, let i be a generic node in the network. Let the interval $[l_i, h_i]$ be the *local* scale of node i. Let r_{ij} be the weight that node i assigns to the arc pointing towards node j. Let out_i (in_i) be the out (in) neighbours of node i. Given that $l_i \leq r_{ij} \leq h_i$, the modified weight $\bar{a}_{ij}$ of the arc that goes from i to j is defined as:

$$\bar{a}_{ij} = \frac{r_{ij} - l_i}{|out_i|(h_i - l_i)} \tag{1}$$

which is significantly different from the normalized arc weight required by PageRank:

$$a_{ij} = \frac{r_{ij}}{\sum_{k \in out_i} r_{ik}} \tag{2}$$

The contribute of the arc from node i to the black hole is given by:

$$b_i = \sum_{j \in out_i} \frac{h_i - r_{ij}}{|out_i|(h_i - l_i)} \tag{3}$$

After the application of the above described transformations, the PageRank computed on the transformed network takes into account the absolute strength of each arc, thus it overcomes some of the mentioned limitations of the original PageRank algorithm, as illustrated in [5].

3 Our Proposal

As previously discussed, in [5] the authors introduce one Black Hole node for the entire network in order to make it consistent with the PageRank dynamics, while in this work we extend that approach by introducing several Black Hole nodes, in order to leverage their PageRank to measure the nodes' cohesion, as detailed in this section.

Generally speaking, arc weights measure the *strength* of each link. For instance, in virtual social network, where nodes are people and links are the relationships among them, they can be easily associated to the *trust* [6–8] a person assigns to his/her acquaintances. Similarly, in peer-to-peer networks they could represent the number of exchanged files among peers, or in power grid networks the amount of energy transmitted and so on; in the following, we continue considering the social network scenario (using the word *trust* to indicate the arc weight) without affecting the generality of our proposal.

Our first research question is the meaning of Black Hole's PageRank. Since each Black Hole node is pointed by weighted arcs whose values depend on how much the source nodes trust their respective neighbours, the PageRank of the black hole node can assume higher values only if these nodes poorly trust each other, whilst a low value of the PageRank of the black hole is a clear symptom that the pointing nodes are trusting each other a lot; this comes from Eqs. 1 and 3, showing that Black Hole nodes gather the residual "amount of trust" not assigned to neighbours.

Based on this reasoning, our idea is to introduce a new metric called C with the aim to measure the level of *cohesiveness* among group of nodes of the network. C ranges from 0 to 1. It assumes a low value in the case of low-trusted groups, – i.e., where each node assigns low values of trust to its neighbours. Conversely, higher values of C means that nodes are highly trusted; therefore, C is an indicator of the level of trustworthiness among the nodes of the considered groups.

To formally introduce the *cohesiveness* metric, let us consider a weighted directed network $G = (V, E)$, where V is the set of nodes, and E is the set of weighted arcs among nodes. Let P be a partition of the network in g disjoint groups of nodes, $P = (G_1, G_2, \ldots, G_g)$. In this general case, we can introduce g Black Hole nodes to measure the internal cohesion of each group and $g(g-1)$ black hole nodes to measure the mutual cohesion between groups. A generic black hole b^{pq} is pointed by those nodes of the group G_p trusting nodes of the group G_q. The contribute of the node $i \in G_p$ to the black hole (i.e. the weight of the arc (i, b^{pq})) is given by:

$$b_i^{pq} = \sum_{j \in (G_q \cap out_i)} \frac{h_i - r_{ij}}{|out_i|(h_i - l_i)} \quad \forall i \in G_p \tag{4}$$

Formally, the cohesiveness between group p and group q is defined as:

$$C_{pq} = 1 - PR'_{b^{pq}} \quad where\, p, q = 1 \ldots g \tag{5}$$

The term $PR'_{b^{pq}}$ indicates the PageRank value of the Black Hole node between group G_p and G_q normalized with respect to the fraction of nodes pointing to it. That is:

$$PR'_{b^{pq}} = \frac{PR'_{b^{pq}} \cdot in_{b_i^{pq}}}{|V|} \tag{6}$$

This step is necessary in order to mitigate the effect of the size of each group on the Black Hole's PageRank value.

Figure 1 shows a toy example network of a network with 9 nodes, split in two groups (4 in the group G_1 and 5 in the group G_2). As discussed above, we transform the network by introducing 4 Black Holes: b^{11} and b^{22} to measure the internal cohesion of respectively G_1 and G_2, and b^{12} and b^{21} to asses the mutual cohesion between groups G_1 and G_2.

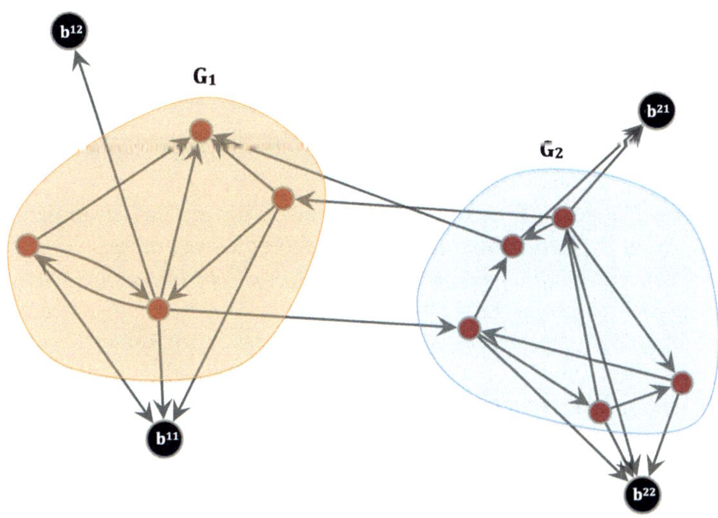

Fig. 1. An example of a network modified in order to measure groups cohesion.

4 Experiments

In our experiments, we study three different real-world scenarios using our proposed metrics. In particular, we analyse the *advogato* trust network [9–11], the friendship *Dutch College* [11–13] temporal network and the *Bitcoin Alpha* [11,14,15] trust temporal network.

For each network we first find the communities to use them as groups, then we find the intra- (internal) and inter- (external) cohesion between them, and in the case of temporal networks we also study their evolution in time. We stress that the first step is not required as groups are arbitrary and that our formulation is general.

Please note that in the following, *cohesiveness* values are scaled to fit the range 0..1 for each network (or layer) because we are interested in the *relative cohesiveness* between groups.

Advogato *Network.* The *Advogato* network is a (monoplex) trust network from the Advogato online community where users (nodes) can assign three different values of trust to each other, each one represented by a link. The network has 6,541 nodes and 51,127 edges, and we find 26 communities using the Louvain algorithm implemented in Pajek [16]. We compute the *cohesiveness* values among those groups and report them in Fig. 2a. As shown in the figure, most have very high intra-group cohesiveness values, which is expected as they are communities, while a few groups tend to be way less cohere (e.g., group 18 and 19). The inter-group values, however, are another story and are very mixed. For instance, some groups (e.g., 2) tend to trust all other groups in the network and, mutually, all other groups in the network tend to trust them, while other groups (e.g., 4) do not enjoy such mutuality. On the other hand, some groups (e.g., 8), do not trust any other group but are trusted nonetheless.

We also show the *cohesiveness* values in Fig. 3 as a function of the size of a group. Moreover, it should be noted that the size of the group and the cohesion values are uncorrelated, as shown in the figure.

Dutch College *N*etwork. The *Dutch College* network contains friendship ratings between 32 university freshmen at seven different time points. Each student is a node and there are 3,062 edges in the network. We pre-process the network by filtering out the nodes that are not present at every timestamp and find four communities at the second one (since the first is too small), using again a Louvain algorithm.

In Fig. 4 we report the evolution of the *cohesiveness* values of those communities in time. While the relative intra-group *cohesiveness* of Groups 1 and 4 are constant at the maximum and minimum values respectively, meaning that they are the most and least cohere communities in the network at each timestamp, their inter-group values change abruptly, showing that the relations among (all) the communities are unstable in time. Even if similar considerations can be made for the other two groups, their relative internal cohesion values show that the friendship among these students is strong at every point in time.

Bitcoin Alpha *N*etwork. The *Bitcoin Alpha* network is a trust/distrust network from the Bitcoin Alpha trade platform. It contains data from 3,783 users (nodes) and 24,186 trust values (edges) collected in six years. We pre-process this network by quantizing this time interval in ten timestamps and find over sixty multi-slice communities using the multi-slice Louvain algorithm implementation in [17], inspired to the paper [18]. We report the *cohesiveness* values in Fig. 5a for each timestamp, and also the multi-slice ones (computed by simple average across slices) in Fig. 5b. Again, the size of the group and the cohesion values are uncorrelated, as shown in the figures.

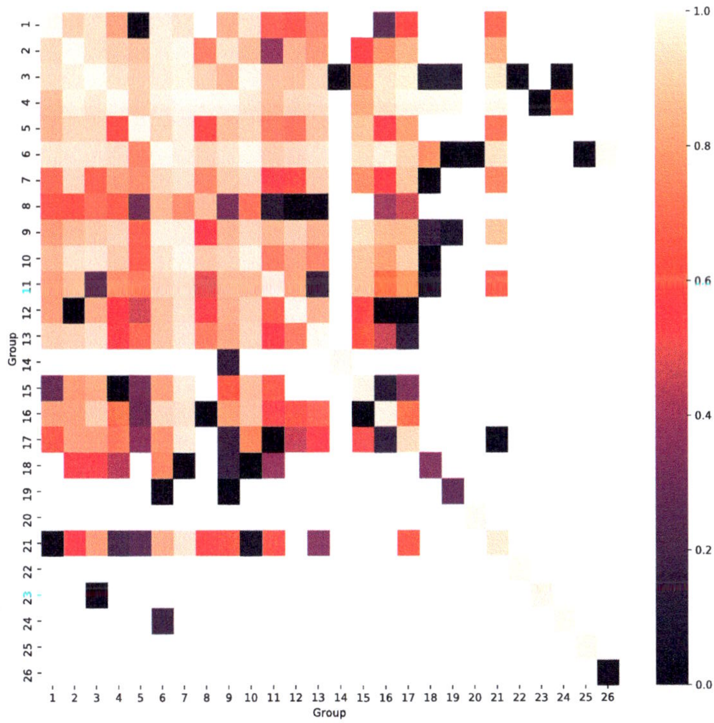

Fig. 2. *Advogato* network inter-group *cohesiveness* values of each group pair (x, y), where x represents the source group and y target. Also, missing values mean that there is no incoming link to the respective Black Hole (i.e., there are no links between the two groups).

4.1 Discussion

By looking at the intra-group cohesion in both Figs. 5a and b we can notice a pattern that is also visible in Fig. 3. In fact, smaller groups exhibit a wide range of internal cohesion values, while larger groups generally have higher internal cohesion. The same pattern is not observable for inter-group relations.

It is interesting to note that in the *Bitcoin Alpha* network the biggest group is that with the highest value of multilayer internal cohesion, but also that with the smallest multilayer inter-group cohesion, meaning that larger groups tends to be closed, probably because members are more focussed on developing intra-group trusted relations than building solid relationships with members of other groups.

Another interesting point to investigate is the correlation between the intra and inter group cohesion. It can shed light on the different role a group has in comparison to the other groups. In Fig. 6 the inter group cohesion is reported as a function of the intra group cohesion. In particular, in Advogato network (Fig. 6a) we can identify two sets of groups, those with high values of intra

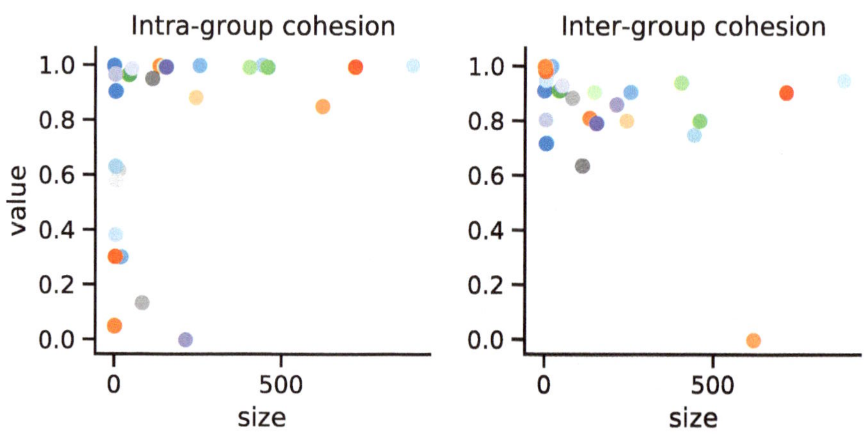

Fig. 3. Advogato (monoplex). Each community is represented by a different color.

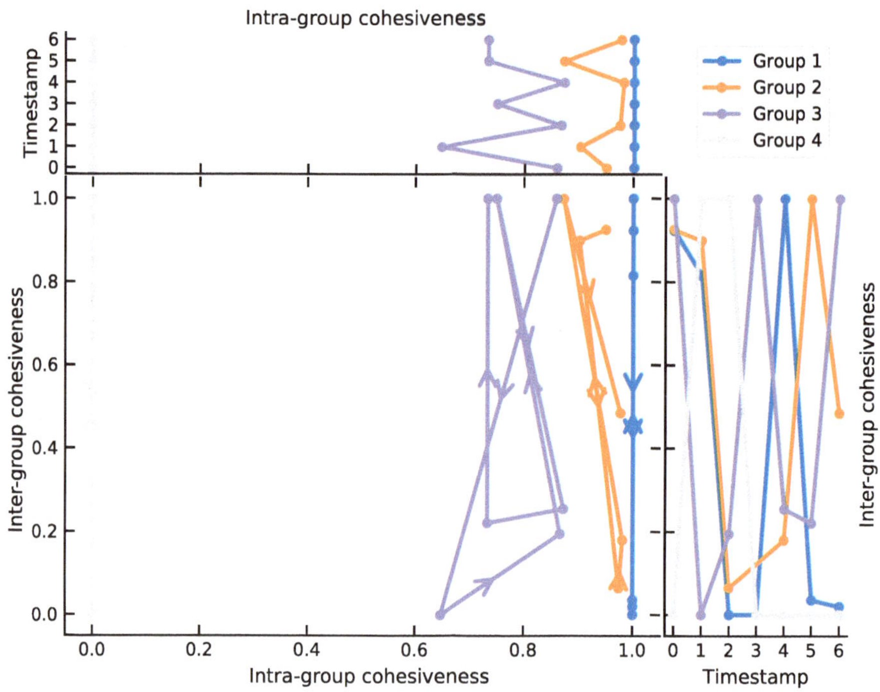

Fig. 4. Inter- and intra- group *cohesiveness* values evolution in Dutch College network. The central pan shows the evolution of both values for each group by connecting subsequent timestamps with a directed line.

and inter groups cohesion (upper-right corner of the graph) and those with low values of intra group cohesion and high values of intra group cohesion. The first set of groups includes groups that are both *open* to the outside and *internally*

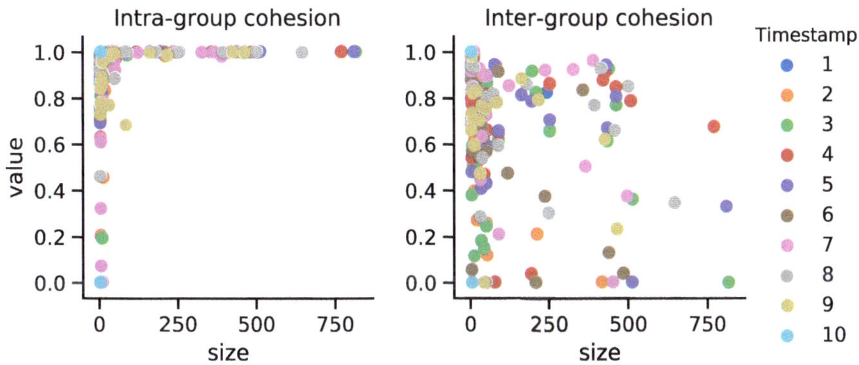

(a) Cohesion values for each timestamp.

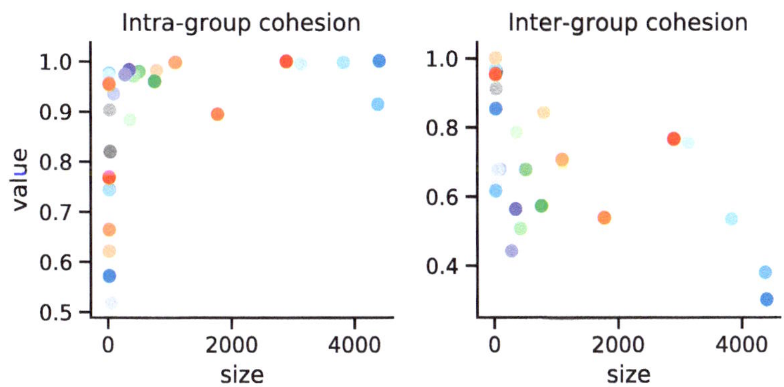

(b) Multilayer cohesion values, computed by averaging the value across timestamps. The size is the multilayer-wise size of the community. Circle colors are used to identify the different groups.

Fig. 5. Bitcoin Alpha network cohesiveness values.

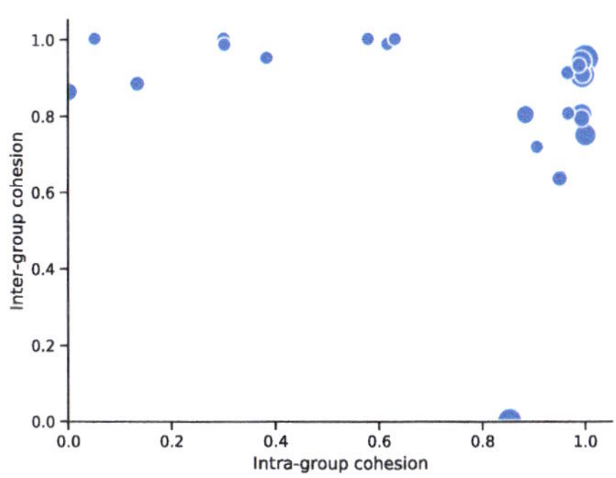

(a) Advogato network

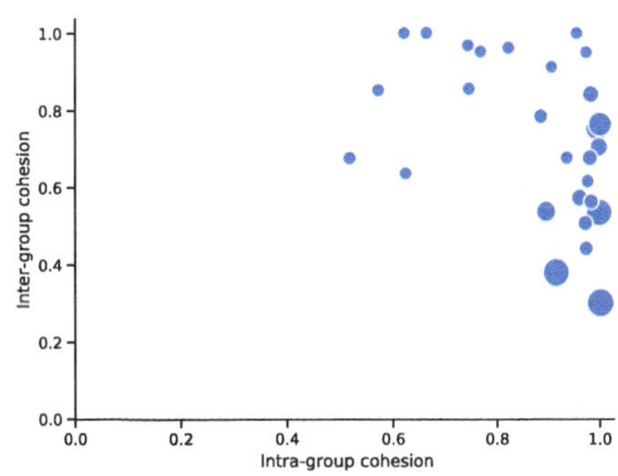

(b) Bitcoin Alpha network

Fig. 6. Inter vs Intra group cohesion. The size of each point is proportional to the size of the corresponding group.

cohesive, so members of these groups are able to establish "good" relationships with any other person. The second set of groups is characterized by a weak internal cohesion, meaning that members of these groups tend to establish better relation with people of other groups.

A different picture appears, however, if we look at the Bitcoin Alpha network (Fig. 6b). Here, in fact, the majority of groups exhibit an high internal cohesion, while the inter group cohesion is high only for low/medium-sized groups. Bigger groups, on the other hand, tend to be *closed* to the outside.

5 Conclusions

In this paper, the cohesiveness of a network has been introduced. First, we presented the origin of such a measure, recalling the Black Hole metric used to address the normalization PageRank problem; then, we added several black hole nodes instead of a single one to assess intra- and inter-group cohesion in partitioned networks.

We considered three different real-world scenarios, namely Advogato, Dutch College and Bitcoin Alpha networks, grouped according to detected communities, discussing relevant results even for what concerns the temporal evolution. Our conclusions are promising and further works will be focused on an in depth analysis of the role of a group in real world networks with ground truth groups.

References

1. Albert, R., Barabasi, A.L.: Statistical mechanics of complex networks. Rev. Mod. Phys. **74**, 47 (2002)
2. Newman, M.: The structure and function of complex networks. SIAM Rev. **45** (2003)
3. Newman, M.: Networks, 2nd edn. Oxford University Press Inc., New York (2018)
4. Latora, V., Nicosia, V.: Complex Networks: Principles, Methods and Applications. Cambridge University Press, Cambridge (2017)
5. Buzzanca, M., Carchiolo, V., Longheu, A., Malgeri, M., Mangioni, G.: Black hole metric: overcoming the pagerank normalization problem. Inf. Sci. **438**, 58–72 (2018)
6. Buzzanca, M., Carchiolo, V., Longheu, A., Malgeri, M., Mangioni, G.: Direct trust assignment using social reputation and aging. J. Ambient Intell. Humaniz. Comput. **8**(2), 167–175 (2017)
7. Carchiolo, V., Longheu, A., Malgeri, M., Mangioni, G.: Trusting evaluation by social reputation. In: Intelligent Distributed Computing, Systems and Applications, pp. 75–84. Springer (2008)
8. Carchiolo, V., Longheu, A., Malgeri, M., Mangioni, G.: Trust assessment: a personalized, distributed, and secure approach. Concurr. Comput.: Practi. Exp. **24**(6), 605–617 (2012)
9. Advogato network dataset – KONECT, April 2017
10. Massa, P., Salvetti, M., Tomasoni, D.: Bowling alone and trust decline in social network sites. In: Proceedings of the International Conference on Dependable, Autonomic and Secure Computing, pp. 658–663 (2009)

11. Kunegis, J.: KONECT – the Koblenz network collection. In: Proceedings of the International Conference on World Wide Web Companion, pp. 1343–1350 (2013)
12. Dutch college network dataset – KONECT, April 2017
13. Van de Bunt, G.G., Van Duijn, M.A.J., Snijders, T.A.B.: Friendship networks through time: an actor-oriented dynamic statistical network model. Comput. Math. Organ. Theory **5**(2), 167–192 (1999)
14. Bitcoin alpha network dataset – KONECT, February 2018
15. Kumar, S., Spezzano, F., Subrahmanian, V.S., Faloutsos, C.: Edge weight prediction in weighted signed networks. In: Proceedings of the International Conference on Data Mining, pp. 221–230 (2016)
16. Batagelj, V., Mrvar, A.: Pajek - program for large network analysis (1999)
17. Traag, V.: vtraag/louvain-igraph: 0.6.1, November 2017
18. Mucha, P.J., Richardson, T., Macon, K., Porter, M.A., Onnela, J.P.: Community structure in time-dependent, multiscale, and multiplex networks. Science **328**(5980), 876–878 (2010)

Node Classification with Bounded Error Rates

Pivithuru Wijegunawardana[1](✉), Ralucca Gera[2], and Sucheta Soundarajan[1]

[1] Department of Electrical Engineering and Computer Science, Syracuse University, Syracuse, USA
{ppwijegu,susounda}@syr.edu

[2] Department of Applied Mathematics, Naval Postgraduate School, Monterey, CA, USA
RGera@nps.edu,
https://www.overleaf.com/project

Abstract. Node classification algorithms are widely used for the task of node label prediction in partially labeled graph data. In many problems, a user may wish to associate a confidence level with a prediction such that the error in the prediction is guaranteed. We propose adopting the Conformal Prediction framework [17] to obtain guaranteed error bounds in node classification problem. We show how this framework can be applied to (1) obtain predictions with guaranteed error bounds, and (2) improve the accuracy of the prediction algorithms. Our experimental results show that the Conformal Prediction framework can provide up to a 30% improvement in node classification algorithm accuracy while maintaining guaranteed error bounds on predictions.

Keywords: Node classification · Conformal prediction · Bounded error rates

1 Introduction

In real world network analysis problems, it is common for data to be incomplete. In such cases, node classification algorithms play an important role: given a partially labeled graph, these algorithms predict labels for unlabeled nodes by using known node's labels and connections between nodes. For example, consider a criminal group hidden inside a general social network. If some criminals and non-criminals are identified, can an algorithm predict whether the unlabeled nodes are criminals? By taking advantage of connections, algorithms specifically designed for node classification generally perform better on semi-supervised graph classification tasks as compared to traditional classification algorithms [11,19].

This material is based upon work supported by the U.S. Army Research Office under grant number W911NF1810047.

H. Barbosa et al. (Eds.): *Complex Networks XI*, SPCOM, pp. 26–38, 2020.
https://doi.org/10.1007/978-3-030-40943-2_3

The user of a node classification algorithm often may wish to associate a confidence with each prediction. For example, when predicting whether a node in a social network is a criminal or not, a prediction may lead to a criminal investigation. In such applications, it is thus essential to have prediction algorithms that can provide guaranteed error rates on unseen data.

The performance of a classification algorithm is generally measured with metrics such as accuracy, precision, and recall. Such metrics describe the algorithm performance in aggregate, but do not measure certainty of individual predictions.

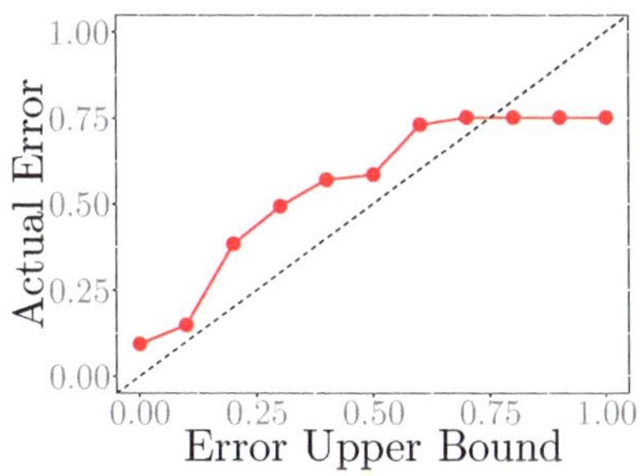

(a) Class Probability

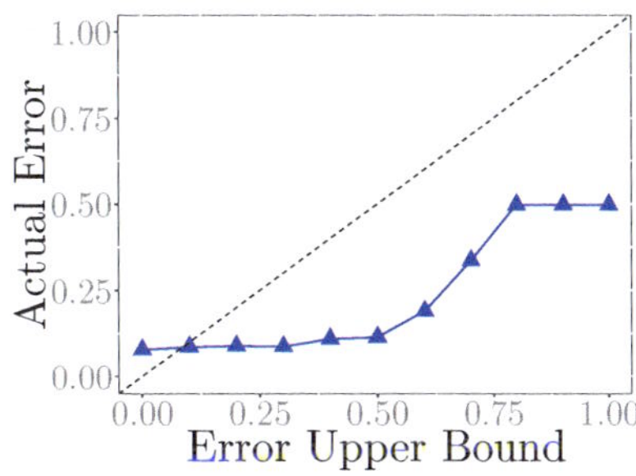

(b) Conformal Prediction

Fig. 1. Comparison of using Conformal Prediction framework vs class probability as upper error bound for Cora citation dataset classification using Iterative Classification Algorithm. Conformal prediction actual error is always no greater than the error bound where as the actual error for class probability does not follow the error bound.

Node classification algorithms can generally output a vector indicating the probability that a node belongs to each class. One can consider the probability of a node belonging to some class as the confidence of the label (1-probability would be the upper error bound). Figure 1a shows that actual errors are larger than the upper error bounds when using the label probability as error bound. Therefore, label probability should not be interpreted as confidence values

In this work, we demonstrate how the *Conformal Prediction* framework can be used to obtain error bounds for the node classification task. Conformal Prediction (CP) is a framework to provide guaranteed error bounds for prediction algorithms [16]. This framework works on top of a prediction algorithm (e.g. SVM, Neural Network) and for a specified error bound, considers how unusual a data instance is in consideration to training data. Since CP framework is a very mathematical framework, application to prediction algorithms require customizing the framework according to the algorithm. The CP framework has been applied to provide guaranteed error bounds for machine learning algorithms [4,7,12]. However, to our knowledge, CP has not been applied to the network setting. Figure 1b shows an example application of the CP framework to node classification problem. The prediction error rate is always lower then the expected error the CP framework suggests.

Our contributions in this paper are: (1) We show conditions under which node classification problem would satisfy the CP framework assumptions to obtain valid error bounds, (2) We show how to apply the CP framework to node classification algorithms from different categories (3) We conduct an experimental analysis over various types of node attributes and graphs and show that the CP framework can improve node classification algorithm accuracy.

2 Background

2.1 Node Classification Algorithms

Node classification algorithms consider both node attributes and node connectivity patterns when making predictions. There are three main categories of node classification algorithms. The first category contains local classifier based algorithms, where a local classifier is iteratively trained using node attributes and network information to predict labels for unlabeled nodes, such as logistic regression local classifier. Iterative Classification Algorithm (ICA) [11] and Link Based Classification algorithm [6] are examples of such algorithms. These algorithms iteratively predict labels for the unlabeled nodes in the graph using predicted labels in the previous round of predictions.

The second category of algorithms are label propagation based algorithms, where the algorithms use random walks to learn a global labeling function across the network [19]. These algorithms predict labels for nodes in the graph by considering hitting probability of each label in a random walk.

The third, and newest, category of node classification algorithms learn a deep representation of the network and labeling function. There are two approaches to learning this representation. The first approach uses network embedding-based algorithms, which generate feature vectors for nodes in a graph in an unsupervised manner. These algorithms use multiple random walks starting at each node, and trains a prediction model based on these features [2,13]. The second approach is learning a labeling function using deep neural networks based on graph representation. Graph Convolutional Neural Networks (Graph CNN) are widely used to conduct node classification under this category [5].

In the current work, we consider node classification algorithms from each category mentioned above and show how the CP framework can be applied to obtain predictions with guaranteed error bounds.

2.2 Conformal Prediction Framework

The CP framework outputs a set of predictions for a given sample with a bounded error rate by comparing "how typical" the sample is as contrasted to other samples [16]. Suppose that we are given a data set $Z = \{z_1, z_2, \ldots, z_n\}$ where $z_i = (x_i, y_i)$; $x_i \in \mathbb{R}^d$ is the feature vector of the sample i, and $y_i \in Y$ is the class label for i^{th} sample. Here, Y is the set of class labels, i.e. $Y = \{y^1, y^2, \ldots, y^\ell\}$.

Given a new sample with feature vector x_{n+1}, the CP framework measures how typical the following sequence is: $(z_1, z_2, \ldots, z_n, (x_{n+1}, y^k))$, where $y^k \in Y$.

Since we already know the labels for $z_1, z_2, \ldots, z_n$, we are in effect measuring how typical the sequence is when label y^k is assigned to the new sample, and how likely is that $n+1$'s true label is y^k [12].

The CP framework uses a test for randomness to measure how likely a sequence is, where the p_{cp}-value for a given sequence is calculated using Eq. (1). A given conformity measure calculates the "typicalness" of a data instance (α values). The α_i is the conformity score for i^{th} data instance [12].[1]

$$p_{cp}(z_1, z_2, \ldots, (x_{n+1}, y^k)) = \frac{|\{i = 1, \ldots, n : \alpha_i \leq \alpha_{n+1}\}|}{n} \tag{1}$$

p-values vs. p_{cp}-values: We adopt notation p_{cp}-value instead of p-value in this paper to avoid any confusion since a higher p_{cp}-value in CP framework means the label in consideration is highly likely. Conversely, a higher p-value in general means that there is stronger evidence towards the alternative hypothesis.

Given some significance value ϵ, CP framework first calculates p_{cp}-values for all sequences considering all possible class labels; then the prediction set of $n+1$ sample at ϵ significance is calculated using Eq. (2).

$$P(n+1, \epsilon) = \{y^k : y^k \in Y \quad \& \quad p_{cp}(z_1, z_2, \ldots, (x_{n+1}, y^k)) > \epsilon\} \tag{2}$$

For example, consider we are predicting hobbies in a social network. A node can have one of the hobbies among the following; {reading, singing, dancing, cooking}. When we use the CP framework, for an unlabeled node v, we observe that the p_{cp}-values for each label are $\{0.2, 0.1, 0.01, 0.02\}$. If we set significance to 0.05, both reading and singing will be predicted as hobbies of node v since both these labels have p_{cp}-values higher than the significance level. This also shows that the chance of generating sequences including dancing and cooking as labels is less than 5%, implying that these labels are highly unlikely.

Note that in Eq. (2), the CP framework outputs the set of labels that satisfy the specified significance rather than a single prediction. Therefore, CP framework predictions can have one prediction, multiple predictions, or zero predictions, in case none of the labels satisfy the significance requirement. The probability of not including true labels in the prediction set is less than the specified threshold, providing an error rate bounded by the significance level. If we are to predict labels at significance ϵ, the probability of not including the correct label in the prediction set is ϵ and the confidence in the prediction is $1 - \epsilon$.

The CP framework provides guaranteed error bounds for predictions under the assumption that the data is exchangeable, meaning any permutation of the sequence $(z_1, z_2, \ldots, z_n, (x_{n+1}, y^k))$ should result in the same p_{cp}-value. This assumption is necessary to obtain the p_{cp}-value using Eq. (1).

The CP framework was originally introduced in the *transductive* setting, where the true label of the current sample is revealed before the arrival of the next sample [16]. In this setting, the given model is trained considering each

[1] Note that the "$\leq$" sign in Eq. 1 changes to "$\geq$" if we are using a non-conformity function instead of conformity.

possible label for new data instance and the framework measures how typical the model is. Since this setting requires training the model for each new data instance and each possible label, applying this in a real world setting would be very inefficient.

The *Inductive Conformal Prediction* (ICP) is an alternative approach which splits the training data into actual training set and a calibration set, and uses the calibration set to conduct CP [12]. The ICP framework uses the training set to train the underlying prediction model, and the calibration set to calculate the p_{cp}-value. In the ICP setting, we only consider the calibration set when calculating the p_{cp}-value of Eq. (1).

2.3 Related Work

Bayesian Framework, Probably Approximately Correct Learning theory (PAC theory) [3] and generalization error bounds [9] are other frameworks that provide bounded error rates in machine learning applications. Bayesian Framework error rates are dependent on the priors that are used in the estimation. Hence, the error bounds are not guaranteed in case priors are wrong. PAC theory and generalization error bounds provides upper bounds on the trained model rather than individual samples. The only assumption that the CP framework makes is that the data is exchangeable, which is valid for most machine learning data. Dashevskiy et al. [1] show that even in cases where exchangeability assumption is violated (e.g., time series data), the CP framework still provides reasonable error bounds. The CP framework, unlike PAC theory and generalization error bounds, can provide error bounds for individual samples rather than the algorithm.

Initial work on the CP framework was primarily theoretical, and focused on proving the error bounds. Applying the CP framework to machine learning algorithms required defining conformity measures specific to algorithms, showing that the data is in fact exchangeable. Research in this area shows how the CP framework can be applied to various algorithms including decision trees [4], neural networks [12], and SVM [7] etc.

To best our knowledge, this is the first work that considers providing guaranteed error bounds for node classification algorithms, and shows how the CP framework can be applied to obtain those error bounds

3 Methodology

We now introduce the details of how the ICP framework can be applied to node classification algorithms. In the $z_n = (x_n, y_n)$ a node classification problem, we have that $x_n \in \mathbb{R}^d$ is the d-dimensional feature vector for node n, and y_n is the label of node n.

To show that the ICP framework applies, we must demonstrate that the data is exchangeable. Note that this does *not* require that the data is i.i.d., simply that all permutations of each sequence of training samples are equally likely. Since we are drawing training samples uniformly at random from the set of nodes, exchangeability holds. We also considered sampling training data using a network crawling algorithm such as random walk or snowball sampling. Resulting error bounds are not valid in these cases since any training node ordering is not equally likely (not exchangeable) for random walk or snowball sampling.

The conformity function is an integral part of the ICP framework, measuring how different the data instance in consideration from the calibration set. Any real valued conformity function that measures how different a sample is can be used to produce valid nested prediction regions [17], but the efficiency (smaller prediction regions) of the algorithm depends on how well the nonconformity function measures differences between data instances. For example, an efficient prediction according to our hobby prediction example in Sect. 2 would be predicting one hobby as the label. An inefficient prediction would have no hobby or more than one hobby in the prediction set.

Consider a prediction algorithm that outputs a vector $\sigma_n \in \mathbb{R}^{|Y|}$ for some unlabeled node n, indicating the probability that n would belong to each class in Y. One possible conformity measure for such an algorithm is the probability margin, which is the difference between the label in consideration and the highest probability of any other label [14]. We can calculate the probability margin conformity score for some label $y^k \in Y$ using Eq. 3.

$$C(n, y^k) = \sigma_n(y^k) - \max_{y^i \in Y : y^i \neq y^k} (\sigma_n(y^i)). \tag{3}$$

Given a node classification algorithm M, a graph G, set L of labeled nodes, set U of unlabeled nodes, a significance level ϵ, and a conformity function C, we introduce Algorithm 1 to show how the ICP framework can be applied to node classification problem.

Algorithm 1 ICP for Node Classification

Input: G= Graph, L = *labeled_nodes*, U = *unlabeled_nodes*, M = prediction algorithm, C= Conformity function, ϵ = significance

Output: Prediction set for each node in U at significance ϵ

procedure ICP
    Divide L into $T = training\_set$ and $S = calibration\_set$
    Train M using G and T ▷ Train prediction model M
    for $s \in S$ **do**
        $\sigma_s = M(s)$ ▷ Get prediction probability vector σ_s for s
        $\alpha_s = C(\sigma_s, y_s)$ ▷ Calculate conformity score for s and s's label y_s
    for $u \in U$ **do**
        $P_u = \{\}$ ▷ u's prediction set at significance ϵ
        for $y^k \in Y$ **do**
            $\sigma_u = M(u)$
            $\alpha_u = C(\sigma_u, y^k)$
            $p = \frac{|\{s \in S : \alpha_s \leq \alpha_u\}|}{|S|}$ ▷ Calculate p-value for label y^k
            if $p > \epsilon$ **then**
                $P_u.add(y^k)$ ▷ Add y^k to u's prediction set

4 Experiments

We conduct experiments to evaluate whether the ICP framework predictions meet the specified error bounds. We consider node classification algorithms from different categories: Iterative Classification Algorithm (ICA), Label Propagation (LP), Graph Convolutional Network (GCN), and Deepwalk (DW). We now introduce the performance metrics and data sets used in our research.

4.1 Performance Metrics

Our evaluation closely follows the evaluation criteria in [4]. We use several measures to evaluate the quality of predictions made by ICP framework for the node classification problem:

1. We check whether the ICP framework predictions meet the specified maximum error bounds. Since node classification graph data meets the exchangeability assumption, the specified error bounds should be met.
2. We evaluate the ICP framework predictions based on their efficiency. Since the ICP framework outputs a set of predictions for a node based on its conformity score, an efficient prediction would have only a single class in the prediction set. We consider the fraction of predictions with only one class (OneC), multiple classes (MultiC) and zero classes (ZeroC) to evaluate the efficiency of the ICP framework.
3. We compare the accuracy of the baseline prediction model (BaselineAcc) with the accuracy of one class predictions (OneAcc) from the ICP framework to show that the ICP framework enhances performance of the baseline prediction model.

4.2 Datasets

Node classification algorithms generally perform well on assortative networks, but less well on nodes are not assortative. Accordingly, we have selected graph datasets with varying levels of assortativity. For each network, we use the largest connected component. Cora [8,15] and PubMed [10] are citation networks, showing citation relationships between papers. Facebook100[2] is the Amhrest college Facebook friendship network. BlogCatalog [18] is a blogger friendship network. The Protein-Protein Interaction network [2] is a subgraph of the PPI Homo Sapiens network. Networks are described in Table 1.

[2] Obtained from https://archive.org/download/oxford-2005-facebook-matrix.

Table 1. Network dataset statistics

Dataset	Type	Nodes	Edges	Label	Classes	Label assortavity
Cora	Citation	2708	5278	Research area	7	0.771
PubMed	Citation	19717	44327	Research area	3	0.686
Blogcatalog	Social	10312	333983	Blogger group	39	0.05
Facebook100	Social	2235	90954	Year	9	0.409
PPI	Biological	3890	38739	Biological state	50	0.05

4.3 Experimental Setup

We run experiments as a multi-class prediction problem where we vary the percentage of labeled nodes in the network from 10% up to 50%. We randomly sample the labeled data from each class proportional to the size of the class and report average performance over 10 runs. We used 25% of the training data as the calibration set to conduct conformal prediction.

For ICA and Deepwalk, we use a multi-class logistic regression classifier as the base classifier. We set Deepwalk hyper parameters for all data sets as follows: 80 walks, 128 dimension representation, window size 10, and walk length 40 according to [13]. GCN hyper parameters are set at 0.5 dropout rate, 5.10^{-4} $L2$ regularization and 16 hidden units, according to [5].

5 Results

Figure 2 shows results of `ICP` using ICA on the Cora citation network with 10%, 30% and 50% of the nodes labeled, using the performance metrics discussed in Sect. 4.1. First, we see that the actual errors in all algorithms are very close to the given significance level, demonstrating that the `ICP` framework in fact provides accurate error bounds for node classification algorithms.

Second, as expected, the percentages of OneC (one-class predictions) and MultiC (multiple-class predictions) increase and decrease as we increase the significance level, respectively. Recall our hobby prediction example in Sect. 2 where p_{cp}-values of node v for labels; reading, singing, dancing and cooking are $\{0.2, 0.1, 0.01, 0.02\}$ respectively. If we set significance to 0.01, all label p_{cp}-values will satisfy the significance requirement and hence will be included in the prediction set. Therefore, we can observe many multiple-class predictions at lower significance values. When we increase significance level to, e.g., 0.15, only one label satisfies the significance requirement, increasing the number of one-class predictions. ZeroC (zero class predictions) slightly increases at higher significance values causing OneC to reduce slightly, because if we set significance to 0.2, none of the labels meet significance.

Finally, we see that the accuracy of the `ICP` framework is higher than the accuracy of the baseline node classification algorithm, showing that the predictions from `ICP` are more reliable than those from the baseline prediction algorithm. Results are consistent across different algorithms and labeled node percentages.

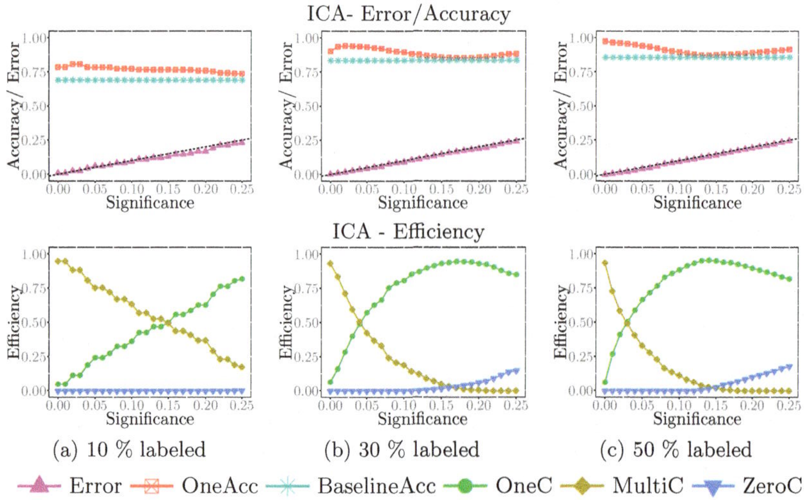

(a) 10 % labeled (b) 30 % labeled (c) 50 % labeled

Fig. 2. Accuracy and efficiency for Cora citation data set using ICA as the baseline algorithm when 10%, 30% and 50% of the nodes are labeled. The actual error is always no greater than the specified significance level.

Table 2. Conformal prediction framework performance using ICA as the baseline algorithm. Average performance over 10 runs where randomly selected 30% of the nodes are labeled in each run.

Dataset	Significance	Error	OneC	MultiC	ZeroC	OneAcc	BaseAcc
Fb100 Amherst	0.05	$\mathbf{0.045 \pm 0.01}$	0.58	0.42	0	0.94	0.82
	0.15	$\mathbf{0.133 \pm 0.02}$	0.88	0.12	0	0.87	
	0.25	$\mathbf{0.25 \pm 0.02}$	0.85	0	0.15	0.88	
BlogCatalog	0.05	$\mathbf{0.05 \pm 0.01}$	0.05	0.95	0	0.52	0.23
	0.15	$\mathbf{0.15 \pm 0.01}$	0.11	0.89	0	0.48	
	0.25	$\mathbf{0.25 \pm 0.01}$	0.15	0.85	0	0.44	
PubMed	0.05	$\mathbf{0.046 \pm 0.01}$	0.52	0.48	0	0.92	0.83
	0.15	0.154 ± 0.01	0.97	0.03	0	0.85	
	0.25	$\mathbf{0.257 \pm 0.01}$	0.84	0	0.16	0.89	
PPI	0.05	$\mathbf{0.051 \pm 0.01}$	0.03	0.97	0	0.21	0.10
	0.15	0.147 ± 0.01	0.08	0.92	0	0.18	
	0.25	$\mathbf{0.248 \pm 0.02}$	0.14	0.86	0	0.17	

Tables 2 and 3 summarize the performance of the `ICP` framework applied to ICA and Label Propagation, respectively. Both algorithms closely maintain the given error bounds. `ICP` framework can cause the prediction errors to be slightly higher than the given error bound since the predictions are based on the calibration set rather than the whole training set.

Table 3. Conformal prediction framework performance using Label Propagation as the baseline algorithm. Average performance over 10 runs where randomly selected 30% of the nodes are labeled in each run.

Dataset	Significance	Error	OneC	MultiC	ZeroC	OneAcc	BaseAcc
Cora	0.05	$\mathbf{0.043 \pm 0.01}$	0.60	0.40	0	0.94	0.83
	0.15	$\mathbf{0.141 \pm 0.03}$	0.89	0.09	0.01	0.87	
	0.25	0.252 ± 0.04	0.85	0	0.15	0.89	
PubMed	0.05	0.052 ± 0.01	0.54	0.46	0	0.91	0.82
	0.15	$\mathbf{0.149 \pm 0.01}$	0.92	0.08	0	0.85	
	0.25	0.253 ± 0.01	0.87	0	0.13	0.86	
Fb100 Amherst	0.05	0.052 ± 0.01	0.38	0.62	0	0.93	0.78
	0.15	$\mathbf{0.144 \pm 0.01}$	0.71	0.29	0	0.88	
	0.25	0.252 ± 0.03	0.92	0.01	0.07	0.81	
BlogCatalog	0.05	$\mathbf{0.049 \pm 0.01}$	0.04	0.96	0	0.36	0.22
	0.15	$\mathbf{0.149 \pm 0.01}$	0.10	0.90	0	0.38	
	0.25	0.253 ± 0.01	0.15	0.85	0	0.39	
PPI	0.05	$\mathbf{0.048 \pm 0.01}$	0.04	0.96	0	0.12	0.10
	0.15	$\mathbf{0.146 \pm 0.01}$	0.10	0.90	0	0.11	
	0.25	$\mathbf{0.239 \pm 0.03}$	0.15	0.85	0	0.11	

Further, applying `ICP` improves baseline accuracy of both algorithms in all data sets. The `ICP` improves ICA accuracy in FB100 data from 0.82 to 0.94, while predicting singleton labels for 58% of the nodes with a guaranteed error rate of 5%. In the Label Propagation algorithm, `ICP` improves accuracy for Facebook100 data from 0.78 to 0.93, while predicting singleton labels for 38% of the nodes with a guaranteed error rate of 5%. When the baseline predictor accuracy is reasonable, `ICP` provides efficient predictions (more singleton predictions). When the baseline predictor does not perform well, conformity scores also become less meaningful leading `ICP` to make more multiple predictions. In blogcatalog, at significance level 0.15, only 11% of the predictions are singletons.

Tables 4 and 5 summarize results for GCN and DeepWalk respectively. GCN algorithm works well when node labels show homophily (Cora, FB100 and PubMed). In the Blogcatalog and PPI data sets, GCN algorithm baseline accuracy is 0.12 and 0.05, making it impractical to get meaningful predictions. The Deepwalk algorithm only considers network structure when predicting labels. If node labels are not correlated with the structure, even if data shows high homophily, Deepwalk baseline accuracy is low. In general, both these algorithms maintain the error bounds but provide inefficient predictions in some cases.

Table 4. Conformal prediction framework performance using GCN as the baseline algorithm. Average performance over 10 runs where randomly selected 30% of the nodes are labeled in each run.

Dataset	Significance	Error	OneC	MultiC	ZeroC	OneAcc	BaseAcc
Cora	0.05	**0.045 ± 0.01**	0.70	0.30	0	0.95	0.83
	0.15	**0.141 ± 0.02**	0.93	0.07	0	0.86	
	0.25	**0.248 ± 0.03**	0.83	0	0.17	0.91	
PubMed	0.05	**0.047 ± 0.01**	0.72	0.28	0	0.94	0.85
	0.15	0.151 ± 0.01	0.98	0.01	0.01	0.86	
	0.25	**0.249 ± 0.01**	0.82	0	0.18	0.92	
Fb100 Amherst	0.05	**0.048 ± 0.01**	0.43	0.57	0	0.96	0.72
	0.15	**0.139 ± 0.02**	0.59	0.41	0	0.9	
	0.25	0.251 ± 0.02	0.90	0.10	0	0.77	
BlogCatalog	0.05	**0.049 ± 0.01**	0.001	0.999	0.05	0	0.12
	0.15	**0.139 ± 0.01**	0.005	0.995	0	0.05	
	0.25	0.238 ± 0.01	0.01	0.99	0	0.11	
PPI	0.05	**0.049 ± 0.01**	0.0002	0.9998	0	0.03	0.05
	0.15	**0.143 ± 0.02**	0.002	0.998	0	0.18	
	0.25	0.239 ± 0.02	0.005	0.995	0	0.11	

Table 5. Conformal prediction framework performance using DeepWalk as the baseline algorithm. Average performance over 10 runs where randomly selected 30% of the nodes are labeled in each run.

Dataset	Significance	Error	OneC	MultiC	ZeroC	OneAcc	BaseAcc
Cora	0.05	**0.048 ± 0.01**	0.59	0.41	0	0.94	0.82
	0.15	**0.150 ± 0.02**	0.90	0.09	0.01	0.86	
	0.25	**0.239 ± 0.03**	0.88	0	0.12	0.87	
PubMed	0.05	**0.048 ± 0.01**	0.53	0.47	0	0.92	0.80
	0.15	**0.149 ± 0.01**	0.89	0.11	0	0.84	
	0.25	**0.245 ± 0.01**	0.90	0	0.10	0.84	
Fb100 Amherst	0.05	**0.047 ± 0.02**	0.001	0.999	0	0.14	0.15
	0.15	0.155 ± 0.02	0.005	0.995	0	0.12	
	0.25	0.268 ± 0.04	0.01	0.99	0	0.13	
BlogCatalog	0.05	0.057 ± 0.01	0.012	0.988	0	0.82	0.28
	0.15	0.153 ± 0.01	0.02	0.98	0	0.79	
	0.25	0.255 ± 0.01	0.03	0.97	0	0.76	
PPI	0.05	0.051 ± 0.01	0.0002	0.9998	0	0.43	0.11
	0.15	0.151 ± 0.02	0.005	0.995	0	0.32	
	0.25	**0.250 ± 0.02**	0.007	0.993	0	0.31	

5.1 Perturbation Analysis

Real world network data collection can be prone to errors. In Fig. 3, we show the effect of mislabeled data on ICP framework predictions. We consider the CORA data set with 30% of the nodes initially labeled and change labels randomly for 10%, 30% and 50% of the nodes in the training data. Figure 3 shows that mislabeled training data does not affect ICP error bounds. As we increase the percentage of mislabeled data, the efficiency of predictions decreases, since the percentage of singleton predictions decreases.

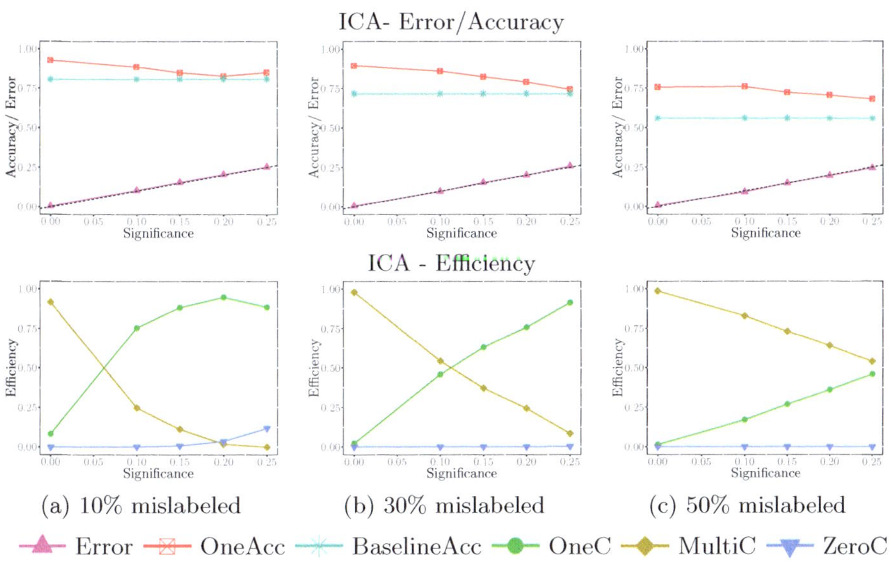

(a) 10% mislabeled (b) 30% mislabeled (c) 50% mislabeled

Fig. 3. Performance of ICP applied to CORA citation data with 30% of nodes initially labeled. ICA is used as the baseline. Note that 10%, 30% and 50% of training data mislabeled. ICP maintains the error bounds even when 50% of the training data is mislabeled. But the efficiency decrease as there are more errors.

6 Discussion and Conclusion

In this work we consider the problem of providing guaranteed error bounds for predictions in node classification algorithms. We use the CP framework, which works with a given prediction model to provide bounded error rates. We use ICP a more efficient variant of the CP framework and show how this can be applied to ICA, Label Propagation, GCN and DeepWalk algorithms to improve prediction accuracy and provide more reliable predictions. We evaluate performance of this framework using citation, social and biological networks and show that (1) Specified significance levels are maintained across all data sets and Algorithms, and (2) ICP can in fact improve accuracy of baseline algorithms. We conduct a perturbation analysis to show that `ICP` framework error bounds are not affected by the perturbations, rather the efficiency is affected.

References

1. Dashevskiy, M., Luo, Z.: Time series prediction with performance guarantee. IET Commun. **5**(8), 1044–1051 (2011)
2. Grover, A., Leskovec, J.: Node2vec: scalable feature learning for networks. In: Proceedings of the 22nd ACM SIGKDD International Conference on Knowledge Discovery and Data Mining, pp. 855–864. ACM (2016)
3. Haussler, D.: Probably approximately correct learning. University of California, Santa Cruz, Computer Research Laboratory (1990)
4. Johansson, U., Boström, H., Löfström, T.: Conformal prediction using decision trees. In: 2013 IEEE 13th International Conference on Data Mining (2013)
5. Kipf, T.N., Welling, M.: Semi-supervised classification with graph convolutional networks. arXiv preprint arXiv:1609.02907 (2016)
6. Lu, Q., Getoor, L.: Link-based classification. In: Proceedings of the 20th International Conference on Machine Learning (ICML 2003), pp. 496–503 (2003)
7. Makili, L., Vega, J., Dormido-Canto, S., Pastor, I., Murari, A.: Computationally efficient SVM multi-class image recognition with confidence measures. Fusion Eng. Des. **86**(6–8), 1213–1216 (2011)
8. Motl, J., Schulte, O.: The CTU Prague relational learning repository. arXiv preprint arXiv:1511.03086 (2015)
9. Nadeau, C., Bengio, Y.: Inference for the generalization error. In: Advances in Neural Information Processing Systems, pp. 307–313 (2000)
10. Namata, G., London, B., Getoor, L., Huang, B.: Query-driven active surveying for collective classification. In: 10th International Workshop on Mining and Learning with Graphs, p. 8 (2012)
11. Neville, J., Jensen, D.: Iterative classification in relational data. In: AAAI-2000 Workshop on Learning Statistical Models from Relational Data (2000)
12. Papadopoulos, H.: Inductive conformal prediction: theory and application to neural networks. In: Tools in artificial intelligence. IntechOpen (2008)
13. Perozzi, B., Al-Rfou, R., Skiena, S.: DeepWalk: online learning of social representations. In: Proceedings of the 20th ACM SIGKDD International Conference on Knowledge Discovery and Data Mining, pp. 701–710. ACM (2014)
14. Schapire, R.E., Freund, Y., Bartlett, P., Lee, W.S., et al.: Boosting the margin: a new explanation for the effectiveness of voting methods. Ann. Stat. **26**(5), 1651–1686 (1998)
15. Sen, P., Namata, G., Bilgic, M., Getoor, L., Galligher, B., Eliassi-Rad, T.: Collective classification in network data. AI Mag. **29**(3), 93–93 (2008)
16. Shafer, G., Vovk, V.: A tutorial on conformal prediction. J. Mach. Learn. Res. **9**(Mar), 371–421 (2008)
17. Vovk, V., Gammerman, A., Shafer, G.: Algorithmic Learning in a Random World. Springer, Heidelberg (2005)
18. Zafarani, R., Liu, H.: Social computing data repository at ASU (2009). http://socialcomputing.asu.edu
19. Zhu, X., Ghahramani, Z., Lafferty, J.D.: Semi-supervised learning using Gaussian fields and harmonic functions. In: Proceedings of the 20th International Conference on Machine Learning (ICML 2003), pp. 912–919 (2003)

Assessment of the Effectiveness of Random and Real-Networks Based on the Asymptotic Entropy

Raihana Mokhlissi[1(✉)], Dounia Lotfi[1], Joyati Debnath[2], and Mohamed El Marraki[1]

[1] LRIT, Rabat IT Center, Faculty of Sciences, Mohammed V University in Rabat, B.P 1014, Rabat, Morocco
mokhlissiraihana@gmail.com, {lotfi,marraki}@fsr.ac.ma
[2] Winona State University, Winona, MN 55987, USA
jdebnath@winona.edu

Abstract. Recently, the analytical study of the structural properties of complex networks has attracted increasing attention due to the growth of these real-world networks. Mathematical graph theory helps to understand and predict their behavior. This paper examines and compares the structural properties such as the small-world effect, the clustering coefficient and the degree distribution of Erdos-Renyi random networks with some real-world networks. Besides, we propose an algorithm to calculate the number and the entropy of spanning trees of these networks by using the electrically equivalent transformations. The result allows us to evaluate the robustness and the homogeneity of their structure. The proposed technique is efficient and more general compared to the classical ones.

Keywords: Spanning trees · Complex network · Erdos-Renyi network · Small-world network · Scale-free network · Electrically equivalent transformations · Complexity · Entropy

1 Introduction

In network analysis, the structural properties are studied to understand the mechanism and the behaviour of real-world networks. The type of a complex network can be defined using the value of some measures such as the average path length, the diameter, the clustering coefficient, the degree distribution, etc. These features play a crucial role in the recent studies of the network theory. In fact, the first attempt of Barabasi and Albert in their research on Scale-Free networks was that the degree distributions of many real-world networks have a power-law form [8]. On the other hand, Watts and Strogatz discovered that the construction of Small-World networks relies on a small average path length or a small diameter as a random network and large clustering coefficient [7]. Whereas, the first model of a random network was introduced by Erdos

H. Barbosa et al. (Eds.): *Complex Networks XI*, SPCOM, pp. 39–50, 2020.
https://doi.org/10.1007/978-3-030-40943-2_4

and Renyi [17,18]. There are two definitions of a random network: the first model is defined by a graph $G(V, E)$ with V nodes, which are connected with E randomly edges and the second model is a graph $G(V,p)$ where each pair of V nodes is connected with the probability $0 \leq p \leq 1$. This second model defines the known Erdos-Renyi network (ER). In Fig. 1, we illustrate some examples of ER networks with the same number of nodes (30 nodes) and different probability values. The expected number of edges for an ER network is $E = p\frac{V(V-1)}{2}$. This model exhibits some important properties shown in real-world networks. For example, we can get the connected random network if the probability $p \sim \frac{(\ln V)}{V}$. The average degree of the random network is $<z> = p(V-1) \simeq pV$. Its clustering coefficient is $C = p \ll 1$. This means that a random network does not have a high clustering coefficient. Its degree distribution follows the binomial distribution (or Poisson if V is large). The ER network is characterized by a small-world effect (Small diameter or small average path length). In this paper, an analytical comparison of the structural properties between real-world networks such as Zachary's karate club [15], Dolphin social network [13], Les Miserables [16], Books about US politics [11], Word adjacencies [14] and American College football [12] and other models of networks known in the literature such as the Flower network [1], the Mosaic network [4], the Koch Network [2], the Small-World Exponential Network [2,3], the Fractal Scale-Free Lattice [5], the Farey network [9], the Watts-Strogatz Network [7], the barabasi-albert Network [8] and the Erdos-Renyi random networks [17,18] is introduced in Sect. 2.

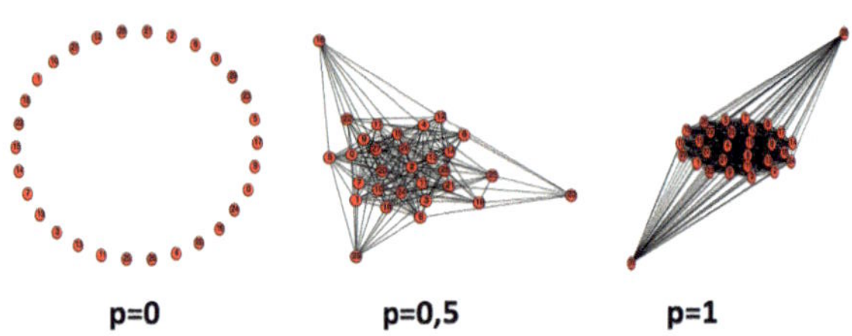

Fig. 1. Erdos-Renyi networks with 30 nodes and the probability values 0, 0.5 and 1

In addition, as a crucial structure invariant, the number of spanning trees of a connected network G or what is called by the complexity of a network [10], denoted by $\tau(G)$, will be helpful to predict the reliability [21] and the robustness of a network [20]. It is known that the problem of calculating the number of spanning trees has been solved by the "Matrix Tree Theorem" [22] of the physicist Kirchhoff. However, this theorem is not efficient for real-world networks having a large value of nodes and links because it will be infeasible to calculate the determinant of this matrix for large networks. Therefore, our motivation is to use a new method named "The electrically equivalent technique" [19] to study the number of spanning trees of real-world networks and random networks without using the Kirchhoff's matrix. This technique includes five transformations:

Parallel and Serial edges, Wye-Delta, Delta-Wye and Star-Mesh transformations in order to get the exact value of the number of spanning trees. In Sect. 3, we propose the algorithms of the electrically equivalent transformations to calculate the number of spanning trees of the examined networks mentioned above. As an application, the entropy of spanning trees of these networks is evaluated and compared in order to estimate which model is more effective.

2 Structural Properties of Complex Networks

In this section, we investigate the structural properties as the clustering coefficient C, the average path length l, the diameter D and the degree distribution of real-world networks mentioned in Table 1 and compare them with those of random networks having a similar number of nodes and edges.

Table 1. Structural properties of real and random networks.

N	Type of network	V	E	C_{real}	C_{random}	D_{real}	D_{random}	l_{real}	l_{random}
1	Zachary's karate club	34	78	0.570638	0.163772	5	5	2.408199	2.399286
2	Dolphin social network	62	159	0.258958	0.087321	8	5	3.356953	2.693283
3	Les Miserables	77	254	0.573136	0.083168	5	4	2.641148	2.488380
4	Watts-Strogatz network	100	500	0.102503	0.098122	4	4	2.235555	2.226060
5	Books about US politics	105	441	0.487526	0.075720	7	4	3.078754	2.394139
6	Word adjacencies	112	425	0.172840	0.046981	5	6	2.535553	2.558236
7	American College football	115	613	0.403216	0.102725	4	4	2.508161	2.250343
8	Fractal scale-free lattice	172	341	0.539960	0.006646	16	7	6.622943	3.777373
9	Barabasi-Albert network	213	1040	0.128816	0.111455	4	4	2.510541	2.595756
10	2-Mosaic networks	343	1024	0	0.011570	32	7	6.240033	3.471007
11	Koch network	513	768	0.818473	0.006092	10	13	5.157894	5.727294
12	Farey network	513	1023	0.692396	0.008893	9	11	5.449211	4.669691
13	2-Flower network	684	1024	0	0.003118	32	15	11.142830	5.845276
14	Small-world exponential	729	1092	0.760371	0.002891	11	14	7.010989	5.919793

From Table 1, we notice that the clustering coefficients of different real networks are larger than that of random networks, except two networks: 2-Flower networks and 2-Mosaic networks, because the neighbors of any node of these networks are never neighbors of one another. In general, we can say that the ER networks do not have high clustering coefficient compared to real networks. We notice also that the values of the diameter and the APL of random networks (ER networks) are almost the same as the real-world networks. Both the real network and the random network exhibit the same behaviour in the small-world property and for the degree distribution, its exact form for random networks is the binomial distribution. Whereas, for the 2-Flower network, Koch Network, 2-Mosaic network, Fractal Scale-Free Lattice and Barabasi-albert Network, their degree distribution follows the power-law form. The Small-World Exponential Network and Farey network, their degree distribution follows an exponential distribution and for the Watts-Strogatz Network, Dolphin social network, Les Miserables, Zachary's karate club, Books about US politics, Word adjacencies and American College football, the shape of their degree distribution is similar to that of a random network.

3 The Number and the Entropy of Spanning Trees

In this section, the proposed methodology for calculating the number of spanning trees, an example and its application for some real-world and random networks are presented.

3.1 Methodology

The technique of the electrically equivalent transformations is a new approach based on the knowledge of electrical networks. It simplifies the structure of a network and changes the weight of its edges [19]. The main objective of using this technique is to get the exact value of the number of spanning trees of complex networks. The electrically equivalent technique includes five transformations: Parallel and Serial edges, Wye–Delta, Delta–Wye and Star-Mesh transformations. Each one has its characteristics and its properties. Based on these five transformations, we propose 5 algorithms that facilitate the computation of the number of spanning trees of a connected network. These transformations must be applied in the following order:

- **Parallel edge:** Two parallel edges having the weights a and b change into a single edge with a new weight $a + b$. The number of spanning trees of the obtained graph remains the same: $\tau(G') = 1 \times \tau(G)$ (See Fig. 2). This transformation reduces the number of edges by deleting all the multiple edges. The Algorithm 1 generates the transformation of parallel edges.

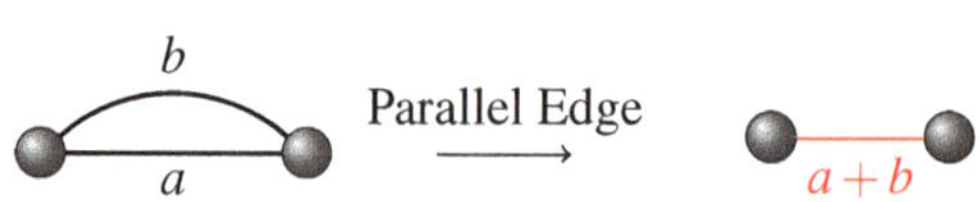

Fig. 2. Parallel edges transformation.

Algorithm 1: The algorithm of the parallel edge transformation

```
Function ParallelEdge(G)
    L1: List of pairs of nodes that are connected by multiple edges;
    n1: The length of L1 ;
    for i = 1 to n1 do
        w: The sum of the weights of all multiple edges between the pair of nodes of L1[i] ;
        Remove all multiple edges between the pair of nodes of L1[i] ;
        Add a new edge between the pair of nodes of L1[i] ;
        Assign the weight w to the new edge created between the pair of nodes of L1[i];
        τ(G) ← 1 * τ(G) ;
    Empty L1;
```

- **Serial edge:** Two serial edges with the weights a and b will be transformed into a single edge with a new weight $\frac{ab}{(a+b)}$. The number of spanning trees of G' will be: $\tau(G') = \frac{1}{a+b}\tau(G)$ (See Fig. 3). This transformation reduces the number of vertices and the edges by deleting the vertices having the degree '2' and the edges attached to it. The Algorithm 2 generates the serial edges transformation.

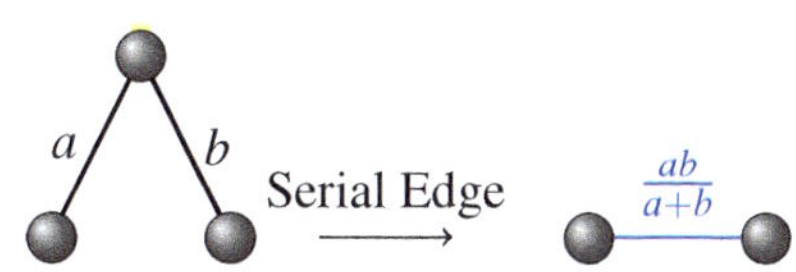

Fig. 3. Serial edges transformation.

Algorithm 2: The algorithm of the serial edge transformation

```
Function SerialEdge(G)
    L2: List of nodes having the degree 2 with their two neighbors;
    n2: The length of L2 ;
    for i = 1 to n2 do
        a, b : Weights of two edges between the node having the degree 2 and its two
          neighbors in L2[i] ;
        c ← (a * b)/(a + b);
        Remove the node having the degree 2 of L2[i];
        Add a new edge between two neighbors of L2[i];
        Assign the weight c to this new edge created between two neighbors of L2[i;
        τ(G) ← (a + b) * τ(G) ;
    Empty L2;
```

- **Wye-Delta transformation:** A wye graph with the weights a, b and c changes into a Delta graph with new weights x, y and z (See Fig. 4). Its number of spanning trees will be $\tau(G') = \frac{1}{a+b+c}\tau(G)$. This transformation reduces the number of vertices by deleting the vertices having the degree '3' and the number of edges does not change. The Algorithm 3 generates the transformation of Wye-Delta.

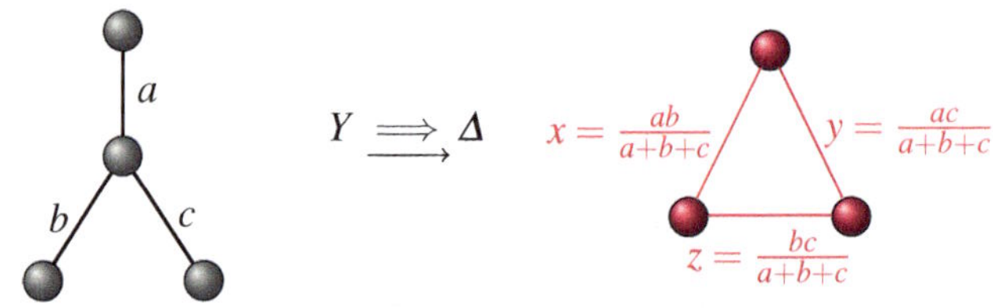

Fig. 4. Wye-Delta transformation.

Algorithm 3: The algorithm of Wye-Delta transformation

```
Function Wye − Delta(G)
    L3: List of nodes having the degree 3 with their three neighbors;
    n3: The length of L3 ;
    for i = 1 to n3 do
        a, b, c : Weights of three edges between the node having the degree 3 and its three
          neighbors in L3[i] ;
        x ← (a ∗ b)/(a + b + c);
        y ← (a ∗ c)/(a + b + c);
        z ← (b ∗ c)/(a + b + c);
        Remove the node having the degree 3 in L3[i];
        Add three new edges between the three neighbors in L3[i];
        Assign the weights x, y and z to the new three edges created between the edges
          having the weight a and b, the weight a and c and the weight b and c,
          respectively;
        τ(G) ← (a + b + c) ∗ τ(G) ;
    Empty L3;
```

- **Delta-Wye transformation:** This transformation is the converse of the Wye–Delta transformation. A Delta graph with the weights a, b and c changes into a wye graph with new weights x, y and z (See Fig. 5). Its number of spanning trees will be $\tau(G') = \frac{(ab+bc+ca)^2}{abc}\tau(G)$. This transformation produces a node of degree '3', while the degrees of Delta nodes will be decreased by 1. The Algorithm 4 presents the transformation of Delta-Wye.

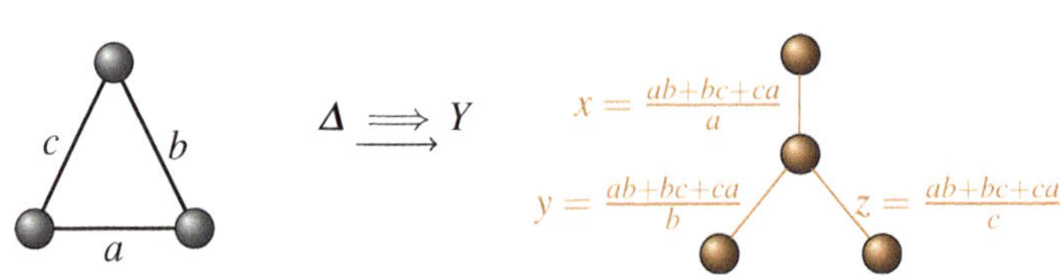

Fig. 5. Delta-Wye transformation.

Algorithm 4: The algorithm of Delta-Wye transformation

```
Function Delta − Wye(G)
    L4: list of triangles in G;
    n4: The length of L4 ;
    for i = 1 to n4 do
        a, b, c : Weights of the three edges of a triangle of L4[i] ;
        x ← (a ∗ b + b ∗ c + a ∗ c)/a;
        y ← (a ∗ b + b ∗ c + a ∗ c)/b;
        z ← (a ∗ b + b ∗ c + a ∗ c)/c;
        Remove all the three edges of a triangle of L4[i];
        Add a new node in G;
        Add three new edges between the new node and three nodes of a triangle of L4[i];
        Assign the weights x, y and z to the new three edges created between the edges
          having the weight a and b, the weight a and c and the weight b and c,
          respectively;
        τ(G) ← (a ∗ b ∗ c)/(a ∗ b + b ∗ c + a ∗ c)^2 ∗ τ(G);
    Empty L4;
```

- **Star-Mesh transformation:** A star graph S_n with n vertices and the weights $a_1, a_2, ..., a_n$ changes into a complete graph K_n with new weight for the edge v_iv_j $(i \neq j)$: $x_{i,j} = \frac{a_i a_j}{\sum_{k=0}^{n} a_k}$. The number of spanning trees will be $\tau(G') = \frac{1}{\sum_{k=0}^{n} a_k}\tau(G)$. As an example, Fig. 6 presents the transformation of Star graph S_4 to the Complete graph K_4. This transformation reduces the number of vertices by deleting the vertices having the degree '$n > 3$'. We consider the transformation of Star-Mesh as a generalization of serial-edge and Wye-Delta transformation. The Algorithm 5 generates the transformation of Star-Mesh.

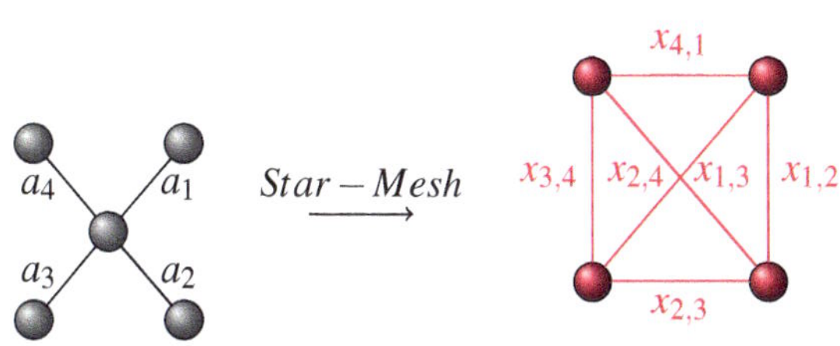

Fig. 6. Star-Mesh transformation.

Algorithm 5: The algorithm of Star-Mesh transformation

```
Function Star − Mesh(G)
    L5: Sorted list of nodes having the degree > 3 with their neighbors;
    n5: The length of L5 ;
    for i = 1 to n5 do
        t: The size of L5[i];
        s ← 0;
        for j = 2 to t do
            a(j) : Weight of the edge between the node having the degree greater than 3
              and the neighbor j;
            s ← s + a(j);                          // Summing all the weights a(j)
        for j = 1 to t − 1 do
            for k = j + 1 to t do
                x(j)(k) ← a(j) ∗ a(k)/s ;
                Add new edge between two edges that have the weight a(j) and a(k);
                Assign the weight x(j)(k) to the new edge;
        Remove the node having the degree greater than 3 in L5[i];
        τ(G) ← s ∗ τ(G) ;
    Empty L5;
```

3.2 Example

As an application of the electrically equivalent technique, we give an example in the Fig. 7. We enumerate the spanning trees of the network G by applying four electrically equivalent transformations. The corresponding computations of the conductances are as follows:

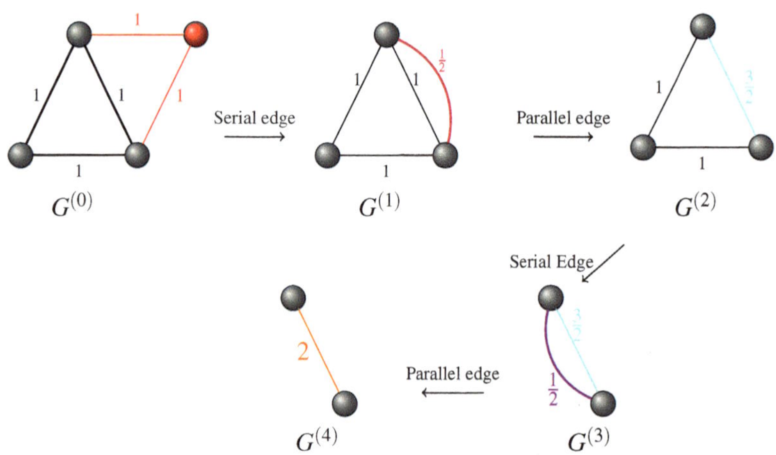

Fig. 7. An example of the electrically equivalent transformations.

1. We start by the conductance "1" for each edge of the original network, then the corresponding conductance of the transformed network for serial edge is $\frac{1.1}{1+1} = \frac{1}{2}$.

2. For two parallel edges with conductances 1 and $\frac{1}{2}$, the conductance of the new edge is the sum of two original conductances, i.e. $1 + \frac{1}{2} = \frac{3}{2}$.
3. When two serial edges with conductances 1 are merged into a new edge, its conductance is $\frac{1.1}{1+1} = \frac{1}{2}$.
4. For two parallel edges, the conductance of the new edge is $\frac{3}{2} + \frac{1}{2} = 2$.

Combining the above four transformations, the weighted number of spanning trees in the network G is calculated as follows:

1. $\tau(G) = \tau(G^{(0)})$.
2. $\tau(G) = 2 \times \tau(G^{(1)}) \rightarrow$ *Serial edge.*
3. $\tau(G) = 2 \times 1 \times \tau(G^{(2)}) \rightarrow$ *Parallel edge.*
4. $\tau(G) = 2 \times 2 \times \tau(G^{(3)}) \rightarrow$ *Serial edge.*
5. $\tau(G) = 4 \times 1 \times \tau(G^{(4)}) \rightarrow$ *Parallel edge.*
6. $\tau(G) = 4 \times 2 = 8$.

Then, the original network G has 8 spanning trees according to the factors of those transformations.

3.3 Application

As an application of the number of spanning trees of a network, there is a measure that characterizes its structure and describes the exponential growth of the number of spanning trees $\tau(G)$ with the number of nodes V_G [20], named by the entropy of spanning trees of a network G or the asymptotic complexity of G, denoted by ρ_G. This constant evaluates the robustness of a network, which is related to its capacity to withstand random changes, failures and perturbations in its structure over evolutionary time. The best known mathematical model of the network robustness is offered by the percolation theory [6]. In this work, we propose a new measure of the entropy to quantify the robustness of a network, which is defined as:

$$\rho_G = \lim_{V_G \to \infty} \frac{\ln |\tau(G)|}{|V_G|} \tag{1}$$

The most robust network is the network that has the highest entropy, because the increase of the number of spanning trees provides more possibilities of connecting two nodes related by defective links, that ensures a good robustness of a network. In Table 2, applying the electrically equivalent technique, we give the results of the number and the entropy of spanning trees for some real-world networks. In Table 3, we give the results of the number and the entropy of spanning trees for Erdos-Renyi random networks having the same number of nodes and edges as the presented networks in Table 2. The only condition for calculating the number of spanning trees is that all these real and random networks must be connected. In Fig. 8, we compare the entropy of spanning trees of some real and random networks. We can see that the entropy of random networks is almost larger than those of real-world networks due to their large number of spanning trees. Consequently, the random networks are robust to random deletion of edges and their structure is more homogeneous than the real-world networks having the same number of nodes and links.

Table 2. The number and the entropy of spanning trees of real networks.

N	Type of network	V	E	τ_{real}	ρ_{real}
1	Zachary's karate club	34	78	$5.09099632302 \times 10^{15}$	0.3616
2	Dolphin social network	62	159	$2.17175713551 \times 10^{32}$	0.7445
3	Les Miserables	77	254	$2.03974706969 \times 10^{42}$	0.9742
4	Watts-Strogatz network	100	500	$9.16480112874 \times 10^{93}$	2.1635
5	Books about US politics	105	441	$5.55633429016 \times 10^{82}$	1.9052
6	Word adjacencies	112	425	$6.8594178138 \times 10^{76}$	1.7692
7	American College football	115	613	$7.74002476226 \times 10^{111}$	2.5763
8	Fractal scale-free lattice	172	341	$5.78960446186 \times 10^{76}$	1.0397
9	Barabasi-Albert network	213	1040	$2.98840370474 \times 10^{186}$	2.0158
10	2-Mosaic networks	343	1024	$2.006582604 \times 10^{205}$	1.3862
11	Koch network	513	768	$1.39008452377 \times 10^{122}$	0.5493
12	Farey network	513	1023	$9.90418468985 \times 10^{209}$	0.9458
13	2-Flower network	684	1024	$2.006582604 \times 10^{205}$	0.6931
14	Small-world exponential	729	1092	$4.7004205677 \times 10^{173}$	0.5493

Table 3. The number and the entropy of spanning trees of random networks.

N	V	E	τ_{random}	ρ_{random}
1	34	78	$1.05860958932 \times 10^{18}$	0.4150
2	62	159	$1.45016355662 \times 10^{37}$	0.8556
3	77	254	$5.99585272075 \times 10^{55}$	1.2843
4	100	500	$7.05245964064 \times 10^{93}$	2.1609
5	105	441	$1.43050550088 \times 10^{89}$	2.0528
6	112	425	$1.65635617438 \times 10^{89}$	2.0543
7	115	613	$4.62483134316 \times 10^{110}$	2.5481
8	172	341	$7.19229226742 \times 10^{78}$	1.0556
9	213	1040	$1.05596179974 \times 10^{196}$	2.1190
10	343	1024	$1.66480921312 \times 10^{234}$	1.5723
11	513	768	$3.94135015033 \times 10^{159}$	0.7163
12	513	1023	$1.26835361025 \times 10^{236}$	1.0597
13	684	1024	$1.33756890122 \times 10^{215}$	0.7241
14	729	1092	$3.68796589012 \times 10^{229}$	0.7250

4 Discussion

According to the above results, we deduce that the random network is the wrong model for most real networks because its structural properties are different from those of the real networks, but it is considered as the most robust model under random attack due to its high entropy and its large number of spanning trees.

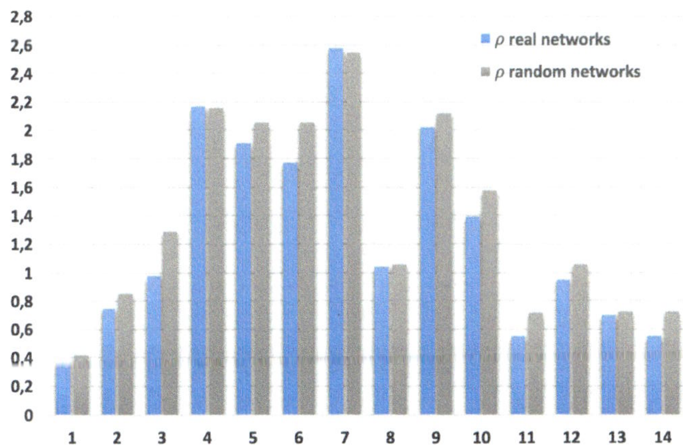

Fig. 8. Comparison between the entropy of spanning trees of real and random networks.

5 Conclusion

In this paper, we have proposed an analytical study of some real-world networks and random networks having the same number of nodes and links. We have discussed their structural properties. Then, we have calculated their number of spanning trees by using the electrically equivalent technique. Finally, we have evaluated and compared their entropy of spanning trees in order to predict which network is more robust. As a perspective, we intend to work on other properties of new complex networks.

References

1. Mokhlissi, R., Lotfi, D., Debnath, J., El Marraki, M.: An innovative combinatorial approach for the spanning tree entropy in Flower Network. In: El Abbadi, A., Garbinato, B. (eds.) International Conference on Networked Systems, pp. 3–14. Springer, Cham (2017)
2. Mokhlissi, R., Lotfi, D., Debnath, J., El Marraki, M., El Khattabi, N.: The evaluation of the number and the entropy of spanning trees on generalized small-world networks. J. Appl. Math. **2018**, 1–7 (2018)

3. Mokhlissi, R., Lotfi, D., Debnath, J., El Marraki, M.: Complexity analysis of "small-world networks" and spanning tree entropy. In: Cherifi, H., Gaito, S., Quattrociocchi, W., Sala, A. (eds.) International Workshop on Complex Networks and their Applications, pp. 197–208. Springer, Cham (2016)
4. Mokhlissi, R., Lotfi, D., El Marraki, M.: A theoretical study of the complexity of complex networks. In: 2016 7th International Conference on Sciences of Electronics, Technologies of Information and Telecommunications (SETIT), pp. 24–28. IEEE (2016)
5. Mokhlissi, R., Lotfi, D., El Marraki, M., Debnath, J.: The structural properties and the spanning trees entropy of the generalized Fractal Scale-Free Lattice. J. Complex Netw. cnz030 (2019)
6. Stauffer, D., Aharony, A., Redner, S.: Introduction to percolation theory. Phys. Today **46**, 64 (1993)
7. Watts, D.J., Strogatz, S.H.: Collective dynamics of 'small-world' networks. Nature **393**(6684), 440 (1998)
8. Barabási, A.L., Albert, R.: Emergence of scaling in random networks. Science **286**(5439), 509–512 (1999)
9. Zhang, Z., Comellas, F.: Farey graphs as models for complex networks. Theor. Comput. Sci. **412**(8–10), 865–875 (2011)
10. Knuth, D.E.: Aztec diamonds, checkerboard graphs, and spanning trees. J. Algebraic Comb. **6**(3), 253–257 (1997)
11. Krebs, V.: Books about US politics (2004, unpublished). http://www.orgnet.com
12. Girvan, M., Newman, M.E.: Community structure in social and biological networks. Proc. Nat. Acad. Sci. **99**(12), 7821–7826 (2002)
13. Lusseau, D., Schneider, K., Boisseau, O.J., Haase, P., Slooten, E., Dawson, S.M.: The bottlenose dolphin community of Doubtful Sound features a large proportion of long-lasting associations. Behav. Ecol. Sociobiol. **54**(4), 396–405 (2003)
14. Newman, M.E.: Finding community structure in networks using the eigenvectors of matrices. Phys. Rev. E **74**(3), 036104 (2006)
15. Zachary, W.W.: An information flow model for conflict and fission in small groups. J. Anthropol. Res. **33**(4), 452–473 (1977)
16. Knuth, D.E.: The Stanford GraphBase: A Platform for Combinatorial Computing, pp. 74–87. ACM Press, New York (1993)
17. Erdős, P.: On the evolution of random graphs. Publ. Math. Inst. Hung. Acad. Sci. **5**, 17–61 (1960)
18. Erdős, P., Rényi, A.: On the strength of connectedness of a random graph. Acta Math. Hung. **12**(1–2), 261–267 (1961)
19. Teufl, E., Wagner, S.: On the number of spanning trees on various lattices. J. Phys. A: Math. Theor. **43**(41), 415001 (2010)
20. Garcia, A., Noy, M., Tejel, J.: The asymptotic number of spanning trees in d-dimensional square lattices. J. Comb. Math. Comb. Comput. **44**, 109–114 (2003)
21. Bistouni, F., Jahanshahi, M.: Reliability analysis of Ethernet ring mesh networks. IEEE Trans. Reliab. **66**(4), 1238–1252 (2017)
22. Chaiken, S., Kleitman, D.J.: Matrix tree theorems. J. Comb. Theory Series A **24**(3), 377–381 (1978)

Unsupervised Strategies to Network Topology Reconfiguration Optimization with Limited Link Addition

William R. Paiva, Paulo S. Martins(✉), and André F. de Angelis

School of Technology, University of Campinas (UNICAMP), Limeira, São Paulo, Brazil
will.unicamp@gmail.com, {paulo,andre}@ft.unicamp.br

Abstract. The evolution of networks may lead to an undesired configuration of its properties due to different forces driving their growth. A planned topology change aiming to bring the properties to an acceptable range is called *Network Topology Reconfiguration Optimization with Limited Link Addition* (NTRLA). We faced an NTRLA problem when we were investigating ways to improve the efficiency of large power grids. In the search for solutions, we developed strategies to add new edges in unsupervised automatic applications. The strategies were tested over thousands of realizations of random and scale-free networks, as well as over a power grid map by means of computer programs that have implemented them and collected the efficiency of the networks along the change processes. An attempt to determine the maximum possible performance provided a comparison reference to the strategies. We show that the best result was obtained by linking the node with larger closeness to the one with the smallest closeness, although this procedure is not scalable. Furthermore, the strategy that links nodes whose betweenness values are near to the median was shown to be scalable and easy to implement, even if it has not delivered the best performance. We have found that the min-cut procedure has improved the results for each strategy. It became apparent how the network topology plays a fundamental role in NTRLA problems, what prompted for new insights and further research work in the field of complex networks.

1 Introduction

The growth of real physical networks is driven by different forces under the constraints of the environment, as discussed by [1]. The networks adapt themselves to the conditions of their domains while evolving. Consequently, the properties of established networks may deviate far from their optimal values, thus harming their ability to efficiently deal with their expected operations as in infrastructure systems where the network features need to be within a desirable range.

In these cases, a planned change can be carried out to bring specific properties to satisfactory ranges. Therefore, edges and nodes are added to or removed

H. Barbosa et al. (Eds.): *Complex Networks XI*, SPCOM, pp. 51–59, 2020.
https://doi.org/10.1007/978-3-030-40943-2_5

from the network and this process leads to a modification in the topology, hopefully keeping the other network properties unchanged. This change, when applied under resource constraints, is called *network topology reconfiguration optimization with limited link addition* (NTRLA) [2].

While looking for ways of improving the power grid infrastructure to reduce or avoid electrical blackouts, we have noticed that NTRLA could be an alternative solution, since we were working with a spatial network where the power plants, stations, or even the existing high voltage lines could not be moved. Thus, we considered the addition of a single redundant line in the grid. The building of such a line is usually a costly alternative that requires a series of considerations about its effects on the environment and people, and may even prove infeasible in some scenarios. However, from the theoretical point of view, the investigation of this course of action is valid and may unveil new possibilities in this field.

Our problem can be stated as the search for ways to choose the end nodes of a new link in a spatial network in order to improve its efficiency. There are many forms to perform this process, but some of them are not suitable for large networks because of the cost, the length of the lines, or other factors.

In this work, we present general strategies to perform NTRLA increasing the network efficiency by inserting one or, at most, a few new edges. We considered a scenario of large networks that are sparse. Indeed, if the network is already dense, the efficiency gains that could be reached by a handful of new edges are very small.

The exhaustive test of all possibilities for the insertion of a set of edges is an NP-hard problem. On the other hand, it is relatively easy to optimize a specific given network by means of guided processes. Thus, we highlight that we have focused on general strategies, that could be automatically applied to a set of network instances without supervision. The strategies can be used when there is no physical network established, as in the model-based research.

There are three particular network configurations that need to be considered because they are efficient but not feasible for large systems: the complete mesh, the star-like, and the onion-like shapes. The first and the second are not scalable, and the third can seldom be built with the inclusion of a few edges into a sparse non-hierarchical network. Therefore, we have avoided strategies that converged towards these shapes.

We proposed a set of strategies and tested them by means of computational simulations over two network models and one real map. The models are very well known in the Complex Networks field: scale-free (SF) and random network (RN) [3,4], whereas the map represents a real country-area power grid (PG).

Within this context, we noticed that the work of Li et al. [2] aimed at improving the robustness and efficiency of SF networks by reconfiguring them, as they state that little attention has been paid to the problem of how to improve the robustness of existing networks. They showed that NTRLA is an NP-hard problem, which led to our decision to avoid methods that test all new edge possibilities in a network. The authors mainly evaluated approaches based on the degree of nodes, even when they tested two specific methods: one based on the degree fitness, a function of the neighbor's degree; others based on the creation of node protection cycles to avoid node isolation during an attack. These two methods outperformed all other degree-based approaches they have tried, including the connection of the high-degree nodes. Their work is narrowed to SF networks and node degree-dependent methods. In our work, we also considered other network models.

Zhao and Xu [5] tested three approaches to add edges to the network, focusing on its robustness: (1) randomly chosen new edges; (2) edges linking the high degree nodes; (3) edges linking the low degree nodes. Their third approach obtained the best results. Also looking for robustness, Zhuo et al. [6] defined redundant allocation of edges based on the degree of nodes. After a series of configurations, they found that linking the lowest-degree nodes offered the best results against network attacks in SF networks. As both works delivered the best results by linking low degree nodes, we also considered this approach in our work.

Concerned with attacks to networks, Louzada et al. [7] have proposed a reconfiguration method that prohibits changes to node degrees and is focused on swaps between edges. They arrived at onion-like structures, a topology where each layer is composed of nodes connected with nodes of the same degree, with few connections between layers. As stated, we avoided methods that could drive the network structure to a star or onion-like topology, so we have not reproduced their work.

The remainder of this paper is organized as follows: Sect. 2 presents the proposed strategies for increasing network efficiency. Section 3 deals with the background information, i.e. the metrics and procedures used to assess the proposed strategies. Section 4 presents and discusses the main findings and Sect. 5 summarises our conclusions.

2 Proposed Approach

In this section, we introduce our proposed strategies to insert a new edge in a network, emphasizing that all the networks are connected, i.e., there are no isolated nodes. The new edge inserted into the network will link two existent nodes and we select them as indicated in Table 1.

The LLo, MLo, NNo, and NTLo strategies consider the betweenness of the nodes or of their neighbors, LDg inspects the node degrees, and HLC strategy takes into account the topological position of the nodes. The RCalc is an exhaustive search over the network, looking for the best pair to connect.

In fact, we doubled the number of assessed strategies because we also test a modification of them where the two nodes to be linked are always chosen from two weakly connected network partitions. These partitions were found using the spectral bisection or minimum cut (*min-cut*) procedure, as proposed by [8]. This algorithm finds the minimum set of edges that could split a graph into two subgraphs of similar size. The original purpose of it was to obtain an improvement in the robustness of the network, but we noticed a consistent gain in the efficiency

Table 1. Strategies to insert new edges: in the first column, each figure is an example of the insertion of a new edge (*bold line*) using the related strategy.

Fig	Strat.	Summary	How to Add a New Edge	Metric
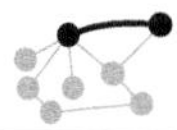	HLC	High to Lower Closeness	link the node with the larger closeness with the one with the smallest closeness	Position
	LDg	Lowest Degree	link the two nodes with lowest degree	Degree
	LLo	Lowest Betweenness	link the two nodes with the lowest betweenness	Betweenness
	NTLo	Non-Terminal Lowest Betweenness	link the two non-terminal nodes with the lowest betweenness	Betweenness
	MLo	Median Betweenness	link the two nodes with betweenness values nearest to the median	Betweenness
	NNo	Neighbor Nodes	link the lowest-degree neighbor of each of the two nodes with bigger betweenness	Betweenness
Not Available	RCalc	Reference Calculus	link any two nodes that cause the largest increase in efficiency	Combination

as well. Using min-cut, each node is selected from each of the sub-graphs found, while the reasoning of each strategy keeps unchanged. From now on, we identify the use of the min-cut procedure with a strategy by "+mc". We assessed all those variants with computational simulations running over network models.

3 Background

In this section, we discuss the models, the measurements and the procedures that were used to assess the strategies.

3.1 Network Models and Measures

We used three network models: (i) a scale-free (SF) model based on the Barabási-Albert proposal [9]; (ii) a random model (RN) following the Erdös-Rényi template [10]; (iii) a real-world sample of a country-area power grid (PG) [11]. As the latter has a fixed topology, we used the same number of nodes and edges of it to built the SF and RN simulations, ending up with 737 nodes and 1123 edges in all experiments. Two measures were used in this work, as follows:

Betweenness ($b(v)$): it indicates the ratio between the number of shortest paths among the network nodes that pass by v and the total of shortest paths in the network, highlighting the importance of the node in the network structure. The betweenness is obtained by Eq. (1) [12]:

$$b(v) = \sum_{s,t \in V} \frac{\sigma(s,t|v)}{\sigma(s,t)} \tag{1}$$

where V is the set of nodes, $\sigma(s,t)$ is the number of shortest (s,t)-paths, and $\sigma(s,t|v)$ is the number of those paths passing through some node v other than s,t. If $s = t$, $\sigma(s,t) = 1$, and if $v \in s,t$, $\sigma(s,t|v) = 0$.

Global Efficiency ($E(G)$): it shows how shorter are all the paths in the networks, i.e., how easy is the flow. This value was recalculated for each topological change of the simulated network, as indicated by Eq. 2 [13]:

$$E(G) = \frac{1}{N(N-1)} \sum_{i \neq j \in G} \frac{1}{d_{ij}}, \tag{2}$$

where G is the graph that represents the network, N the number of nodes, and d_{ij} is the shortest path between edges i and j.

3.2 Assessment Procedures

We developed a computer program to create realizations of the SF and RN models and to load the PG map in the computer memory. All networks were represented as undirected unweighted connected graphs. The program assessed the efficiency of the networks as the new edges were added using the proposed strategies. In all cases, we did such evaluation with and without the min-cut procedure. We configured the software to generate one thousand realizations of the networks for each SF and RN model. As PG is a map of an existent network, we used just one graph mapping the real-world sample. The tests were conducted as follows.

The software creates or loads a graph of the network and does incremental changes in its topology, adding a new edge between strategy-dependent chosen pair of nodes, in order to assess the variation of network efficiency. Then, this edge is fixed to the network and the procedure is repeated until the graph ends up with 50 new edges. When there was more than one candidate pair of nodes to be connected, the program randomly selects one of the pairs.

The LLo strategy takes into account all the nodes of the network and connects the lowest betweenness pair. We assign *betweenness zero* to terminal nodes (i.e. those with degree 1). Consequently, this strategy tends to first connect terminal nodes if they are present in the network. Thus, we also tested the NTLo strategy, that has the same reasoning as LLo, but is restricted to non-terminal nodes.

The PG graph received an extra evaluation process due to the uniqueness of the sample. This was called the RCalc strategy. In this case, the program places a new edge into the graph and calculates the efficiency increase; then, the edge is removed and the procedure is repeated for all possible pairs. The edge that

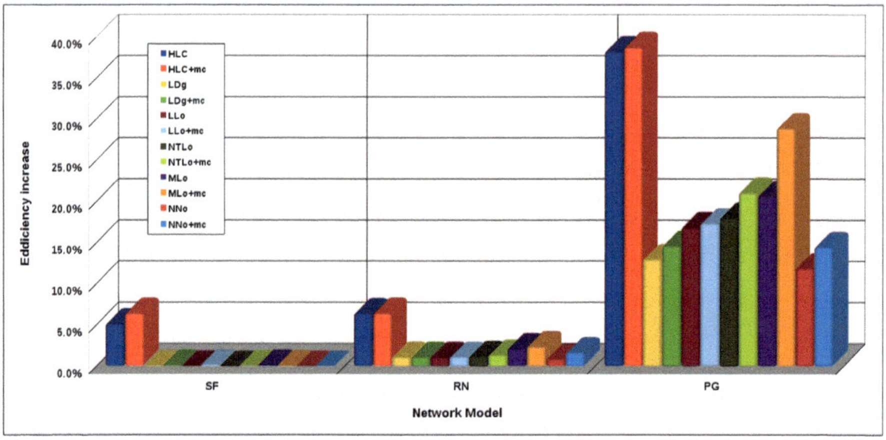

Fig. 1. Network efficiency increase after inserting 50 edges according to each proposed strategy; all networks started with 737 nodes and 1123 edges; the results for SF and RN models are the average of 1000 realizations, whereas the results for PG network are from the unique instance.

brings the larger gain is fixed to the graph and the procedure is repeated over the new graph until 50 new edges are inserted into the network. If two or more edges bring the same gain, we randomly choose one of them. The goal of this strategy was find an upper bound reference for our work. The strategy itself is very CPU-consuming and does not scale. Furthermore, each edge is independently added to the graph and, thus, this procedure can lead to wrong conclusions about optimal solutions because the algorithm explores local peaks of efficiency.

4 Results and Discussion

The efficiency gain differs significantly for each network, being the largest in the PG model (Fig. 1 and Table 2). The relatively satisfactory performance of the

Table 2. Summary of strategies and ranking of maximum performance

Model	Strategy	Gain %	Remarks
PG	HLC + mc	39	Limited number of edges
	MLo + mc	29	Sparser distribution of edges
	NTLo + mc	21	No limitations on # repetitions
RN	HLC + mc	6.5	Same as above
	MLo + mc	2.5	
	NTLo + mc	1.4	
SF	HLC + mc	6.5	Same as above
	MLo + mc	0.1	
	NTLo + mc	0.1	

strategies in the PG model was not observed in the other network models. The use of the min-cut procedure increased the efficiency of all strategies. Following, we discuss the four strategies that offered the best performances:

- **HLC**: its largest gains were verified in the PG model; HLC outperformed all other strategies for the two models and for the map; it achieved its best result in the PG with min-cut, increasing the efficiency up to 39%. The HLC strategy cannot be applied for a large number of repetitions, as it tends to form a star-like topology; however, for a limited amount of new edges, it is an attractive and feasible option.
- **MLo**: it was the second-best strategy, as it yielded gains around 29% in the PG, 2.5% in the RN and 0.1% in the SF, thus reinforcing the notion that strategies are model-dependent; MLo and MLo + mc have resulted in a sparser distribution of the new edges in the graph, which avoids unfeasible topologies and thus may be regarded as valuable alternatives.
- **NTLo**: it presented a 21% efficiency gain in PG and therefore it was the third-best result; furthermore, NTLo has no severe limitations on scalability and terminal nodes are not affected. These results dropped to 1.4% for the RN and 0.1% for the SF.
- **NNo**: it had the smallest gain in efficiency; its graph presented no slope (Fig. 2) for the first 17 new edges; on the other hand, NNo and NNo+mc create a series of redundant paths in the network.

Each other strategies allowed a better result in PG, a smaller gain in RN, and a very poor performance in SF networks. In any case, their performance was much lower than the ones just discussed.

The **RCalc** is not a general strategy for edges addition. It is a sort of brute force scam of the combinatorial explosion of probabilities to insert an edge. It is a NP-hard problem, so the scalability of RCal is limited. Furthermore, it demands the analysis of an identifiable graph, as the results for one network are not directly comparable to any other. There is no guarantee that RCalc finds the optimal global efficiency because the incremental process of fixing edges may lead to optimal local values.

However, it provided us a comparison reference to the other strategies. In Fig. 2, we plotted the increase in efficiency for each strategy as new edges were added to the PG network. RCalc, as expected, outperformed the other methods with a gain around 9% for the first edge and ending up with about 76% for the set of 50 edges. As RCalc also suffered the trend to form star-like topologies that are very efficient, it can only be used with limited scalability.

All strategies, but RCalc, presented a low computational cost because they have required not high-performance computers nor long-time processing. The use of RCalc to study thousands of realizations of networks is unfeasible given the time demanded. Nevertheless, as the PG is a single map, we could examine it in more detail by applying the RCalc to it.

We already saw that HLC presented the best performance for PG over all other strategies, except RCalc. But HLC and HLC + mc are not quite scalable, leading us to consider **MLo + mc** a good choice for a generic strategy. MLo + mc

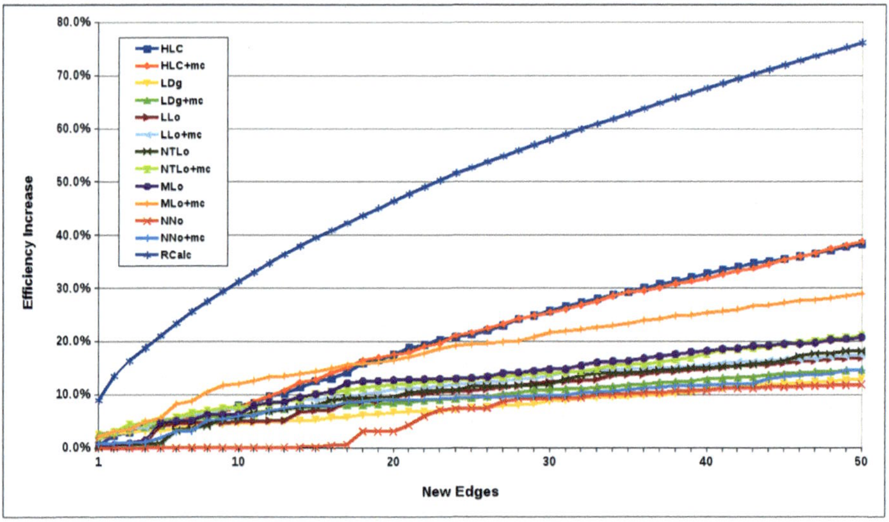

Fig. 2. Network efficiency increase for PG model due the insertion of 50 edges according to each proposed strategy; the RCalc line was added as a reference; the network started with 737 nodes and 1123 edges

reached a gain around 29% with the whole set of 50 new edges and it does not have the tendency to centralization presented by the other methods.

5 Summary and Conclusions

We were looking for ways to improve large power grids and we faced an NTRLA problem that is NP-hard if we try methods that explore all possible combinations of nodes to be linked. We have proposed a number of general unsupervised strategies to improve the efficiency of networks, subject to the constraint of keeping the original topology in place and only adding new edges to them.

After a number of tests using two well-known models and a real map, we were able to rank the strategies based on their performance. We have employed a reference calculation to obtain an insight into the maximum possible performance, but it has two strong limitations: it is not a scalable procedure and it is not guaranteed to find global optimal values. As we have noticed that the min-cut procedure effectively improved each of the strategies, we suggest that attempts to rise efficiency ought to consider its use.

The strategy that links the node with larger closeness with the one with the smallest closeness (HLC) has gotten the best overall results and is suitable for limited-scale insertion of edges. The linking nodes with near-to-median betweenness (MLo) appeared to be a good choice because is significantly scalable and easy to be calculated, although it has not presented the best performance.

This work has shown that the underlying network topology has a strong influence on the behavior of strategies and it provided tools and clues to researchers interested in NTRLA.

Acknowledgments. W. Paiva thanks CAPES for the financial support. The authors also thanks Professor Vitor Rafael Coluci for the use of the *Beowulf Cluster* acquired by FAPESP grant 2010/50646-6.

References

1. Pereira, V.H., Martins, P., de Angelis, A.F., Timóteo, V.S.: A simple model for cascading failures in scale-free networks. In: SBAI-DINCON (2013)
2. Li, L., Jia, Q.-S., Guan, X., Wang, H.: Enhancing the robustness and efficiency of scale-free network with limited link addition. KSII Trans. Internet Inf. Syst. **06**, 1333–1353 (2012)
3. Newman, M.E.J.: The structure and function of complex networks. SIAM Rev. **45**, 167–256 (2003)
4. Boccaletti, S., Latora, V., Moreno, Y., Chavez, M., Hwang, D.: Complex networks: structure and dynamics. Phys. Rep. **424**, 175–308 (2006)
5. Zhao, J., Xu, K.: Enhancing the robustness of scale-free networks, CoRR, vol. abs/0905.2227 (2009)
6. Zhuo, Y., Peng, Y., Long, K., Liu, Y.: On allocating redundancy links to improve robusteness of complex communication network. In: Proceedings of the SPIE, vol. 7633 (2009)
7. Louzada, V.H.P., Daolio, F., Herrmann, H.J., Tomassini, M.: Smart rewiring for network robustness. J. Complex Netw. **1**(2), 150–159 (2013)
8. Rosato, V., Bologna, S., Tiriticco, F.: Topological properties of high-voltage electrical transmission networks. Electr. Power Syst. Res. **77**(2), 99–105 (2006)
9. Barabási, A.-L., Albert, R.: Emergence of scaling in random networks. Science **286**, 509–512 (1999)
10. Erdos, P., Renyi, A.: On random graphs. Publ. Math. Debrecen **6**, 290–297 (1959)
11. Paiva, W.R., Pedro, P.S.M., Angelis, A.F.: Increasing the efficiency of the Brazilian power grid. In: SBAI/DINCON 2015, p. 7 (2015)
12. Brandes, U.: On variants of shortest-path betweenness centrality and their generic computation. Soc. Netw. **30**, 136–145 (2008)
13. Crucitti, P., Latora, V., Marchiori, M.: Model for cascading failures in complex networks. Phys. Rev. E **69**, 045104 (2004)

Embedding of Signed Networks Focusing on Both Structure and Relation

Tsuyoshi Murata(✉) and Hiroki Arihara

Department of Computer Science, School of Computing,
Tokyo Institute of Technology,
W8-59 2-12-1 Ookayama, Meguro, Tokyo 152-8552, Japan
murata@c.titech.ac.jp, arihara@net.c.titech.ac.jp
http://www.net.c.titech.ac.jp/

Abstract. Network embedding is a method for learning representation of nodes in given networks as low-dimensional vectors which preserves the proximity in the networks. Most of the research on network embedding are for simple networks without edge signs. However, relations among users in real social media are often represented as signed networks composed of positive and negative relationships. In this paper, we propose a new method for embedding signed networks focusing on both network structure and node relation (sign of connecting edges). Experimental results show that our method outperforms previous methods for the tasks of relation prediction and link prediction.

Keywords: Signed network · Network embedding · Link prediction · Relation prediction

1 Introduction

Social media is often represented as a network of its users (nodes) connected with their relations (edges). Network embedding is a method for learning representation of nodes in a given network as low-dimensional vectors which preserves the proximity in the given networks. DeepWalk [1], LINE [3] and node2vec [7] are famous examples of network embedding methods. For some tasks on networks such as node classification and link prediction, the methods based on network embedding often outperform those based on traditional methods based on network topology. Many machine learning researchers have been focusing on network embedding recently.

As is often the case with real society, relations between users in social media can be friendly or hostile. Signed networks represent such relations as nodes connected with positive edges or negative edges. Analyzing signed network is important for the recommendation of potential friends and for modeling the social media in order to predicting its future structure. Both of network structure and node relation (sign of connecting edges) are important for the embedding of signed networks.

H. Barbosa et al. (Eds.): *Complex Networks XI*, SPCOM, pp. 60–69, 2020.
https://doi.org/10.1007/978-3-030-40943-2_6

Previous methods for the embedding of signed network are classified into the following two approaches: (i) methods focusing on network structures and (ii) methods focusing on node relation (sign of connecting edges). In this paper, we propose a new method of combining both of these approaches for better embedding. Experimental results show that our method outperforms other state-of-the-art methods for the tasks of relation prediction and link prediction using the signed networks of real Web users.

2 Problem Definition

A signed network is composed of nodes and edges with positive or negative signs. We focus on directed signed networks since real signed networks are often directed. Signed network G is represented as $G = (V, E)$, where V is a set of nodes and E is a set of edges. An edge from node $u \in V$ to node $v \in V$ with sign $r \in \{positive, negative\}$ is represented as (u, v, r).

Network embedding of signed networks is a task of finding d-dimensional vector $\overrightarrow{u}$ corresponding to each of the nodes $u \in V$ when signed network G, dimension of vectors d $(d \ll |V|)$ and hyperparameters for embedding (explained later) are given. Vector representation should reflect the proximity of the nodes based on both of network structure and node relation (sign of connecting edges).

3 Related Work

3.1 Skip-gram

Skip-gram is a model for representation learning of words in the field of natural language processing. DeepWalk [1] applies Skip-gram to networks and it performs well for some tasks on networks such as node classification. Skip-gram is widely used as a method for network embedding recently. Skip-gram performs representation learning by minimizing the following loss function:

$$Loss = -\frac{1}{|D|} \sum_{(u,v) \in D} \log P(v|u), \tag{1}$$

where D is sampled node pairs, $P(v|u)$ is the posterior probability defined with vector $\overrightarrow{u} \in \mathbb{R}^d$ of node u and context node vector $\overrightarrow{c_v} \in \mathbb{R}^d$ as follows:

$$P(v|u) = \frac{\exp(\overrightarrow{u}^T \cdot \overrightarrow{c_v})}{\sum_{i \in V} \exp(\overrightarrow{u}^T \cdot \overrightarrow{c_i})}. \tag{2}$$

Since the denominator of $P(v|u)$ is computationally expensive, the loss function is approximated using negative sampling [1] as follows:

$$Loss = -\frac{1}{|D|} \sum_{(u,v) \in D} \left\{ \psi^+_{u,v} + \sum_{i \in N_s(v)} \psi^-_{u,i} \right\}, \tag{3}$$

where $\psi^{+}_{u,v} = \log\sigma(\overrightarrow{u} \cdot \overrightarrow{c_v})$, $\psi^{-}_{u,i} = \log\sigma(-\overrightarrow{u} \cdot \overrightarrow{c_i})$, and $\sigma()$ is sigmoid function. $N_s(v)$ a set of nodes collected by negative sampling.

The probability $P(v|u)$ depends on the sampling method of D. For example, LINE [3] samples D by edge sampling, and $P(v|u)$ represents "the probability that an edge from node u is connected to node v". Since DeepWalk [1] and node2vec [7] perform sampling by random walk, $P(v|u)$ represents "the probability that node v appears in the sequences of nodes obtained by random walk containing node u".

As shown in Eq. (2), when Skip-gram performs learning of node pair (u, v), it performs learning of other node pairs not contained in D because the learning is done for all nodes in the network other than u and v. For example, the value of $P(v|u)$ for distant node pair (u, v) will be small because such node pair does not appear often in D.

3.2 Network Embedding for Signed Networks

Methods Based on Network Structures. SIDE [2] is an extended model of Skip-gram, and it performs learning by the maximizing likelihood $P(v, u)$ for the co-occurring node pairs in random walk corpus. Although SIDE performs network embedding efficiently with Skip-gram, relations among nodes (such as positiveness and negativeness of edges) are not fully utilized.

Methods Based on Node Relations. StEM [4] performs learning by predicting the relation between given node pair (u, v). StEM minimizes the following loss function using neural networks for obtaining embedding:

$$Loss = -\sum_{(u,v,r)\in E^{+}} \log P(r|u, v). \tag{4}$$

SiNE [6] performs vector learning for nodes x, y and z so that $d(\overrightarrow{x}, \overrightarrow{y}) < d(\overrightarrow{x}, \overrightarrow{z})$ when there is positive relations between nodes x and y, and negative relation between nodes x and z. $d(\overrightarrow{u}, \overrightarrow{v})$ is the distance between the vectors of nodes u and v, and it is represented with neural networks in SiNE.

These methods learn relations between nodes with expressive neural network, so they are good at learning signs of edges between nodes. On the other hand, these method update the values of sampled node pairs only, so they often perform learning without using overall network structure.

4 Proposed Method

4.1 Combination of Structure and Relation

For the embedding of signed networks, it is important to use both (i) network structure (which node pair is connected) and (ii) node relation (whether node pair is connected with positive edge or negative edge). However, most of the previous methods focus on either, not both. We therefore propose a new network embedding method focusing on both.

One thing we have to care about is asymmetry of relation between nodes. Suppose there are a positive directed edge from node u to node v (u likes v) and a negative directed edge from node v to node u (v hates u). In such case, it is not easy to learn unique vector representation $\overrightarrow{u}$ of node u.

In our proposed method, two vectors are learned for each node u independently: $\overrightarrow{y_u} \in \mathbb{R}^{\frac{d}{2}}$ (a feature vector as start node) and $\overrightarrow{x_u} \in \mathbb{R}^{\frac{d}{2}}$ (a feature vector as end node). After learning, these two vectors are concatenated as the vector representation $\overrightarrow{u}$ of node u.

4.2 Loss Function

Our proposed method introduces two loss functions: $Loss_{structure}$ for learning structure and $Loss_{relation}$ for learning relation. The overall loss function $Loss$ is defined as follows:

$$Loss = (1-\alpha)Loss_{structure} + \alpha Loss_{relation} + \lambda Loss_{reg}, \tag{5}$$

where $Loss_{reg}$ is a regularization term and λ is a hyperparameter for weighting the regularization term. α is a hyperparameter within the range $0 \leq \alpha \leq 1$ for balancing $Loss_{structure}$ and $Loss_{relation}$. When α is close to 0, learning focusing on network structure is performed. When α is close to 1, learning focusing on node relation is performed.

4.3 Learning Using Network Structure

Research on network embedding on early stage are for simple networks without signs on edges, and Skip-gram is an efficient method for such networks. However, normal Skip-gram does not take edge signs into consideration. We therefore propose an extension of Skip-gram model.

In our proposed method, learning based on network structure is performed by minimizing the following loss function $Loss_{structure}$:

$$Loss_{structure} = -\frac{1}{|E|}\sum_{(u,v,r)\in E} \log P(v|u,r), \tag{6}$$

where $P(v|u,r)$ is the probability that an edge from node u with sign r is connected to node v, which is defined as follows:

$$P(v|u,r) = \frac{\exp\{f(u,r)^T \cdot \overrightarrow{x_v}\}}{\sum_{i\in V}\exp\{f(u,r)^T \cdot \overrightarrow{x_i}\}}. \tag{7}$$

$f(u,r)$ is a function from node u and edge sign r to corresponding $d/2$-dimensional vector as follows:

$$f(u,r) = \begin{cases} \overrightarrow{y_u}^T W_{positive} + b & (\text{if } r = positive) \\ \overrightarrow{y_u}^T W_{negative} + b & (\text{if } r = negative) \end{cases} \tag{8}$$

where $W_{positive}, W_{negative} \in \mathbb{R}^{\frac{d}{2} \times \frac{d}{2}}$ and $b \in \mathbb{R}^{\frac{d}{2}}$ are the parameters to be learned. Computation of the denominator of Eq. (7) is expensive because it requires the computation for all nodes. We therefore approximate it using negative sampling as follows:

$$Loss_{structure} = -\frac{1}{|E|} \sum_{(u,v,r) \in E} \left\{ \psi^{+}_{u,v,r} + \sum_{i \in N_s(u)} \psi^{-}_{u,i,r} \right\}, \tag{9}$$

where $\psi^{+}_{u,v,r} = \log \sigma(f(u,r)^T \cdot \vec{x_v})$ and $\psi^{-}_{u,i,r} = \log \sigma(-f(u,r)^T \cdot \vec{x_i})$.

4.4 Learning Using Node Relation

Real signed networks often represent the interactions of many nodes. In order to obtain embedding of signed networks, expressive model is required. SiNE and StEM use expressive neural networks for learning. We therefore use neural network for our method as well.

Our proposed method learns node relation with the following two-layer neural network.

$$\begin{aligned} h_0 &= [\vec{y_u}, \vec{x_v}] \\ h_1 &= \sigma(W_1 h_0^T + b_1) \\ g(u,v) &= \sigma(W_2 h_1^T + b_2) \end{aligned} \tag{10}$$

where [ ] represents concatenation, $W_1 \in \mathbb{R}^{d \times d}, W_2 \in \mathbb{R}^{d}, b_1 \in \mathbb{R}^{1 \times d}$ and $b_2 \in \mathbb{R}$ are the parameters to be learned. $\sigma()$ is sigmoid function.

In our method, the probability $P(r|u,v)$ is defined based on the output of neural network $g(u,v)$ as follows.

$$P(positive|u,v) = g(u,v) \tag{11}$$

$$P(negative|u,v) = 1 - g(u,v) \tag{12}$$

With these probability functions, learning is performed by minimizing the following loss function.

$$\begin{aligned} &Loss_{relation} \\ &= -\frac{1}{|E^+|} \sum_{(u,v,positive) \in E^+} \log P(positive|u,v) \\ &\quad -\frac{1}{|E^-|} \sum_{(u,v,negative) \in E^-} \log P(negative|u,v), \end{aligned} \tag{13}$$

where E^+ is the set of edges with positive sign, and E^- is the set of edges with negative sign.

5 Experiments

This section first shows the experiments for the comparison of proposed method and previous methods. Relation prediction task and link prediction task are performed in order to show the effectiveness of the proposed method. The effect of balancing parameter α in the loss function is also investigated in our experiments.

5.1 Datasets

We use the datasets of Slashdot and Epinions available at SNAP[1] site in Stanford University. Both of these signed networks are extracted from real Web sites.

Slashdot is the dataset of the relations of the users of Slashdot, a site for the discussion of science and technology. The site can register the relation with other users as "friend" or "foe". We regard the dataset as signed network of users connected with "friend" as positive edges, and "foe" as negative edges.

Epinions is the dataset of the relations of the users of Epinions, a site for reviewing products by consumers. The site can register the relation with other users as "trust" and "distrust". We regard the dataset as signed network of users connected with "trust" as positive edges, and "distrust" as negative edges. Table 1 shows the number of nodes $|V|$, the number of edges $|E|$, and the ratio of positive edges in each dataset, respectively.

Table 1. Datasets

	$\|V\|$	$\|E\|$	Ratio of Positive Edges(%)
Slashdot	82,140	549,202	76.1
Epinions	131,828	841,371	85.3

5.2 Comparison with Previous Methods

Methods to Be Compared. We compare the proposed method with SiNE [6], StEM [4] and SIDE [2]. SiNE and StEM are the methods for learning focusing on node relations using neural networks. SIDE is a method for learning focusing on network structure using extended Skip-gram. For the experiments of StEM and SIDE, we use the implementations by original authors of the papers, and for the experiments of SiNE and our proposed method, we use the implementation by the authors of this paper. Implementation of the proposed method is available in GitHub[2].

[1] https://snap.stanford.edu/.
[2] https://github.com/ari1219/sne_open.

Table 2. Accuracy of relation prediction

Dataset	Measure	SiNE	StEM	SIDE	Proposed method
Slashdot	AUC	0.863	<u>0.893</u>	0.780	**0.904**
	F1	0.879	**0.890**	0.807	<u>0.881</u>
	Macro-F1	0.755	<u>0.769</u>	0.731	**0.784**
Epinions	AUC	0.882	<u>0.931</u>	0.851	**0.938**
	F1	0.932	<u>0.940</u>	0.901	**0.952**
	Macro-F1	0.782	<u>0.838</u>	0.789	**0.841**

Table 3. Accuracy of link prediction

Dataset	Measure	SiNE	StEM	SIDE	Proposed method
Slashdot	Micro-F1	0.648	<u>0.688</u>	0.651	**0.706**
	Macro-F1	0.601	<u>0.649</u>	0.611	**0.688**
Epinions	Micro-F1	0.721	0.766	<u>0.779</u>	**0.794**
	Macro-F1	0.639	0.679	<u>0.718</u>	**0.728**

Parameter Tuning. The dimension of vectors d is set to $d = 60$ for all of the methods. Since the proposed method learn two different vectors for each node corresponding to incoming direction and outgoing direction, the dimensions of incoming/outgoing vectors are set to 30 each, and concatenated vector is used as the vector representation of the node.

For our proposed method, we set $\alpha = 0.5$, and other parameter values are selected from the best one based on the grid-search of parameters using training data. For other methods, parameter values mentioned in the original papers are used in our experiments. For the parameters not mentioned in the papers, we perform grid-search for the parameters and choose the best ones.

Task 1: Relation Prediction. Relation prediction is the classification task of predicting the sign of an edge (positive or negative) in an assumption that an edge exists between given two nodes. We use 80% of the edges in the dataset as the training data for network embedding. After the network embedding is performed, for each of the connected node pairs, vectors of two end nodes are concatenated and labeled with the sign of the edge (1: positive edges, 0: negative edges). The concatenated vectors with the labels of edge signs are used for the training of logistic regression. Finally, the remaining 20% edges are used for the test of the logistic regression. As the metrics of evaluation, we use AUC, F1, and Macro-F1. Values of these metrics are within the range from 0 to 1, and values close to 1 mean high accuracy.

Experimental results are shown in Table 2. Values shown in bold font are the best ones, and underlined values are the second best ones. Based on the experiments, we can conclude as follows.

- Except F1 value of Slashdot, the proposed method outperforms other methods from the viewpoint of accuracy of relation prediction task. This shows that the proposed embedding method is better.
- Both of the proposed method and StEM are accurate. This shows that node relation is important for the task of relation prediction. On the other hand, the accuracy of SIDE is worse compared with the proposed method and StEM. This shows that the learning based on network structure is less effective because the existence of an edge between given nodes is already assumed for the task of relation prediction.

Task 2: Link Prediction. Link prediction is the task of predicting links between two nodes. Link prediction for simple network (without edge signs) is a two-class classification problem whether an edge exists or not between given two nodes. Since we are focusing on signed networks, signs should also be considered for the task of link prediction. We perform the experiments of three-class classification (positive edge, negative edge or no edge) as the task of link prediction in signed networks. We use 80% of the edges in the dataset as the training data for network embedding. After the network embedding, the same numbers of node pairs that are (i) connected with positive edges, (ii) connected with negative edges, and (iii) not connected are sampled. For each of the node pairs, vectors of two end nodes are concatenated and labeled with the sign or non-existence of the edge (2: no edge, 1: positive edges, 0: negative edge). The concatenated vectors with the labels are used for the training of logistic regression. Three classifiers (positive edge or others, negative edges or others, and no edges or others) are obtained after training. Finally, the remaining 20% edges are used for the test of these three classifiers based on one-versus-the-rest method. As the metrics of evaluation, we use Micro-F1 and Macro-F1 for this multi-class classification. Values of these metrics are within the range from 0 to 1, and values close to 1 mean high accuracy.

Experimental results are shown in Table 3. Values shown in bold font are the best one, and underlined values are the second best one. Based on the experiments, we can conclude as follows.

- The proposed method outperforms other methods from the viewpoint of accuracy of link prediction. This shows that the proposed embedding method is better.
- For the above three-class classification task, edge signs and the existance of edges should be learned. The proposed method learns both relation and structure, and its performance is better than those of other methods.

5.3 Effect of α

We investigate the effect of changing the value of balancing parameter α in the loss function of the proposed method. For the tasks of relation prediction and link prediction, the value of α is set to 0, 0.2, 0.4, 0.6, 0.8, and 1.0, and the accuracy of each case is investigated.

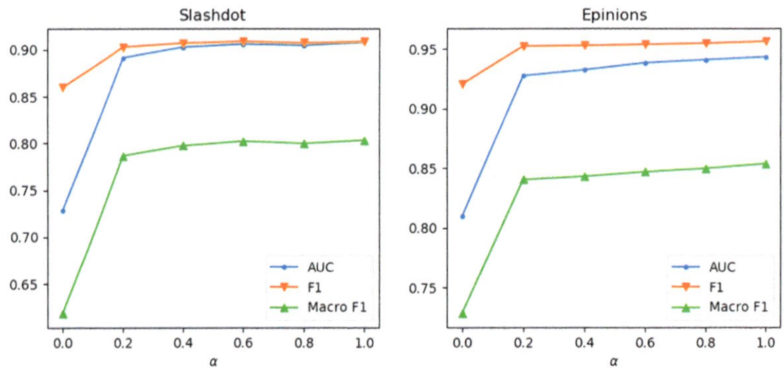

Fig. 1. α and the accuracy of relation prediction

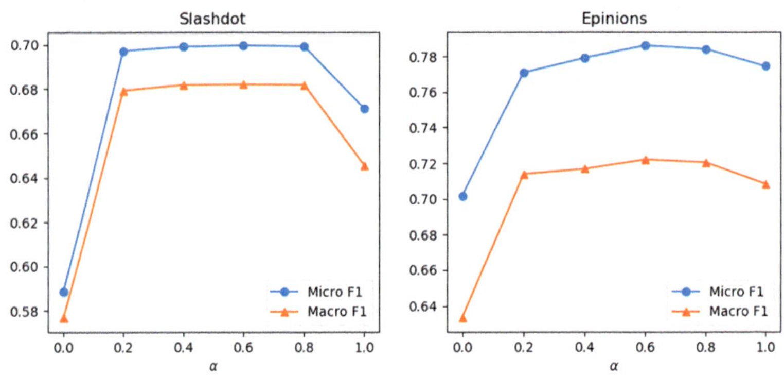

Fig. 2. α and the accuracy of link prediction

Figure 1 shows the relation of α and the accuracy of relation prediction, and Fig. 2 shows the relation of α and the accuracy of link prediction. The x-axis is the value of α, and the y-axis is the value of each measure (such as AUC and F1).

For the task of relation prediction in both datasets, the accuracy is the best when $\alpha = 1.0$. This means that the loss function focusing on node relations with bigger α is effective for this task. On the other hand, for the task of link prediction, the accuracy is high when α is between 0.2 and 0.8. This means that the loss function focusing on both node relations and network structure is effective for this task.

6 Conclusion

We propose a new method for embedding signed networks. The proposed method utilizes both of network structure and node relations for learning. Experimental results show that the proposed method outperforms previous method from the viewpoint of accuracy of relation prediction task and link prediction task. Especially, both of nodes relation and network structure are important for learning for the task of link prediction.

Acknowledgement. This work was supported by JSPS Grant-in-Aid for Scientific Research (B)(Grant Number 17H01785) and JST CREST (Grant Number JPMJCR1687).

References

1. Perozzi, B., Al-Rfou, R., Skiena, S.: Deepwalk: online learning of social representations. In: Proceedings of the 20th ACM SIGKDD International Conference on Knowledge Discovery and Data Mining, pp. 701–710 (2014)
2. Kim, J., Park, H., Lee, J., Kang, U.: Side: representation learning in signed directed networks. In: Proceedings of the 2018 World Wide Web Conference, pp. 509–518 (2018)
3. Tang, J., Qu, M., Wang, M., Zhang, M., Yan, J., Mei, Q.: Line: large-scale information network embedding. In: Proceedings of the 24th International Conference on World Wide Web, pp. 1067–1077 (2015)
4. Inzamam, R., Patrick, H.: A method for learning representations of signed networks. In: Proceedings of the 14th International Workshop on Mining and Learning with Graphs (MLG) (2018)
5. Mikolov, T., Sutskever, I., Chen, K., Corrado, G.S., Dean, J.: Distributed representations of words and phrases and their compositionality. In: Advances in Neural Information Processing Systems, pp. 3111–3119 (2013)
6. Wang, S., Tang, J., Aggarwal, C., Chang, Y., Liu, H.: Signed network embedding in social media. In: Proceedings of the 2017 SIAM International Conference on Data Mining, pp. 327–335 (2017)
7. Grover, A., Leskovec, J.: node2vec: scalable feature learning for networks. In: Proceedings of the 22nd ACM SIGKDD International Conference on Knowledge Discovery and Data Mining, pp. 855–864 (2016)

Power of Nodes Based on Their Interdependence

Sergey Shvydun[1,2(✉)]

[1] National Research University Higher School of Economics, Myasnitskaya Str. 20, 101000 Moscow, Russia
shvydun@hse.ru
[2] V.A. Trapeznikov Institute of Control Sciences of Russian Academy of Science, Profsoyuznaya Str. 65, 117342 Moscow, Russia

Abstract. Power of nodes has been studied in many works, in particular, using centrality concepts. However, in some applications, a large flow between two nodes implies that these nodes become too interdependent on each other. For instance, in trade networks, the possible shortage of flow between two countries may lead to the deficit of goods in the importing country but, on the other hand, it may also affect the financial stability of the exporting country. This feature is not captured by existing centrality measures. Thus, we propose an approach that takes into account interdependence of nodes. First, we evaluate how nodes influence and depend on each other via the same flow based on their individual attributes and a possibility of their group influence. Second, we present several models that transform information about direct influence to a single vector with respect to the network structure. Finally, we compare our models with centrality measures on artificial and real networks.

Keywords: Influence in networks · Interdependence · Individual attributes · Group influence · Trade networks

1 Introduction

Identification of influential nodes in network structures is crucial for the understanding of the associated real-world systems. However, in many applications critical nodes detection is not a straightforward process as connections in a network may have several meanings when we consider the influence. For instance, in some networks (e.g. citation networks) the edge from node A to node B means that node B influences node A. In some other networks (e.g. flow networks) the edge from node A to node B means that node A influences node B.

There are also some cases where nodes influence each other using the same edge. For instance, trade relations make countries increasingly economically interdependent and highly integrated [1]. On the one hand, importers are

H. Barbosa et al. (Eds.): *Complex Networks XI*, SPCOM, pp. 70–82, 2020.
https://doi.org/10.1007/978-3-030-40943-2_7

dependent on exporters in the purchase of products which are not manufactured nationwide. The increase of trade relations and specialization of countries on particular products lead to the fact that exporters may use the trade as a political tool. The power of a country increases with the size of exports from an exporting country. On the other hand, importers may also possess certain power against exporters using trade relations as a potential leverage [1]. The importer may influence the exporter if it suddenly uses the refusal of purchases as a threat for political concession. The potential import ban may lead to economic losses of exporter which, consequently, affect its level of economic growth, employment and the financial stability of the major industries. The importer has also influence over the exporter as the exporter seeks importer's market and investment. Therefore, the edge from node A to node B means that both nodes are highly interdependent. Since trade flows does not represent countries interactions, there have been proposed several measures how to evaluate trade interdependence taking into account the importance of export/import for each country [2–4]. Unfortunately, the network approach was not applied to the problem.

Power of agents in networks is a subject of study in many works, in particular, using centrality concept [5,6]. However, as it was discussed above these measures cannot be applied to some networks because they do not take into that some edges make nodes interdependent on each other (see Fig. 1).

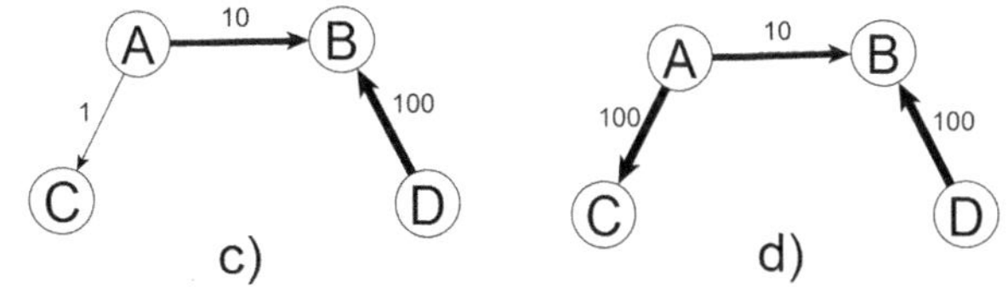

Fig. 1. Interdependence of nodes in networks.

Suppose that connections in a network correspond to the trade flows. According to Fig. 1a, node A influence node B as it is a major exporter of goods for node B. However, node B is the largest importer of goods from node A, consequently, node A may also depend on node B. Thus, nodes A and B are interdependent. On the other hand, if node B is not the largest importer for node A (see Fig. 1b), the influence of node B to node A is insignificant. Similarly, if node B is the major importer for node A while node A is not the major exporter for node B (see Fig. 1c), there is only one-sided influence between nodes (node B influences node A). Finally, if nodes A and B are not the largest exporters and importers for each other (see Fig. 1d), there is no interdependence between nodes A and B.

Although influence in Fig. 1 can be evaluated using existing centrality concepts, – by considering separately how nodes influence each other as exporters and as importers (one-sided nodes influence) – we did not find any measure that takes into account all the cases from Fig. 1 simultaneously. The most relevant measure is the hyperlink-induced topic search (HITS) algorithm which was originally introduced in [7]. The algorithm assigns two scores – the hub score and the authority score – and, according to the model, the node has a high hub score if it is connected to high authority nodes. Similarly, node has a high authority score if it is pointed by nodes with high hubs scores. The HITS algorithm was applied to the global trade network in [8]; however, the measure still does not capture the nodes interdependence. Moreover, the HITS algorithm also does not consider heterogeneity of nodes and their potential group influence on other nodes.

Thus, we propose a novel approach that takes into account the main features of interdependence, which are discussed in Fig. 1. Our main goal is to identify nodes that have high power and at the same time low dependence on other nodes. We analyze how nodes depend on each other and present several models how to aggregate such information taking into account the structure of the network.

The paper is organized as follows. First, we give some notions to define influence in networks. Second, we propose several models that consider interdependence of nodes. Third, we apply the proposed models to an artificial network as well as to world trade network. The final section concludes.

2 Influence in Networks

To define the level of nodes interdependence, we firstly need to understand how nodes influence each other. As connections in a network do not represent the real influence among nodes, we need to transform the initial network to the network of influence.

We operate with a graph $G = (V, E)$, where $V = \{1, ..., n\}$ is a set of nodes, $E \subseteq V \times V$ is a set of edges. We consider directed graphs, i.e., a graph where the existence of edge (i, j) does not imply the existence of the edge (j, i). To describe a graph we also use the adjacency matrix $W = [w_{ij}]$ with parameter w_{ij} defined on each edge (i, j), i.e., we consider directed weighted graph.

As it was illustrated in Fig. 1, one needs to understand how nodes influence each other directly. For instance, in trade networks node i influences node j if its flow w_{ij} is critical for node j. To understand the importance of incoming flows for node j we follow our previous works [9,10] and assume that node j has some threshold of influence q_j^{in} which indicates the minimal level when this node becomes affected. The threshold value can be calculated with respect to a network structure (e.g. the total import of node j) or it may depend on some external parameters (e.g. GDP value of node j). The threshold value appears in many networks. For instance, in trade networks, countries have some minimal requirements for the amount of goods needed to meet their requirements.

If $w_{ij} \geq q_j^{in}$, a flow w_{ij} is critical for node j, consequently, node i can use the flow to influence node j. In trade networks it means that country i is important for country j as it can potentially use the shortage of its export as a political tool. However, if $w_{ij} < q_j^{in}$ it does not mean that node i has no influence on node j. If node i collaborates with some other nodes that have flows to node j (e.g. with nodes from the same alliance), node i becomes also important for node j. Thus, there is a need to consider the group influence of nodes. Following [9] we provide the definition of critical groups and its pivotal members and define the direct influence.

Definition 1. *A critical group for node j is a subset of nodes whose group influence exceeds the threshold value q_j^{in}. More formally, $\Omega(j) \subseteq V$ is a critical group for node j if*

$$\sum_{k \in \Omega(j)} w_{kj} \geq q_j^{in} \tag{1}$$

Definition 2. *Node i is pivotal for some group $\Omega(j)$ if its exclusion from this group makes the group non-critical. Formally, $\Omega^p(j) \subseteq \Omega(j)$ is a subset of pivotal nodes of group $\Omega(j)$ if $\forall k \in \Omega^p(j)$*

$$\sum_{k \in \Omega(j) \setminus \{i\}} w_{kj} < q_j^{in} \tag{2}$$

Definition 3. *If node i is pivotal for node j, the direct influence can be evaluated as*

$$c_{ij}^{in} = max_{\Omega_k(j): i \in \Omega_k^p(j)} \frac{w_{ij}}{\sum_{h \in \Omega_k(j)} w_{hj}} \tag{3}$$

According to formula (3), $0 \leq c_{ij}^{in} \leq 1$; $c_{ij}^{in} = 0$ implies that node i does not influence on node j as it is not connected to the node or it is not pivotal in all critical groups. On the other hand, $c_{ij}^{in} = 1$ implies that node i individually may exceed the threshold of node j. The intermediate values $(0 < c_{ij}^{in} < 1)$ corresponds the maximal impact of node i in a critical group $\Omega(j)$ where node i is pivotal.

We illustrate the idea of direct influence evaluation in Fig. 2. Suppose a simple trade network where node E has a threshold of influence equal to 40% of incoming edges (40% loss of the total import is critical for a node), i.e., $q_E^{in} = 40$. In that case $c_{AE}^{in} = 1$ because the trade flow from node A is higher than the threshold. As for the other nodes, there are only 2 critical groups without node A: $\{B, C\}$, $\{B, C, D\}$. Nodes C and D are both pivotal in these groups, however, the impact of these nodes is higher in a smaller group. Thus, $c_{CE}^{in} = \frac{25}{45} = 0.56$ and $c_{DE}^{in} = \frac{20}{45} = 0.44$. Node B is not pivotal in any group, hence, $c_{BE}^{in} = 0$.

Similarly, to understand how flow w_{ij} is important for node i we assume that node i has some threshold of influence q_i^{out} which indicates the minimal level when this node becomes affected. For instance, in trade networks, as node i also benefits from connections to other nodes, q_i^{out} corresponds to the minimal export

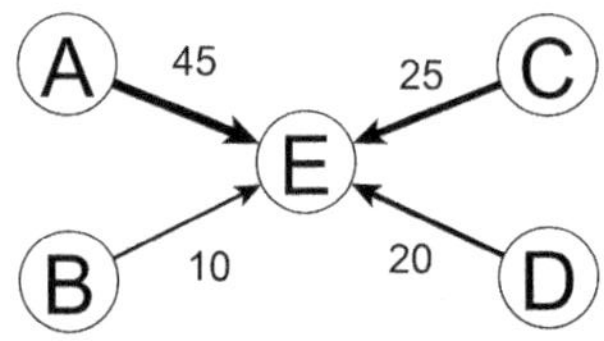

Fig. 2. Direct influence evaluation based on group influence.

value which will be critical (e.g. economically) for node i to loose. As it was discussed above, q_i^{out} can be defined using the network structure or externally.

We can adapt the idea of direct influence calculation (c_{ij}^{in}) to calculation of c_{ij}^{out} which represents the importance of flow w_{ij} for node i. Therefore, one can evaluate the importance of flow w_{ij} for both nodes i and j and present such information as two matrices of pairwise influence $C^{in} = [c_{ij}^{in}]_{n\times n}$ and $C^{out} = [c_{ij}^{out}]_{n\times n}$. One can represent such information as a multiplex network that consists of two layers: the layer of nodes influence and the layer of nodes dependence (see Fig. 3).

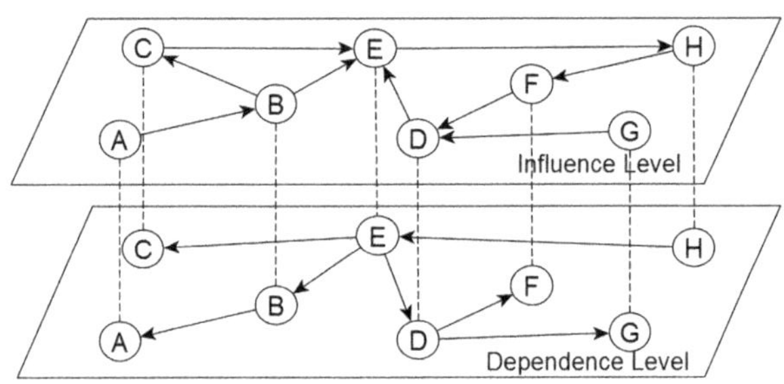

Fig. 3. Multiplex representation of nodes interdependence.

3 Models of Nodes Interdependence

In previous section we defined how nodes influence each other through the adjacent edge. If $c_{ij}^{in} \gg c_{ij}^{out}$, node i influences node j and not vice versa. Similarly, $c_{ij}^{in} \ll c_{ij}^{out}$ means that node i is highly dependent on node j while node j cannot be directly influenced by node i. Finally, if $c_{ij}^{in} \approx c_{ij}^{out} \gg 0$, the nodes are equally dependent on each other. Such information defines the level of their direct interdependence.

However, such approach does not consider the possibility of indirect influence of nodes. Indeed, nodes may affect each other via some intermediate members. Thus, one needs to identify how nodes are interdependent indirectly.

We consider several models how to evaluate nodes interdependence with respect to the network structure.

3.1 Model 1: Nodes Interdependence as Aggregation of Indirect Influences

Since nodes influence each other indirectly, one can extend matrices of pairwise influence C^{in} and C^{out} to matrices of indirect influence. This step can be done with respect to considering different paths or random walks in a graph of direct influence which was performed in [9,10] to evaluate the long-range interaction centrality (LRIC). For instance, one can consider various simple paths in a network and evaluate the strength of each path as a joint 'probabilities'. In other words, if there is a sequence of edges $(i,k_1),(k_1,k_2),(k_2,k_3),...,(k_{s-1},j)$ which can be denoted as $P_{i\to j}$, we can evaluate the path strength as

$$f^{in}(P_{i\to j}) = c^{in}_{ik_1} \times c^{in}_{k_1k_2} \times \cdots \times c^{in}_{k_{s-1}j}. \tag{4}$$

In Fig. 4 we present an example of the path strength calculation according to formula (4). The indirect influence of node A on node C is equal to $0.5\times 0.2 = 0.1$

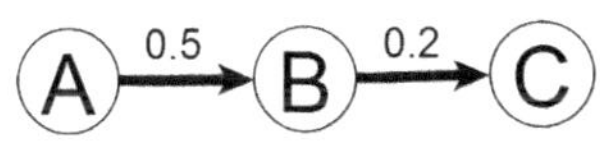

Fig. 4. Indirect influence in networks.

Hence, we can construct matrices $\tilde{C}^{in}$ and $\tilde{C}^{out}$ which correspond to the level of indirect influence/dependence of nodes. As $\tilde{c}^{in}_{ij}$ defines how node i influences node j directly or indirectly, $\tilde{c}^{out}_{ij}$ evaluates how node is dependent on node j, one can aggregate this information into a single matrix $\tilde{C}$. For instance, $\tilde{c}_{ij} = \tilde{c}^{in}_{ij} - \tilde{c}^{out}_{ij}$ can be used as interdependence value. If $\tilde{c}_{ij} > 0$ node i have more influence on node j. On the other hand, if $\tilde{c}_{ij} < 0$, node i is dependent on node j. Finally, $\tilde{c}_{ij} \approx 0$ means that nodes have not enough power to influence each other.

Overall, one should mention that model 1 evaluates power of nodes based on their maximal possible influence and dependence while the relation between layers is not taken into account.

3.2 Model 2: Nodes Interdependence Based on Difference of Direct Influences

Similarly to model 1, one needs to evaluate indirect influence of nodes. However, contrary to the previous model, we firstly propose to analyze how nodes influence each other directly. If we consider trade networks and $\tilde{c}^{in}_{ij} = 1$, node i is the major exporter of goods for node j. Hence, it is highly influential for node j if it uses the shortage of the export as a political tool. However, if $\tilde{c}^{out}_{ij} = 1$, node j is the major of importer for node i and, consequently, the shortage of the export also affects node i itself. Thus, one can conclude that nodes i and j are too interdependent so they cannot use the flow w_{ij} to influence each other (contrary to model 1). This case can be observed in real trade networks. For instance, some countries of the European Union are too interdependent on each other in terms of traded goods and flows of money for them, so they cannot use the flows to influence each other. This idea also has a good correspondence with recent studies which prove that strong trade relations reduce the probability of conflict among the states [11, 12].

Our second model attempts to consider this feature. The main idea is that before considering how nodes influence each other we aggregate matrices of pairwise influence C^{in} and C^{out} into a single interdependence matrix C. We calculate the level of interdependence c_{ij} via flow w_{ij} as

$$c_{ij} = c_{ij}^{in} - c_{ij}^{out} \tag{5}$$

According to formula (4), negative values of c_{ij} means that node i is more dependent on node j than vice versa. Similarly, $c_{ij} > 0$ means that node i influences node j, while $c_{ij} \approx 0$ means that nodes do not influence each other.

The matrix C can be further transformed into matrix $\tilde{C}$ which takes into account indirect connections among nodes. Similarly to the first model, one can perform this step by considering different paths or random walks in a graph based on matrix C (see model 1).

Similarly to model 1, the second model evaluates power of nodes based on their maximal possible influence and dependence, however, it also takes into account that some flows cannot be exploited by its adjacent nodes to influence each other.

3.3 Model 3: Nodes Interdependence as a Search for Influential Paths

In the third model we combine the ideas of two previous models. The first model does not take into account that nodes actually do not influence each other if they are interdependent. The second model does not distinguish the case when nodes are not connected ($c_{ij}^{in} = c_{ij}^{out} = 0$) and the case when nodes are highly interdependent ($c_{ij}^{in} = c_{ij}^{out} = 1$). However, these cases are completely different: in the first case node i has no power on node j and its adjacent nodes while in the second case node i influences node j and, consequently, its neighbors, however, it may not be reasonable to do it as node i suffers some losses.

The third model attempts to consider all these features. The model is based on the following principle: in order to understand whether node i influences node j or not we should consider all possible paths between nodes i and j and choose such path $P_{i\to j}$ that maximizes the influence of node i and, at the same time, minimizes its losses through this path. In other words, interdependence between nodes i and j is defined as

$$\tilde{c}_{ij} = max_{P_{i\to j}}(f^{in}(P_{i\to j}) - f^{out}(P_{i\to j})), \tag{6}$$

where $f^{in}(P_{i\to j})$ and $f^{out}(P_{i\to j})$ are path strengths which can be calculated by formula (4). Note that $f^{in}(P_{i\to j})$ and $f^{out}(P_{i\to j})$ show how node i is critical and dependent for node j through some path $P_{i\to j}$.

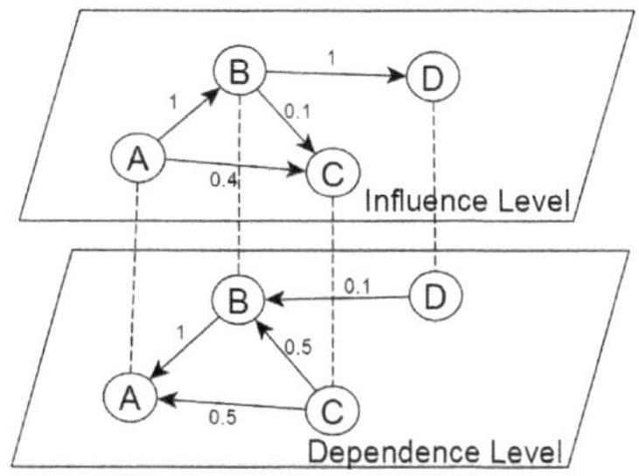

Fig. 5. Illustration of model 3.

The idea of the model can be also illustrated in Fig. 5. According to model 2, $c_{AB} = 0$ if we use formula (5) to aggregate networks; hence, there are no connections from node A to node D, i.e. $\tilde{c}_{AD} = 0$. However, according to this model, $\tilde{c}_{AD} = 1 - 0.1 = 0.9$. There is only one path among nodes A and D in the layer of influence (via node C), consequently, $f^{in}(P_{i \to j}) = 1 \times 1 = 1$ and $f^{out}(P_{i \to j}) = 0.1 \times 1 = 0.1$. Similarly, there are two paths between nodes A and C (direct or via node B), hence, $\tilde{c}_{AC} = max(-0.1, -0.4) = -0.1$, i.e., node A is slightly dependent on node C. Model 1 also does not capture this feature as it maximizes the influence with no respect to the other layer.

Information about aggregated node-to-node interdependence level for all three models can be converted into a single index c_i that corresponds to the overall level of influence in a network. For example, the index for node i may be obtained as a normalized total influence of node i on other nodes in a network, i.e., $\tilde{c}_i = \frac{\sum_{i \neq j} \tilde{c}_{ij}}{\sum_k \sum_{j \neq k} \tilde{c}_{kj}}$.

4 Applications

In this Section we compare our models with existing centrality concepts using a small artificial network and a trade network. All proposed models were implemented in Python while centrality measures were calculated using 'networkx' package.

4.1 Application to Artificial Network

Consider the following example of an artificial network (see Fig. 6), which was taken from our previous work [13].

To calculate the proposed models, one need to define the threshold of influence. We assume that node j influences node i through its flow w_{ji} if it exceeds a certain percentage (75%) of in/out-degree of node i, i.e.,

$$q_i^{out} = q_i^{in} = 0.75 \times max(\sum_k w_{ik}, \sum_k w_{ki}) \tag{7}$$

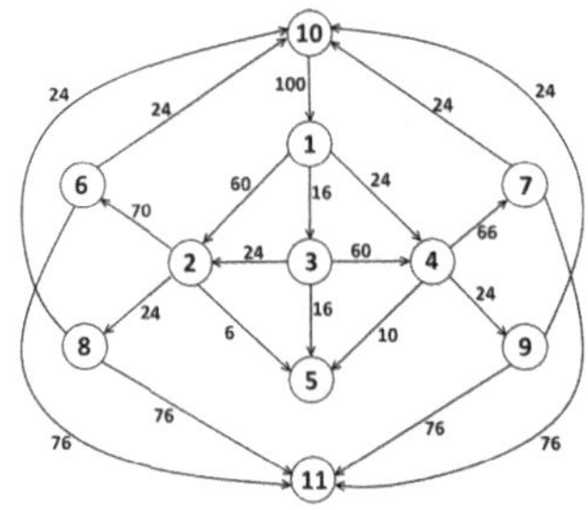

Fig. 6. An example of an artificial network.

The interpretation of formula (7) is the following: if the out-degree of node i is greater than its in-degree, incoming flows to node i will have lower importance. In other words, the importance of flow w_{ji} depends not only on other flows to node i but also on whether node i is an 'exporter' or 'importer'.

The results for classical measures and our models are provided in Table 1.

Table 1. Centrality measures for an artificial network.

Centrality	Node										
	1	2	3	4	5	6	7	8	9	10	11
PageRank	0.11	0.10	0.05	0.08	0.10	0.06	0.05	0.08	0.05	0.09	0.22
Eigenvector	0.28	0.28	0.17	0.28	0.18	0.18	0.46	0.18	0.18	0.44	0.44
Hubs	0	0	0	0	0.25	0.25	0	0.25	0.25	0	0
Authorities	0	0	0	0	0	0	0	0	0	0.24	0.76
Model 1	0.11	0.10	0.05	0.02	0.09	0.08	0.01	0.09	0.06	0.17	0.22
Model 2	0	0.01	0.06	0.04	0.03	0.03	0	0.18	0.08	0	0.58
Model 3	0.05	0.10	0.04	0.08	0.15	0.02	0.01	0.12	0.05	0	0.40

We can observe that node 11 that has strong in-coming edges and no outgoing edges is the most important element in the network by the majority of measures. According to our models, the interpretation is the following: node 11 is the major importer for nodes 6–9, consequently, they are vulnerable to potential refusal of purchases from node 11. Node 10 is the second most influential element as it is the major exporter for node 1. However, node 1 is also the major importer for node 10, hence, these nodes can be considered as interdependent and have no influence on each other. Contrary to models 2–3, this feature is not taken into account by model 1 and classical measures. We should also note that node 8 is the second influential element by models 2–3. Node 8 exports more than imports. Therefore, node 8 has certain power not only against its importers (e.g. node 2) but also against its exporters (e.g. node 10).

Overall, the results of our models are different from classical measures.

4.2 Application to Trade Network

One can also apply the proposed measures to the real trade network. As the input data, we used the STAN Bilateral Trade Database by Industry and End-use category (Revision 4) from OECD [14]. The database presents estimates of bilateral flows between countries from 1990 to 2018 (the current version was published on April-May 2019). According to OECD, the latest year shown is subject to the availability of underlying product-based annual trade statistics, thus, we consider the trade network in 2017. The network is directed and weighted and contains 198 nodes and 27264 edges.

To define how nodes influence each other, we used the gross domestic product (GDP) of countries. We assume that country i influences country j through its export if it exceeds a certain percentage of GDP of country j. Similarly, country i is dependent on the export to country j if it exceeds some level of its own GDP. In other words, we calculated the threshold of influence (q_i^{out}) or dependence (q_i^{in}) of country i as

$$q_i^{out} = q_i^{in} = \lambda \times GDP_i \tag{8}$$

As for λ value, we considered different levels $\lambda = 0.1$ (low influence threshold), $\lambda = 0.2$ (medium influence threshold), $\lambda = 0.3$ (high influence threshold). As a result, we compared the proposed models with some classical centrality measures: PageRank, eigenvector and HITS centralities. In Table 2 we provide a list of countries which were ranked in the TOP-5 by at least one centrality measure.

Table 2. Global trade network in 2017, classical centrality measures (TOP-5, rank).

Country	PageRank	Eigenvector	Hubs	Authorities	Weighted out-degree
USA	1	17	5	1	2
CHN	2	1	1	3	1
DEU	3	16	4	5	3
FRA	4	2	10	10	6
GBR	5	3	12	9	10
JPN	6	21	6	4	4
CAN	7	19	3	7	12
HKG	11	40	8	2	7
MEX	12	13	2	6	11
KOR	13	10	7	8	5
CZE	28	4	33	30	29

According to Table 2, the United States (USA), China (CHN) and Germany (DEU) are the most central elements by most measures. France (FRA) and the United Kingdom (GBR) get high scores by PageRank and eigenvector centrality while Japan (JPN), Canada (CAN), Mexico (MEX) and Hong Kong, China (HKG) are highly ranked by hubs and authorities. If we compare the overall results, the lowest correlation is between eigenvector and all other measures (≈0.3) and between hubs and authorities (≈0.67) while the correlation of other measures is in range [0.82–0.92]. However, if we compare rankings, the correlation is high between all measures: the lowest is between hubs and eigenvector (≈0.78); the highest is between PageRank and authorities (≈0.96).

In Table 3 we provide the results of our models for different influence levels.

Table 3. Global trade network in 2017, classical centrality measures (TOP-5, rank).

Country	$\lambda = 0.1$			$\lambda = 0.2$			$\lambda = 0.3$		
	Model 1	Model 2	Model 3	Model 1	Model 2	Model 3	Model 1	Model 2	Model 3
CHN	1	1	1	2	2	2	3	2	2
USA	2	2	2	1	1	1	1	1	1
DEU	3	3	3	3	3	3	2	3	3
FRA	4	6	4	5	6	5	4	4	4
GBR	5	7	6	6	5	6	5	7	7
JPN	7	5	5	4	4	4	6	5	5
ITA	8	4	7	13	10	10	11	8	10

According to Table 3, China, the United States and Germany are TOP-3 countries with the highest influence levels. China has higher scores than the United States for a low threshold (10% of GPD) which can be explained by the fact that China influences individually ($c^{in}_{CHN,j} = 1$, i.e., $w_{CHN,j} \geq 10\% \times GDP_j$) 46 other countries while the United States influences individually only 36 countries. As for dependence layer, trade flows to China have a high importance ($c^{out}_{j,CHN} = 1$, i.e., $w_{j,CHN} \geq 10\% \times GDP_j$) for 17 countries (mostly from Asia) while trade flows to the United States are important for 15 countries (Mexico, Canada, etc.). If we also take into account the size of the countries (e.g. their GDP) which China or the USA influence, China will have highest score.

For a high threshold (30% of GPD), the United States influences individually only 5 countries while China influences 15 countries. However, if we analyze countries that China or the USA influence in a group with other countries, the total number will be equal to 51 for the USA and 55 for China. As for dependence layer, trade flows to the USA and China are have an importance for 22 and 23 countries respectively. We take into account size of the countries, which China and the USA influence, as well as indirect connections, the United States, on the contrary, will have highest score by our models.

Overall, the correlation coefficient between all presented models is in range [0.96–1.00] for the scores and [0.8–0.99] for the rankings. The highest correlation is between models 1 and 3 because model 2 eliminates many edges.

We can also observe that our models have a good correspondence with classical centrality measures on a trade network. The correlation coefficient with

PageRank is from 0.94 to 0.97 for the scores and from 0.87 to 0.95 for the rankings. Other classical measures have lower correlation: ≈ 0.29 for eigenvector, ≈ 0.79 for hubs and ≈ 0.85 for authorities. However, the proposed models allow to achieve more accurate results, as they takes into account individual attributes of nodes (size, threshold of influence, etc.), a possibility of their group influence as well as nodes interdependence. Moreover, our models provides a more detailed analysis of connections in a network: one can use our approach to identify flows, which are important only for target/source node or for both nodes simultaneously.

5 Conclusion

We have proposed several methods for the power estimation in networks based on interdependence of nodes. The presented models are mostly designed for applications where a flow in a network may result that nodes become too interdependent on each other and, consequently, have some power against each other using the same flow. Our main goal is to identify nodes with high influence and low dependence on other nodes.

Trade relations are an example of such networks: in some cases the potential leverage of the trade flow may affect both exporter and importer. The importer is dependent on exporter in the purchase of goods, which are not produced nationwide. The flow reduction (e.g. due to sanctions or other reasons) affects the importer as it has not enough products. However, the importer may also affect the exporters: if it suddenly uses the refusal of purchases (e.g. import ban), the exporter also suffers economic losses which may lead to financial instability. Thus, one needs to consider nodes interdependence more deeply.

Our models consist of the following steps. First, we evaluate the importance of each connection for the source and target nodes. An important feature of the method is that it takes into account individual attributes of nodes and a possibility of their group influence. Such information allows to analyze influence more accurately. We also do not consider connections in a network, which do not have any impact in terms of the influence. Second, we consider three models that aggregates information about nodes influence and dependence taking into account the structure of the network. Finally, we aggregate information about nodes interdependence to the power index.

We have applied the power indices to the artificial small network as well as to the trade network. The results are distinct from classical measures.

The presented models are implemented to SLRIC library, which can be downloaded from this site: https://github.com/SergSHV/slric.

Acknowledgments. The article was prepared within the framework of the Basic Research Program at the National Research University Higher School of Economics (HSE) and supported within the framework of a subsidy by the Russian Academic Excellence Project '5–100'. This work is also supported by the Russian Foundation for Basic Research under grant No. 18-01-00804a Power of countries in the food security problem.

References

1. Mancheri, N.: China and its neighbors: trade leverage, interdependence and conflict. Contemp. East Asia Stud. **4**(1), 75–94 (2015). https://doi.org/10.1080/24761028.2015.11869082
2. Kojima, K.: The pattern of international trade among many countries. Hitotsubashi J. Econ. **5**(1), 19–36 (1964)
3. Drysdale, P., Garnaut, R.: Trade intensities and the analysis of bilateral trade flows in a many-country world: a survey. Hitotsubashi J. Econ. **22**(2), 62–84 (1982)
4. Frankel, J., Rose, A.: The endogeneity of the optimum currency area criteria. Econ. J. **108**(449), 1009–1025 (1998)
5. Freeman, L.: Centrality in social networks: conceptual clarification. Soc. Netw. **1**, 215–2391 (1979)
6. Bonacich, P., Lloyd, P.: Eigenvector-like measures of centrality for asymmetric relations. Soc. Netw. **23**, 191–201 (2001)
7. Kleinberg, J.: Authoritative sources in a hyperlinked environment. J. ACM **46**, 604–632 (1999)
8. Wei, W., Liu, G.: Bringing order to the world trade network. In: IPEDR Proceedings. IACSIT Press, Singapore, vol. 28, p. 88 (2012)
9. Aleskerov, F., Meshcheryakova, N., Shvydun, S.: Centrality measures in networks based on nodes attributes, long-range interactions and group influence. arXiv preprint arXiv:1610.05892 (2016)
10. Aleskerov, F., Meshcheryakova, N., Shvydun, S.: Power in network structures. In: Models, Algorithms, and Technologies for Network Analysis. Springer Proceedings in Mathematics and Statistics, vol. 197, pp. 79–85. Springer, Heidelberg (2017)
11. Martin, P., Mayer, T., Thoenig, M.: Make trade not war? CEPREMAP Working Papers (Docweb) 0515, CEPREMAP (2005)
12. Tanious, M.: The impact of economic interdependence on the probability of conflict between states. Rev. Econ. Polit. Sci. **4**(1), 38–53 (2019). https://doi.org/10.1108/REPS-10-2018-010
13. Aleskerov, F., Meshcheryakova, N., Nikitina, A., Shvydun, S.: Key borrowers detection by long-range interactions. arXiv preprint arXiv:1807.10115 (2016)
14. OECD Bilateral Trade in Goods by Industry and End-use (BTDIxE), ISIC Rev.4. https://stats.oecd.org/Index.aspx?DataSetCode=BTDIXE. Accessed 1 Dec 2019

Asymmetric Node Similarity Embedding for Directed Graphs

Stefan Dernbach(✉) and Don Towsley

College of Information and Computer Sciences,
University of Massachusetts Amherst, Amherst, USA
{dernbach,towsley}@cs.umass.edu

Abstract. Node embedding is the process of mapping a set of vertices from a graph onto a vector space. Modern deep learning embedding methods use random walks on the graph to sample relationships between vertices. These methods rely on symmetric affinities between nodes and do not translate well to directed graphs. We propose a method to learn vector embeddings of nodes in a graph as well as the parameters of an asymmetric similarity function that can be used to retain the direction of relationships in the embedding space. The effectiveness of our approach is illustrated visually by the 2D embedding of a lattice graph as well quantitatively in multiple link prediction experiments on real world datasets.

Keywords: Node embedding · Skip-gram · Digraphs

1 Introduction

Networks and graphs are ubiquitous in modern information settings. A graph's representational power of objects and relationships make them an essential tool in data visualization and processing. To further aid in many data processing tasks, we seek to learn a vector representation of the objects in a network. Many embedding schemes utilize the spectra of an affinity matrix of the graph to form vector representations of nodes [1,2,15]. These approaches, while effective, require eigen decompositions of large matrices and so do not scale well as the number of nodes in the graph increases. Recent embedding methods sample random walks from the graph and use a stochastic gradient descent process to learn vector representations for the nodes [3,10]. These methods have been shown to be efficient on graphs scaling up to millions of nodes.

Implementing most undirected network embedding methods (such as those above) on a directed network requires making sacrifices to the network structure because these methods rely on symmetric affinities between nodes. This leads to the unrecoverable loss of the asymmetric relationships between nodes. In this work we propose a directed random walk based approach to learning node embeddings that does not sacrifice the asymmetries of the directed graph. Our approach, asymmetric node similarity embedding (ANSE), simultaneously learns

H. Barbosa et al. (Eds.): *Complex Networks XI*, SPCOM, pp. 83–91, 2020.
https://doi.org/10.1007/978-3-030-40943-2_8

the node embedding vectors and the parameters of an asymmetric similarity function on the embeddings. This function allows the direction of edges to be recovered from the embeddings. Additionally we provide an adaptation to our method which embeds the graph onto a hypersphere. Like other random walk approaches, our method scales linearly with the size of the graph. We provide an illustration of a 2-dimensional embedding of a small lattice graph as well as demonstrate the effectiveness of our technique in several link prediction tasks on real world directed networks.

2 Problem Definition

A network, or graph, is used to represent objects and relationships. We define an undirected graph $G = (V, E)$ to be a set of vertices (nodes), $V = \{v_0, v_1, ..., v_{N-1}\}$, and a set of edges, $E = \{(v_i, v_j) : v_i, v_j \in V\}$, representing the objects and relationships respectively. A directed graph (digraph) imposes an ordering on the vertices of each edge such that (v_i, v_j) denotes an edge pointing from node v_i to node v_j. A random walk on a graph is a sequence of nodes $\left(v^{(0)}, v^{(1)}, ..., v^{(K)}\right)$ ordered such that $v^{(k+1)}$ is selected at random from the (outgoing) neighbors of $v^{(k)}$.

A node embedding is a function on a graph that maps each node in the graph to a d-dimensional vector $f_G : V \to \mathbb{R}^d$. The rows of the matrix $\boldsymbol{\Phi} \in \mathbb{R}^{|V| \times d}$ correspond to the vector embeddings of each node, i.e. $\boldsymbol{\Phi}_i = f_G(v_i)$. Define the similarity between two nodes $S(v_i, v_j)$ as proportional to the probability of visiting node v_j within k steps of a random walk beginning at node v_i. This function is asymmetric, $S(v_i, v_j) \neq S(v_j, v_i)$ for most complex networks. This is especially true for a directed graph in which $S(v_i, v_j) > 0 \not\Rightarrow S(v_j, v_i) > 0$.

Our goal is to preserve the asymmetric similarity between nodes in the embedded space. To this end, we aim to learn an embedding matrix $\boldsymbol{\Phi}$ as well as a similarity measure $K : \boldsymbol{\Phi} \times \boldsymbol{\Phi} \to \mathbb{R}$ such that $K(\boldsymbol{\Phi}_i, \boldsymbol{\Phi}_j) \underset{\sim}{\propto} S(v_i, v_j)$.

3 Skip-Gram Embedding

This section describes the skip-gram embedding model that forms the basis for random walk embedding methods including our approach and describes two algorithms that use skip-gram for embedding nodes in undirected graphs. The skip-gram method, first proposed for word embedding [6,7] in natural language processing, extracts sentences from a document corpus and embeds each word to maximize the probability of predicting surrounding words. DeepWalk [10] and subsequent algorithms such Node2Vec [3] adapt the skip-gram model to node embedding for undirected graphs by substituting random walks for sentences.

Algorithm 1 outlines a model for skip-gram style node embedding methods. Lines 4 and 7 are the two key steps in the algorithm and consequently are where many node embedding methods differ from one another. Function $RandomWalk(G, v_i, k)$ on line 4 collects a sequence of k nodes on the graph from a random walk originating at vertex v_i. DeepWalk samples classical walks while Node2Vec uses a random walk that can be weighted to remain close to the initial node or explore further away in the graph. In both methods, every pair of nodes in a walk within k-steps of one another are collected as positive samples.

The probability in line 7 of Algorithm 1 is calculated using a softmax:

$$P(v_k|\boldsymbol{\phi}_j) = \frac{\exp\langle\boldsymbol{\phi}_j, \boldsymbol{\phi}_k\rangle}{\sum_l \exp\langle\boldsymbol{\phi}_j, \boldsymbol{\phi}_l\rangle}. \tag{1}$$

In practice this calculation is prohibitively expensive because it requires an inner product between the embeddings of the source node and each other node in the graph. Many algorithms employ alternative, less computationally expensive, methods of approximating the conditional probabilities. DeepWalk utilizes a hierarchical softmax [8] for computing probabilities while Node2Vec uses a negative sampling approach [7]. Negative sampling samples N node pairs from a noise distribution $P(v)$ as negative samples to approximates the log softmax as:

$$\log P(v_k|\boldsymbol{\phi}_j) \approx \log\left(\sigma\langle\boldsymbol{\phi}_j, \boldsymbol{\phi}_k\rangle\right) + \sum_{n=1}^{N} \mathbb{E}_{v_n \sim P(v)} \log(\sigma\langle -\boldsymbol{\phi}_j, \boldsymbol{\phi}_n\rangle) \tag{2}$$

where σ is the sigmoid function. The hierarchical softmax in DeepWalk reduces the complexity of each softmax calculation from $O(|V|)$ to $O(\log|V|)$ by using a binary tree to calculate the conditional probabilities. The negative sampling approach of Node2Vec reduces the complexity further to $O(N)$ where N is the number of negative samples.

Algorithm 1: Skip-Gram Model of Node Embedding

```
input : graph: G=(V,E)
        window size: w
        embedding size: d
        walks per vertex: γ
        walk length: k
output: node embeddings: φ
1  Initialize Φ ∈ ℝ^{|V|×d}
2  for i = 1 to γ do
3      for All nodes v_i ∈ V do
4          W_{v_i} = RandomWalk(G, v_i, k)
5          for v_j ∈ W_{v_i} do
6              for v_k ∈ W_{v_i}[j − w : j + w] do
7                  J(Φ) = − log P(v_k|φ_j)
8                  Φ = Φ − α * ∂J/∂Φ
9              end
10         end
11     end
12 end
```

4 Method

Many symmetric properties of undirected graphs are asymmetric on digraphs. For example the graph geodesic, the length of the shortest path between nodes, is not symmetric for a directed network. The existence of a path from v_i to v_j does not even imply that a path exists for v_j to v_i.

To learn an embedding that can retain the asymmetric relationships between nodes, we replace the standard (symmetric) inner product used in the softmax (1) and negative sampling (2) equations with an asymmetric bilinear product defined by a matrix $\mathbf{A}$:

$$k_{\mathbf{A}}(v_i, v_j) = \langle \boldsymbol{\phi}_i, \boldsymbol{\phi}_j \rangle_{\mathbf{A}} = \boldsymbol{\phi}_i^T \mathbf{A} \boldsymbol{\phi}_j. \tag{3}$$

If $\mathbf{A}$ is not symmetric then in general $k_{\mathbf{A}}(v_i, v_j) \neq k_{\mathbf{A}}(v_j, v_i)$. Matrix $\mathbf{A} \in \mathbb{R}^{d \times d}$ can be learned in tandem with the embedding matrix $\boldsymbol{\Phi}$ through stochastic gradient descent. Geometrically, $\mathbf{A}$ can be viewed as defining a vector field that determines the direction of the most similar embedding vectors at any point.

Several challenges in sampling random walks on a digraph must be addressed that aren't present in undirected graphs. First, due to the directed nature of edges, positive node pair samples must remain ordered as they appear in the random walk. Second, walks in directed graphs may dead end. In a connected undirected graph a random walk can continue indefinitely by retracing the last edge the walk took to return from an otherwise terminal node. This is not the case for a connected digraph in which nodes may not have any outgoing edges. In such an event the walk is forced to terminate early.

A consequence of these two issues is that nodes without outgoing edges will never be sampled first in a pair. In these cases there is an increased importance in reaching the node from random walks beginning at other nodes so that positive samples containing the node are still collected. In scenarios in which a node has an arbitrarily small likelihood of being reached in a random walk, e.g. the only incoming edge to a node comes from another node with a high out-degree, the node is unlikely to appear as the second node in any positive pair. To address these issues we introduce a reverse random walk sampling method in addition to regular random walk sampling. Reverse walks are sampled from a dual graph, $G^* = (V, (v_j, v_i) : (v_i, v_j) \in E)$, where all edges are reversed. The sequence of nodes in the walk is then reversed again to provide a random walk on G whose transition probabilities are proportional to the in-degrees of nodes rather than their out-degrees. This guarantees that there are sample pairs containing each node with at least one incoming edge as the second node.

We use negative sampling to approximate the softmax function. Nodes are randomly sampled from the graph and used as negative samples as in (2). Combining the positive and negative sampling methods with the asymmetric similarity function produces our method given in Algorithm 2. We split the negative sampling into randomly sampling either the first or second nodes in the negative pairs, line 7 and line 9 respectively.

Algorithm 2: Asymmetric Node Similarity Embedding

input : graph: G(V,E)
embedding dimesnion: d
walks per vertex: γ
walk length: k
output: embeddings matrix: $\boldsymbol{\Phi}$
similarity matrix: $\mathbf{A}$

Initialize $\boldsymbol{\Phi} \in \mathbb{R}^{|V| \times d}, \ \mathbf{A} \in \mathbb{R}^{d \times d}$
for *All nodes* $v_i \in V$ **do**
  for $w = 1$ *to* γ **do**
    $W_{v_i} = RandomWalk(G, v_i, k)$
    $W_{v_i}^{rev} = ReverseRandomWalk(G^*, v_i, k)$
    $J_1(\boldsymbol{\Phi}) = -\sum_{v_j \in W_{v_i}} \log(\sigma \langle \boldsymbol{\phi}_i, \boldsymbol{\phi}_j \rangle_{\mathbf{A}})$
    $J_2(\boldsymbol{\Phi}) = -\sum_{n=1}^{N} \mathbb{E}_{v_n \sim P(v)} \log(\sigma \langle -\boldsymbol{\phi}_i, \boldsymbol{\phi}_n \rangle_{\mathbf{A}})$
    $J_3(\boldsymbol{\Phi}) = -\sum_{v_j \in W_{v_i}^{rev}} \log(\sigma \langle \boldsymbol{\phi}_j, \boldsymbol{\phi}_i \rangle_{\mathbf{A}})$
    $J_4(\boldsymbol{\Phi}) = -\sum_{n=1}^{N} \mathbb{E}_{v_n \sim P(v)} \log(\sigma \langle -\boldsymbol{\phi}_n, \boldsymbol{\phi}_i \rangle_{\mathbf{A}})$
    $J(\boldsymbol{\Phi}) = \sum_{n=1}^{4} J_n(\boldsymbol{\Phi})$
    $\boldsymbol{\Phi} = \boldsymbol{\Phi} - \alpha * \frac{\partial J}{\partial \boldsymbol{\Phi}}$
  end
end

4.1 Hypersphere Embedding

The embedding architecture can be adapted to embed nodes of a (di)graph onto the unit hypersphere. To do so we constrain the node embedding vectors to have unit length: $||\boldsymbol{\phi}_i||_2^2 = 1$. This is accomplished by renormalizing the length of the vector following each backpropagation update.

Matrix $\mathbf{A}$ should also be constrained such that $\forall v_i, v_j \in V : -1 \leq k_{\mathbf{A}}(v_i, v_j) \leq 1$. This constraint allows the similarity function to match the range of a standard inner product between two points on the unit hypersphere. If $\mathbf{A}$ is a unitary matrix, the product $\boldsymbol{\phi}_i^T \mathbf{A}$ will also have unit length and thus $-1 \leq \boldsymbol{\phi}_i^T \mathbf{A} \boldsymbol{\phi}_j \leq 1$ guaranteeing the constraint will hold.

We can project $\mathbf{A}$ onto the set of unitary matrices whenever it diverges during learning similar to renormalizing the embedding vectors. Unfortunately, this projection is computationally costly, requiring a singular value decomposition of the matrix. Alternatively, we compose $\mathbf{A}$ as the product of a set of elementary reflector matrices of the form $\mathbf{A}^{(m)} = \mathbf{I} - 2 * \frac{\mathbf{v}_m \mathbf{v}_m^T}{\mathbf{v}_m^T \mathbf{v}_m}$ where $\mathbf{v}_m$ is any vector and $\mathbf{I}$ is the identity matrix. Any unitary matrix can be decomposed into a product of elementary reflectors. We use this decomposition to efficiently construct a unitary matrix $\mathbf{A} = \prod_{m=1}^{M} \mathbf{A}^{(m)}$ where M can chosen from 1 to the embedding dimension d. Smaller values of M restrict the space of possible unitary matrices but also reduces both the computational cost to calculate $\mathbf{A}$ and the number of parameters for the model to learn. The vectors $\mathbf{v}_k$ are learned by backpropagating the loss through $\mathbf{A}$.

5 Experiments

In this section we conduct multiple experiments to demonstrate both quantitatively and qualitatively the effectiveness of our approach.

5.1 Lattice Example

We use a 2D lattice graph to provide a visual representation of our embedding scheme. The lattice is composed of 10 rows and 12 columns of vertices. All lateral edges in the graph are oriented to point right and all vertical edges to point up. Eight walks of length three are sampled from every node in the graph. The lattice is shown in Fig. 1a and the learned embedding of the vertices in shown in Fig. 1b. Additionally the effect of the matrix **A** in the similarity function is drawn as a vector field in the background such that a source node is most similar to target nodes that lie in the direction of the local arrows. The embedded nodes form a spiral pattern with the bottom-left-most node of the original lattice innermost in the spiral and the top-right-most one outermost. Diagonal sets of nodes in the original lattice representation are structurally similar in the graph and are roughly clustered together along the spiral. Edges are also oriented along the direction of the field induced by **A**.

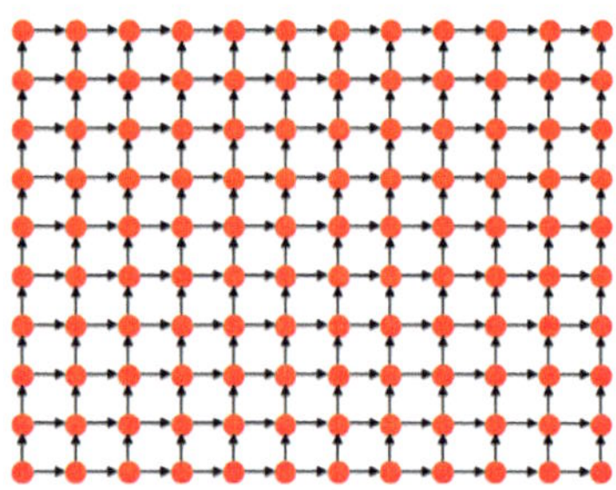

(a) Original Lattice Graph

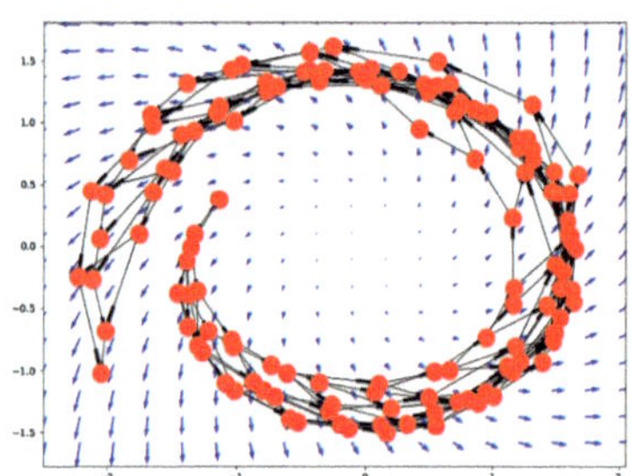

(b) Embedded Lattice Graph

Fig. 1. The original lattice (a) and vector embeddings (b). The direction and magnitude of the vector field illustrates the bias of the similarity measure across the space.

5.2 Link Prediction

Node embeddings can be used to predict missing or future edges in a graph by measuring pairwise similarities. Node pairs with a high similarity score are more likely to form an edge. We evaluate the area under the receiver-operator curve (AUC) for several real world networks to evaluate our method and compare to several other skip-gram algorithms.

Arxiv [5] is a co-authorship network consisting of 5242 nodes representing authors and 28980 edges linking authors who have co-authored a paper together.

Arxiv is the only undirected network in this set of experiments. **Cora** [12] is a citation network where the 23166 nodes in the graph are papers and the 91500 edges points from one paper to another if the first paper cites the second. **Epinions** [11] is a social network dataset with 75879 users (nodes) and 508837 edges indicating trust placed by one user in another.

The reciprocity of a graph is the ratio of the number of bidirectional edges to the total number of edges. The three datasets we evaluate on vary wildly in reciprocity from Arxiv whose reciprocity as an undirected graph is 1.00 to Cora with nearly 0 bidirectional edges and a reciprocity of 0.05. Epinions sits in the middle at 0.40. Together, the three graphs provide a range of node-pair relations from bidirectional, to one-directional, to unrelated.

We use the same hyper-parameters for asymmetric node similarity embeddings (ANSE) and the hypersphere variant (ANSE-H) across all three experiments. Nodes are embedded into a 32-dimensional space and 16 total walks (8 forward, 8 backward) are collected at each node. Each walk continues for 3 steps. For ANSE, we also clip the length of the embedding vectors to be less than 1. In ANSE-H, we use a single elementary reflector matrix for $\mathbf{A}$. We evaluate our method against several modern skip-gram style embedding methods. Deepwalk [10] and Node2Vec [3] are methods developed for undirected graphs. Line [13] and asymmetric proximity preserving embeddings (APP) [16] are methods for embedding nodes of either undirected or directed graphs. We compare ANSE against the scores for the other methods reported in [16]. We also compare against HOPE [9] a recent spectral method for directed graphs that learns both a source and target embedding for each node. We also using a 32 dimensional embedding for HOPE.

The embedding methods are trained using 70% of the edges of the graph while the remaining 30% appear as positive examples in the test set along with an equal number of random node pairs without edges to form the negative examples. The pairwise node score is used to predict the existence of an edge or not between each pair of nodes in the test set. The AUC for each method is given in Table 1. On two of the three graph domains ANSE and ANSE-H demonstrates a significant improvement over all the other methods tested. On Cora, APP joins ANSE and ANSE-H in outperforming the other methods tested. Despite Arxiv being an undirected graph our method learns the asymmetric random walk similarities between pairs of nodes resulting in the high AUC.

6 Related Work

Several node embedding methods for digraphs embed the nodes twice, once into a source space and a second time into a target space [4,9,16]. APP [16] is a skip-gram model that learns dual embeddings for each node in the graph. Node pair samples are collected from the endpoints of (directed) random walks. These pairs are ordered so that the similarity score between nodes is calculated using the source embedding of the first node and the target embedding of the second.

Our embedding technique can also be viewed as learning a dual embedding where the source embedding for node v_i is $\boldsymbol{\phi}_i$ and the target embedding is $\mathbf{A}\boldsymbol{\phi}_i$.

Table 1. Link prediction Area Under Curve (AUC)

Network	Arxiv	Cora	Epinions
DeepWalk	0.887	0.936	0.823
Node2Vec	0.810	0.734	0.865
Line	0.750	0.694	0.867
APP	0.887	**0.944**	0.926
HOPE	0.596	0.874	0.629
ANSE	0.902	**0.941**	**0.948**
ANSE-H	**0.920**	**0.942**	0.924

Compared to learning two embeddings, our approach reduces the number of embedding parameters from $2|V|d$ to $|V|d + d^2$, as typically $d \ll |V|$. Additionally, tying the source and target embeddings together in our approach overcomes a potential issue in [16] in which the source or target embeddings of a node in a digraph may have no positive samples if the node has no outgoing or incoming edges respectively.

A related asymmetric bilinear product is used in [14] for text retrieval. The asymmetric Hermitian inner product is used to score co-attention between complex valued word vectors. The bilinear product $s_{ij} = Re(a_i^T \mathbf{M} b_j)$ is also studied. Unlike our work, however, the matrix $\mathbf{M}$ was tuned as a hyperparameter rather than allowed to change during learning.

7 Conclusion

In this paper, we proposed ANSE, a scalable method to embed a digraph into a vector space, and ANSE-H, a variant of ANSE that embeds the graph onto a hypersphere. ANSE simultaneously learns a vector representations of nodes and an asymmetric similarity function for embedding directed networks. Learning both the embedding and the similarity function offers the ability to recover the direction of edges from the embedded nodes. Additionally we proposed a random walk sampling method to improve learning for nodes without either incoming or outgoing edges. On multiple real world datasets, ANSE and ANSE-H outperforms other skip-gram embedding schemes for link prediction.

References

1. Belkin, M., Niyogi, P.: Laplacian eigenmaps for dimensionality reduction and data representation. Neural Comput. **15**(6), 1373–1396 (2003). https://doi.org/10.1162/089976603321780317
2. Coifman, R.R., Lafon, S.: Diffusion maps. Appl. Comput. Harmonic Anal. **21**(1), 5–30 (2006)

3. Grover, A., Leskovec, J.: Node2vec: scalable feature learning for networks. In: Proceedings of the 22nd ACM SIGKDD International Conference on Knowledge Discovery and Data Mining, KDD 2016, pp. 855–864. ACM, New York, NY, USA (2016). https://doi.org/10.1145/2939672.2939754. http://doi.acm.org/10.1145/2939672.2939754
4. Khosla, M., Leonhardt, J., Nejdl, W., Anand, A.: Node representation learning for directed graphs. ArXiv **abs/1810.09176** (2018)
5. Leskovec, J., Kleinberg, J., Faloutsos, C.: Graph evolution: densification and shrinking diameters. ACM Trans. Knowl. Discov. Data (TKDD) **1**(1), 2 (2007)
6. Mikolov, T., Chen, K., Corrado, G., Dean, J.: Efficient estimation of word representations in vector space. In: 1st International Conference on Learning Representations, ICLR 2013, Workshop Track Proceedings, Scottsdale, Arizona, USA, 2–4 May 2013 (2013)
7. Mikolov, T., Sutskever, I., Chen, K., Corrado, G.S., Dean, J.: Distributed representations of words and phrases and their compositionality. In: Advances in Neural Information Processing Systems, pp. 3111–3119 (2013)
8. Mnih, A., Hinton, G.E.: A scalable hierarchical distributed language model. In: Advances in Neural Information Processing Systems, pp. 1081–1088 (2009)
9. Ou, M., Cui, P., Pei, J., Zhang, Z., Zhu, W.: Asymmetric transitivity preserving graph embedding. In: Proceedings of the 22nd ACM SIGKDD International Conference on Knowledge Discovery and Data Mining, KDD 2016, pp. 1105–1114. ACM, New York, NY, USA (2016). https://doi.org/10.1145/2939672.2939751. http://doi.acm.org/10.1145/2939672.2939751
10. Perozzi, B., Al-Rfou, R., Skiena, S.: Deepwalk: online learning of social representations. In: Proceedings of the 20th ACM SIGKDD International Conference on Knowledge Discovery and Data Mining, KDD 2014, pp. 701–710. ACM, New York, NY, USA (2014). https://doi.org/10.1145/2623330.2623732. http://doi.acm.org/10.1145/2623330.2623732
11. Richardson, M., Agrawal, R., Domingos, P.: Trust management for the semantic web. In: International Semantic Web Conference, pp. 351–368. Springer (2003)
12. Šubelj, L., Bajec, M.: Model of complex networks based on citation dynamics. In: Proceedings of the 22nd International Conference on World Wide Web, pp. 527–530. ACM (2013)
13. Tang, J., Qu, M., Wang, M., Zhang, M., Yan, J., Mei, Q.: Line: Large-scale information network embedding. In: Proceedings of the 24th International Conference on World Wide Web, pp. 1067–1077. International World Wide Web Conferences Steering Committee (2015)
14. Tay, Y., Luu, A.T., Hui, S.C.: Hermitian co-attention networks for text matching in asymmetrical domains. In: IJCAI, pp. 4425–4431 (2018)
15. Tenenbaum, J.B., De Silva, V., Langford, J.C.: A global geometric framework for nonlinear dimensionality reduction. Science **290**(290), 2319–2323 (2000)
16. Zhou, C., Liu, Y., Liu, X., Liu, Z., Gao, J.: Scalable graph embedding for asymmetric proximity. In: Thirty-First AAAI Conference on Artificial Intelligence (2017)

Consistent Recovery of Communities from Sparse Multi-relational Networks: A Scalable Algorithm with Optimal Recovery Conditions

Sharmodeep Bhattacharyya[1] and Shirshendu Chatterjee[2](✉)

[1] Oregon State University, Corvallis, OR 97330, USA
bhattash@science.oregonstate.edu
[2] City University of New York, New York, NY 10031, USA
shirshendu@ccny.cuny.edu

Abstract. *Multi-layer* and *multiplex* networks show up frequently in many recent network datasets. We consider the problem of identifying the common *community membership structure* of a finite sequence of networks, called *multi-relational networks*, which can be considered a particular case of multiplex and multi-layer networks. We propose two *scalable* spectral methods for identifying communities within a finite sequence of networks. We provide theoretical results to quantify the performance of the proposed methods when individual networks are generated from either the *stochastic block model* or the *degree-corrected block model*. The methods are guaranteed to recover communities consistently when either the number of networks goes to infinity arbitrarily slowly, or the expected degree of a typical node goes to infinity arbitrarily slowly, even if all the individual networks have fixed size and are sparse. This condition on the parameters of the network models mentioned above is both sufficient for consistent community recovery using our methods and also necessary to have any consistent community detection procedure. We also give some simulation results to demonstrate the efficacy of the proposed methods.

Keywords: Spectral clustering · Community detection · Multi-relational networks · Multi-layer networks · Stochastic block model · Degree-corrected block model

1 Introduction

In this paper, we focus on the problem of identifying common community structure present in a finite sequence of (possibly incredibly sparse) networks. The

S. Chatterjee was partially supported by PSC-CUNY Cycle 50 Enhanced Research Award # 62781-00 50 while writing this paper.

Electronic supplementary material The online version of this chapter (https://doi.org/10.1007/978-3-030-40943-2_9) contains supplementary material, which is available to authorized users.

H. Barbosa et al. (Eds.): *Complex Networks XI*, SPCOM, pp. 92–103, 2020.
https://doi.org/10.1007/978-3-030-40943-2_9

community detection problem can be thought of as a particular case of vertex clustering problem, in which the goal is to divide the set of vertices of a given network (or a finite sequence of networks) into groups based on some common properties of the vertices. The primary objective in the community detection problem is to identify groups of vertices of a given network (or a finite sequence of networks) so that the average number of connections within the groups are *significantly more* than those between the groups. Several random graph models have been proposed in the literature for generating networks with community structure. Examples of random graph models for a single network with community structure include stochastic block model [11], degree-corrected block model [13] and random dot product model [30].

1.1 Community Detection Methods for a Single Network

Many methods have been proposed in the statistics and machine learning literature to identify the community structure (see [8] for a review) within a given single network. An important class of methods for detecting communities within a given network, which we refer to as *spectral methods*, involve the spectrum of various matrices (e.g. the adjacency matrix, the laplacian matrix) associated with the network. Spectral methods for community detection was introduced in [7], and analyzed in many subsequent papers (see [1,17,19,21,25,26]). In addition to being model agnostic, the spectral methods are highly scalable, as the main numerical procedure involved in these methods is matrix factorization, and many scalable implementations of matrix factorization algorithms have been developed in the numerical analysis literature. The accuracy of some spectral methods in recovering communities within a given single network has been proven theoretically if the network is dense and is generated from some form of exchangeable random graph models [25,27]. But, to the best of our knowledge, no known community detection algorithm is scalable and has been proved to perform consistently to identify communities within several kinds of sparse network.

1.2 Existing Community Detection Methods from Multiple Networks

Several approaches have been put forward to develop statistical frameworks for inference on temporal and multi-layer network models. Although most of such methods have not been developed with the goal of community detection, many of them can be used for such a purpose. For example, the methods developed in [20,29,32] can be used to perform model-based community detection, and the authors of [10] and [23] use likelihood-based methods (e.g., profile-likelihood) to identify communities in networks generated from multi-layer network models. Various other authors have proposed model agnostic procedures (see, e.g., [2,3,14,28]) for detecting communities in multi-layer networks. Spectral algorithms have also been used to find communities from a finite sequence of networks [22,24]. However, most of these works lack quantitative estimates evaluating the performances of the proposed methods and theoretical results which guarantee the consistent recovery of communities. Also, most of these methods including the existing spectral methods do not work when individual networks as well as an aggregated version of the multi-relational networks, are both sparse.

1.3 Our Contribution

Realizing the above limitations of the existing approaches for performing community detection on a single (resp. multiple) network (resp. networks), and recognizing the advantages of using spectral methods (e.g., scalability and model agnostic nature) for a given single network, we propose and analyze two spectral algorithms for finding the common community structure within a given finite sequence of networks.

The main contributions of our work can be summarized as follows.

(a) We propose and analyze two scalable and model agnostic methods, for identifying communities within a multi-relational network having a common community structure. These methods
  - can be used to identify communities within a single network too.
  - are flexible enough to accommodate both sparse and dense networks.

(b) We prove theoretically that our methods outperform existing methods when the given network is generated from either the stochastic block model or the degree-corrected block model or their extensions in a multi-relational setup.

(c) We also prove analytically that, under the mildest (necessary) parametric condition, the proposed methods identify communities in the networks generated from single or multi-layer stochastic block models and degree-corrected block models consistently. We show that in the multi-relational network setting, our spectral clustering methods can recover the common community structure consistently even if each of the individual networks has fixed size and is highly sparse (e.g., has a constant average degree) and has connectivity below the community detectability threshold.

2 Community Detection Algorithms

A *multi-relational network* can be considered as an *edge-colored multi-graph*, where different colors correspond to edge sets of different network layers. The t-th layer $G_n^{(t)}$ is represented by the corresponding adjacency matrix $\mathbf{A}_{n\times n}^{(t)}$. *Let* $\mathbf{Z}_{n\times K}$ *denote the actual common community membership matrix of the nodes in each of the graphs* $G_n^{(t)}$, *where,* $\mathbf{Z}_{ik} = 1$ *if the* i*-th node belongs to the* k*-th community for all* $G_n^{(t)}$. *The goal is to estimate* $\mathbf{Z}$. The algorithms are given in Algorithms 1 and 2.

Let $[n] := \{1, 2, \ldots, n\}$ for $n \in \mathbb{N}$, $\mathscr{M}_{m,n}$ be the set of all $m \times n$ matrices which have exactly one 1 and $n-1$ 0's in each row. $||\cdot||_2, ||\cdot||, ||\cdot||_F$ denote Euclidean ℓ_2-norm, operator norm and Frobenius norm respectively. $\lambda_i(\cdot)$ denotes the i-th largest eigenvalue. For the truncation parameter δ in Algorithm 1, any small positive value is a good choice. In our implementation, we used $\delta = 0.01$.

As mentioned above, the primary goal is to estimate $\mathbf{Z}$ based on the data consisting of $T \geqslant 1$ adjacency matrices $\mathbf{A}^{(1)}, \ldots, \mathbf{A}^{(T)} \in \mathbb{R}^{n\times n}$. In Algorithm 1, we consider the **sum of adjacency matrices** $\mathbf{A}_0 = \sum_{t=1}^{T} \mathbf{A}^{(t)}$. Let $\delta, \varepsilon > 0$ be the truncation and approximation parameters respectively. Let $\bar{d} = \frac{1}{nT}\mathbf{1}_n^T \mathbf{A}_0 \mathbf{1}_n$ be the average degree of a node and n' be the number of rows (with indices $1 \leqslant k_1 < \cdots < k_{n'} \leqslant n$) of $\mathbf{A}_0$ having row sum at most $e(T\bar{d})^{1+\delta}$. In order to deal with the sparse case, we consider a truncated form of the matrix $\mathbf{A}_0$ by removing specific high-degree nodes. Let $\mathbf{A} \in \mathbb{R}^{n'\times n'}$ be the submatrix of $\mathbf{A}_0$ such that $A_{ij} := (A_0)_{k_i k_j}$ for $i, j \in [n']$. We then perform spectral clustering of the nodes of the network using top K orthogonal eigenvectors corresponding to the top K absolute eigenvalues of $\mathbf{A}$. Finally, $\hat{\mathbf{Z}}$ is extended to $\hat{\mathbf{Z}}_0 \in \mathscr{M}_{n,K}$ by taking $(\hat{\mathbf{Z}}_0)_{k_j,*} := \hat{\mathbf{Z}}_{j,*}, j \in [n']$, and filling the other rows in arbitrarily.

Algorithm 1: Spectral Clustering of the Sum of the Adjacency Matrices
Input: Adjacency matrices $\mathbf{A}^{(1)}, \mathbf{A}^{(2)}, \ldots, \mathbf{A}^{(T)}$; number of communities K; approximation parameter ε, truncation parameter δ.
Output: Membership matrix $\hat{\mathbf{Z}}_0$.

1. Obtain the **sum of the adjacency matrices**, $\mathbf{A}_0 = \sum_{t=1}^{T} \mathbf{A}^{(t)}$.
2. Get $\bar{d} := \frac{1}{nT}\mathbf{1}_n^T \mathbf{A}_0 \mathbf{1}_n$. Let n' be the number of rows (having indices $1 \leqslant k_1 < k_2 < \cdots < k_{n'} \leqslant n$) of $\mathbf{A}_0$ having row sum at most $e(T\bar{d})^{1+\delta}$.
3. Let $\mathbf{A} \in \mathbb{R}^{n'\times n'}$ be the submatrix of $\mathbf{A}_0$: $A_{i,j} = (A_0)_{k_i,k_j}, i, j \in [n']$.
4. Obtain $\hat{U} \in \mathbb{R}^{n'\times K}$ consisting of the leading K eigenvectors of $\mathbf{A}$ corresponding to its largest absolute eigenvalues.
5. Use $(1+\varepsilon)$ approximate K-means clustering algorithm on the row vectors of $\hat{\mathbf{U}}$ to obtain $\hat{\mathbf{Z}} \in \mathscr{M}_{n',K}$ and $\hat{\mathbf{X}} \in \mathbb{R}^{K\times K}$ satisfying

$$||\hat{\mathbf{Z}}\hat{\mathbf{X}} - \hat{\mathbf{U}}||_F^2 \leqslant (1+\varepsilon) \min_{\mathscr{V}\Gamma \in \mathscr{M}_{n'\times K}, \mathbf{X} \in \mathbb{R}^{K\times K}} ||\mathscr{V}\Gamma\mathbf{X} - \hat{\mathbf{U}}||_F^2. \quad (1)$$

6. Extend $\hat{\mathbf{Z}}$ to obtain $\hat{\mathbf{Z}}_0 \in \mathscr{M}_{n,K}$ as follows. $(\hat{\mathbf{Z}}_0)_{j,*} = \hat{\mathbf{Z}}_{i,*}$ (resp. $(1, 0, \ldots, 0)$) for $j = k_i$ (resp. $j \notin \{k_1, \ldots, k_{n'}\}$).
7. $\hat{\mathbf{Z}}_0$ is the estimate of $\mathbf{Z}$.

The reason for using an $(1+\varepsilon)$-approximate K-means clustering algorithm is completely theoretical. K-means clustering is originally an NP-hard problem with any K-means clustering algorithm generating an approximate solution. However, we need a guarantee on the error of K-means clustering algorithm. So, we choose to use the K-means algorithms that can give us a guarantee on the error of the optimized objective function like algorithms proposed in [6,15].

In Algorithm 2, we formulate the spherical spectral clustering method, which is a modification (proposed by [12] and used by [17]) of Algorithm 1 on $\mathbf{A}$, which is a submatrix of $\mathbf{A}_0 = \sum_{t\in[T]} \mathbf{A}^{(t)}$ as described in Algorithm 1. Let $\mathbf{A}$ and $\hat{\mathbf{U}}$ be defined as in Algorithm 1. Let n'' be the number of nonzero rows (with indices $1 \leqslant l_1 < l_2 < \cdots < l_{n''} \leqslant n'$) of $\hat{\mathbf{U}}$. Let $\hat{\mathbf{U}}^+ \in \mathbb{R}^{n'' \times K}$ consist of the normalized nonzero rows of $\hat{\mathbf{U}}$, i.e. $\hat{\mathbf{U}}^+_{i,*} = \hat{\mathbf{U}}_{l_i,*}/||\hat{\mathbf{U}}_{l_i,*}||_2$ for $i \in [n'']$. We then perform clustering of the rows of $\hat{\mathbf{U}}^+$ to get $\check{\mathbf{Z}}^+$. Finally, $\check{\mathbf{Z}}^+$ is extended to $\check{\mathbf{Z}} \in \mathscr{M}_{n',K}$, and then $\check{\mathbf{Z}}$ is extended to $\check{\mathbf{Z}}_0 \in \mathscr{M}_{n,K}$ by taking $\check{\mathbf{Z}}_{l_j,*} := \check{\mathbf{Z}}^+_{j,*}, j \in [n'']$, and $(\check{\mathbf{Z}}_0)_{k_j,*} := \check{\mathbf{Z}}_{j,*}, j \in [n']$, and filling in the remaining rows arbitrarily. $\check{\mathbf{Z}}_0$ is the estimate of $\mathbf{Z}$ from this method. Unlike in Algorithm 1, we use a K-medians clustering algorithm in Algorithm 2 since the objective function in Eq. (2) is in ℓ_2-norm form in stead of sum of squared form as in Eq. (1) of Algorithm 1. However, like in Algorithm 1, the reason for using an $(1+\varepsilon)$ approximate K-medians clustering algorithm in Algorithm 2 is also purely theoretical as we need guarantee for the heuristic algorithms used to solve the K-medians problem as given in works like [6,15]. The proposed method is highly scalable as $\mathbf{A}$ is a sparse matrix and eigen-decomposition with top K eigenvalues of a sparse matrix is highly scalable [4,9,16].

Algorithm 2: Spherical Spectral Clustering of the Sum of the Adjacency Matrices
Input: Adjacency matrices $A^{(1)}, A^{(2)}, \ldots, A^{(T)}$; number of communities K; approximation parameter ε, truncation parameter δ.
Output: Membership matrix $\check{\mathbf{Z}}$.

1. Perform till Step 4 of Algorithm 1.
2. Let n_+ be the number of nonzero rows of $\hat{\mathbf{U}}$. Obtain $\hat{\mathbf{U}}^+ \in \mathbb{R}^{n_+ \times K}$ consisting of normalized nonzero rows of $\hat{\mathbf{U}}$, i.e. $\hat{\mathbf{U}}^+_{i,*} = \hat{\mathbf{U}}_{i,*}/\left\|\hat{\mathbf{U}}_{i,*}\right\|_2$ for i such that $\left\|\hat{\mathbf{U}}_{i,*}\right\|_2 > 0$.
3. Use $(1+\varepsilon)$ approximate K-median clustering algorithm on the row vectors of $\hat{\mathbf{U}}^+$ to obtain $\check{\mathbf{Z}}^+ \in \mathscr{M}_{n_+,K}$ and $\check{X} \in \mathbb{R}^{K \times K}$ satisfying

$$\left\|\check{\mathbf{Z}}^+\check{\mathbf{X}} - \hat{\mathbf{U}}^+\right\|_F \leqslant (1+\varepsilon) \min_{\mathscr{V}\Gamma \in \mathscr{M}_{n'' \times K}, \mathbf{X} \in \mathbb{R}^{K \times K}} \left\|\mathscr{V}\Gamma\mathbf{X} - \hat{\mathbf{U}}^+\right\|_F. \quad (2)$$

4. Extend $\check{\mathbf{Z}}^+$ to obtain $\check{\mathbf{Z}}$ by (arbitrarily) adding $n' - n_+$ many canonical unit row vectors at the end, like in Step 6 of Algorithm 1.
5. Extend $\check{\mathbf{Z}}$ to obtain $\check{\mathbf{Z}}_0 \in \mathscr{M}_{n,K}$ as follows. $(\hat{\mathbf{Z}}_0)_{j,*} = \hat{\mathbf{Z}}_{i,*}$ (resp. $(1, 0, \ldots, 0)$) for $j = k_i$ (resp. $j \notin \{k_1, \ldots, k_{n'}\}$).
6. $\check{\mathbf{Z}}_0$ is the estimate of $\mathscr{V}Z$.

3 Theoretical Results About the Performance of the Algorithms

We consider two different models for a *multi-relational network* generation. The first one is *Multi-layer stochastic block model* with (i) the latent membership vector $\mathscr{V}z = (z_1, \dots, z_n)$, where each $z_i \in [K]$, (ii) the set of T, $K \times K$ connectivity probability matrices $\{\mathbf{B}^{(t)}\}_{t=1}^T$ and (iii) the $K \times 1$ probability vector of allocation in each community, $\mathscr{V}\pi = (\pi_1, \dots, \pi_K)$.

$$\mathbf{z}_1, \dots, \mathbf{z}_n \overset{iid}{\sim} \text{Mult}(1; (\pi_1, \dots, \pi_K)), \mathbb{P}\left(A_{ij}^{(t)} = 1 \middle| \mathbf{z}_i, \mathbf{z}_j\right) = B_{\mathbf{z}_i \mathbf{z}_j}^{(t)}, \text{ so} \tag{3}$$

$$A_{ij}^{(t)} \sim Bernoulli(P_{ij}^{(t)}), \text{ where } \mathbf{P}^{(t)} := \mathbf{Z}\mathbf{B}^{(t)}\mathbf{Z}^T. \tag{4}$$

Let d be the maximum expected degree of a node, λ be the average of the smallest eigenvalues of normalized probability matrices $\{\mathbf{B}^{(t)}\}_{t=1}^T$ and

$$\lambda = \frac{n}{Td} \sum_{t \in [T]} \lambda_K(\mathbf{B}^{(t)}) > 0 \tag{5}$$

Theorem 1. *For any $\varepsilon, \eta, \delta > 0$ and $c \in (0, 1)$, there are constants $C_1 = C_1(\varepsilon, c, \delta), C_2 = C_2(c, \delta) > 0$ such that if $Td \geqslant C_2(K/\lambda)^{1+\delta}, n \geqslant 3K$ and if $n_{min} > 2/\lambda$, then the proportion of misclassified nodes in Algorithm 1 is*

$$\leqslant \left(\frac{n_{min}}{n}\right)^{-1} e^{-(1-c)Td} + C_1 \left(\frac{n_{min}}{n} - e^{-(1-c)Td}\right)^{-2} K\lambda^{-2}(Td)^{-1+2\eta+2\delta}$$

with probability $\geqslant 1 - 5\exp(-\min\{cTd\lambda, \frac{1}{5}(Td)^{2\eta}\log n\})$. n_{min} = smallest community size. Therefore, in the special case, when (i) K is a constant and (ii) the community sizes are balanced, i.e. $n_{max}/n_{min} = O(1)$, then the proportion of misclassified nodes in $\hat{\mathbf{Z}}_0$ goes to zero with probability $1 - o(1)$ if $Td\lambda \to \infty$.

The other model is *multi-layer degree-corrected block model* with (i) the latent membership vector $\mathscr{V}z = (z_1, \dots, z_n)$, where each $z_i \in [K]$, (ii) the set of T, $K \times K$ connectivity probability matrices $\{\mathbf{B}^{(t)}\}_{t=1}^T$, (iii) a set of *degree parameters*

$$\mathscr{V}\psi = (\psi_1, \dots, \psi_n) \text{ satisfying } \max_{i \in \mathscr{C}_k} \psi_i = 1 \text{ for all } k \in \{1, 2, \dots, K\} \tag{6}$$

and (iv) the $K \times 1$ probability vector of community allocation $\mathscr{V}\pi = (\pi_1, \dots, \pi_K)$.

$$\mathbf{z}_1, \dots, \mathbf{z}_n \overset{iid}{\sim} \text{Mult}(1; (\pi_1, \dots, \pi_K)), \mathbb{P}\left(A_{ij}^{(t)} = 1 \middle| \mathbf{z}_i, \mathbf{z}_j\right) = \psi_i\psi_j B_{\mathbf{z}_i \mathbf{z}_j}^{(t)}, \text{ so} \tag{7}$$

$$A_{ij}^{(t)} \sim Bernoulli(\tilde{P}_{ij}^{(t)}), \text{ where } \tilde{\mathbf{P}}^{(t)} := Diag(\mathscr{V}\psi)\mathbf{Z}\mathbf{B}^{(t)}\mathbf{Z}^T Diag(\mathscr{V}\psi). \tag{8}$$

For $k \in [K]$, let $\tilde{n}'_k := \sum_{i \in \mathscr{C}_k \cap \{k_1, \dots, k_{n'}\}} \psi_i^2$ and $\tau_k := \sum_{i \in \mathscr{C}_k} \psi_i^2 \sum_{i \in \mathscr{C}_k} \psi_i^{-2}$ be a measure of heterogeneity of $\mathscr{V}\psi$.

Theorem 2. *For any $\varepsilon, \eta, \delta > 0$ and $c \in (0, 1)$, there are constants $C_1(\varepsilon, c, \delta), C_2(c, \delta) > 0$ such that if $Td \geqslant C_2(K/\lambda)^{1+1/\delta}$ and $n \geqslant 3K$ is large enough, then the total number of misclassified nodes in Algorithm 2 is*

$$\leqslant \frac{n}{e^{(1-c)Td}} + C_1 \left[\frac{K\tilde{n}'_{max}}{(\psi_{min}\lambda\tilde{n}'_{min})^2} + \frac{\sqrt{K\sum_{k \in [K]} \tau_k n (Td)^{-(1/2)+\delta+\eta}}}{\lambda\tilde{n}'_{min}}\right] \tag{9}$$

with probability at least $1 - 5\exp(-\min\{cTd\lambda, \frac{1}{5}(Td)^{2\eta}\log n\})$.

Therefore, in the special case, when (i) K is a constant, (ii) the community sizes are balanced, i.e. $n_{max}/n_{min} = O(1)$ and (iii) $\psi_i = \alpha_i / \max\{\alpha_j : z_i = z_j\}$, where $(\alpha_i)_{i=1}^n$ are i.i.d. positive weights, then consistency holds for $\mathbf{Z}_0$ with probability $1 - o(1)$ if $\mathscr{E}[\max\{\alpha_1^2, \alpha_1^{-2}\}] < \infty$ and $Td\lambda \to \infty$.

Remark 3. *The condition "$Td\lambda \to \infty$" is necessary and sufficient in order to have a consistent estimator of $\mathbf{Z}$. Theorems 1 and 2 proves the sufficiency. The necessity of the condition follows from the work of* [31].

Remark 4. *Note that the assertion of Theorems 1 and 2 are non-asymptotic results, so, the asymptotic result on consistent label recovery can hold for different conditions like - (i) constant T and $n \to \infty$; (ii) constant n and $T \to \infty$; (iii) $K \to \infty$ and suitable conditions on n, d and T and so on.*

4 Simulation Results

We simulate multiple stochastic block model with $n = 40,000$, $K = 4$ and $T = 10$, but varying $\{\mathbf{B}^{(t)}\}_{t=1}^T$ such that $Td\lambda$ increases. We simulate multiple degree-corrected block model with $n = 20,000$, $K = 4$ and $T = 10$, but varying $\{\mathbf{B}^{(t)}\}_{t=1}^T$ such that $Td\lambda$ increases. The degree parameters in multiple degree-corrected block model are generated from $U(0.5, 1)$.

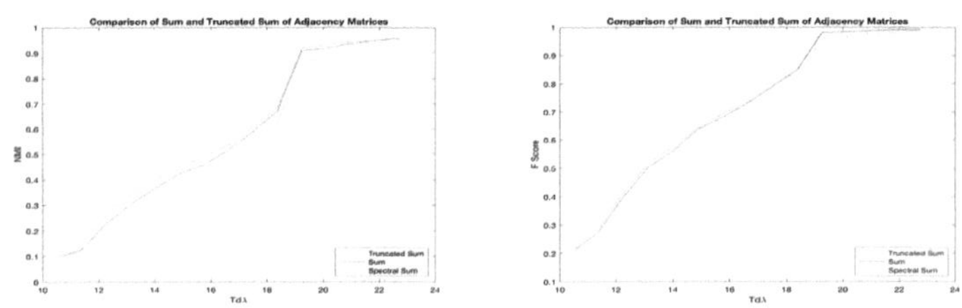

(a) (b)

Fig. 1. Comparison of (i), (ii) and Algorithm 1 (Truncated Sum) (a) using normalized mutual information and (b) using F-score.

We implement five different algorithms - (i) *Sum*: spectral clustering with sum of adjacency matrices without truncation; (ii) *Spectral sum*: clustering the rows of sum of eigen-spaces $\sum_{t=1}^T \hat{\mathbf{U}}^{(t)}$ of each network snapshot (where, $\hat{\mathbf{U}}_{n\times K}^{(t)}$ is the matrix formed by the eigenvectors of top K eigenvectors of $A^{(t)}$); (iii) *Sum (Spherical)*: spherical spectral clustering with sum of adjacency matrices without truncation; (iv) Algorithm 1; (v) Algorithm 2. For models generated under multiple stochastic block model, we compare the algorithms (i), (ii) and (iv). For models generated under multiple degree-corrected block model, we compare the algorithms (iii) and (v). For metric of success, we use normalized mutual information and F-score. We can see from Figs. 1 and 2 that for $Td\lambda$ between

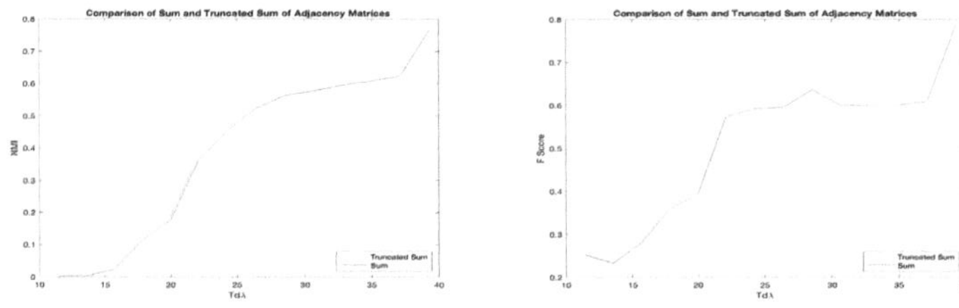

(a) (b)

Fig. 2. (a) Comparison of (iii) and Algorithm 2 (Truncated Sum) (a) using normalized mutual information and (b) using F-score.

$(10, 20)$ for multiple stochastic block model and $(10, 40)$ for multiple degree-corrected block model, Algorithms 1 and 2 out-performs all other algorithms. The simulation results are in concert with the theoretical results in Theorems 1 and 2.

5 Discussion

In this paper, we consider the problem of community detection in a given multi-level network having a common community structure and (possibly) varying connectivity among the network nodes. We propose a spectral clustering and a spherical spectral clustering algorithm on a suitably truncated version of the sum of adjacency matrices. We show theoretically that our spectral (resp. spherical spectral) method is guaranteed to recover communities of a single network and a multi-level network consistently if the networks are generated from stochastic block model (resp. degree-corrected block model) irrespective of the sparsity of the aggregated network. Naturally, our spherical spectral method also works for the multi-level stochastic block model too, as it is a particular case of multi-level degree-corrected block model.

As mentioned in the Introduction, the community detection problem is a particular case of the vertex clustering problem. We discuss the general vertex clustering problem on multi-level networks in a separate paper, which will appear somewhere else. Next, we will consider the case of sparse dynamic networks, where the community structure evolves with time, and the edge structures and community structures of individual networks are correlated. Finding the number of communities for community detection in multi-level networks is another potential future problem.

Finding the number of communities for community detection in multi-level networks is another potential future problem.

6 Proof of Theorems 1 and 2

The proof of Theorem 1 follows from Lemma 5 and Theorem 6. Without loss of generality, we can assume $k_i = i, i \in [n']$. Let $n'_a := ||\mathbf{Z}_{[n'],a}||_2^2, a \in [K]$,

$\Delta := Diag(\sqrt{n'_1}, \ldots, \sqrt{n'_K}), \mathbf{B} := \sum_{t=1}^{T} \mathbf{B}^{(t)}$. and $\mathbf{P} = \mathbf{Z}_{[n'],*}\mathbf{B}\mathbf{Z}_{[n'],*}^T$. Also, let $\mathbf{R}\mathbf{D}\mathbf{R}^T$ be the spectral decomposition of $\Delta\mathbf{B}\Delta - Diag(\mathbf{B})$ and $\mathbf{U} := \mathbf{Z}_{[n'],*}\Delta^{-1}\mathbf{R}$.

Lemma 5. *There is an orthogonal matrix* $\mathbf{Q} \in \mathbb{R}^{K\times K}$ *such that if* $S_a := \{i \in [n'] : Z_{ia} = 1, 2\sqrt{n'_a}||(\hat{\mathbf{Z}}\hat{\mathbf{X}} - \mathbf{U}\mathbf{Q})_{i,*}||_2 > 1\}, a \in [K]$, *then all nodes in* $[n'] \setminus \cup_{a\in[K]} S_a$ *are correctly classified in Algorithm 1 and* $\sum_{a\in[K]} n_a^{-1}|S_a| \leqslant 32K(4 + 2\varepsilon)(n/n'_{min})^2(Td\lambda)^{-2}||\mathbf{A} - \mathscr{E}(\mathbf{A})||^2$.

Theorem 6. *For any* $\eta, \delta > 0$ *and* $c \in (0,1)$ *there are constants* C, $C' > 0$ *(depending on* c, δ*) such that if* $Td \geqslant C'(K/\lambda)^{1+1/\delta}$ *and* $\mathscr{A} := \{||\mathbf{A} - \mathscr{E}(\mathbf{A})|| \leqslant C(Td)^{1/2+\delta+\eta}\} \cap \{(T\bar{d})^{1+\delta} \geqslant Td\}$, *then* $\mathbb{P}(\mathscr{A}) \geqslant 1 - 4\exp[-\min\{cTd\lambda, (Td)^{2\eta}\log n\}]$.

The proof of Lemma 5 uses some known techniques but involves many additional technical details. We present the proof in the supplement.

Remark 7. *Theorem 6 cannot be proved by only using the conventional matrix concentration inequalities,* e.g., *the matrix Bernstein inequality, which would provide suboptimal bound for the spectral norm (with an extra* $\log(n)$ *factor).*

Remark 8. *There are some methods (in case of single networks) available in the literature for bounding the spectral norm of the centered adjacency matrix, but when these methods are applied on the sum of adjacency matrices, they produce suboptimal bounds. For example,* [18] *use a path counting technique in random matrix theory, but their method would require the condition* $Td \geqslant c(\log(n))^4$ *in order to obtain a similar bound for the spectral norm. In* [17] *the authors use the Bernstein inequality and a combinatorial argument to bound the spectral norm of the entire adjacency matrix (in the single network case), but they need the maximal expected degree to be* $\geqslant c\log(n)$. *So if we adopt that method in our setting, we would need the condition* $Td \geqslant c\log(n)$.

The proof of Theorem 6 involves intricate technical details, as it uses crucial large deviation estimates and combinatorial arguments. Our proof develops on the techniques used in [17] (originally developed by [5] for bounding the second largest eigenvalue of an Erdós-Rényi random graph).

Proof (Theorem 1). First we will bound n'. Let $\mathscr{A}' := \mathscr{A} \cap \{n' \geqslant n(1 - e^{-(1-c)Td})\}$. Note that for any $i \in [n]$, $\sum_{j\in[n],t\in[T]} A_{ij}^{(t)}$ is stochastically dominated by $X \sim Binomial(Tn, d/n)$. So, using standard large deviation argument and the properties of $\mathscr{A}$ described in Theorem 6.

$$\mathscr{E}(n - n'; \mathscr{A}) = \sum_{i\in[n]} \mathbb{P}\left(\sum_{j\in[n],t\in[T]} A_{ij}^{(t)} > e(T\bar{d})^{1+\delta}; \mathscr{A}\right) \leqslant n\mathbb{P}(X \geqslant eTd) \leqslant ne^{-Td}.$$

By the above bound and Markov inequality,

$$\mathbb{P}(\mathscr{A}'^c) \leqslant \mathbb{P}(\mathscr{A}^c) + \frac{\mathscr{E}(n - n'; \mathscr{A})}{ne^{-(1-c)Td}} \leqslant 5\exp\left(-\min\left\{\frac{1}{5}(Td)^{2\eta}\log n, cTd\lambda\right\}\right).$$

Using Lemma 5 and Theorem 6, there is a constant C such that

$$\sum_{a\in[K]} f_a \leqslant e^{-(1-c)Td}\frac{n}{n_{min}} + 32K(4+2\varepsilon)\frac{(Td)^{1+2\delta+2\eta}}{(Td\lambda n'_{\min}/n)^2} \text{ on the event } \mathscr{A}'.$$

This completes the proof as $n_{\min} - n'_{\min} \leqslant n - n' \leqslant ne^{-(1-c)Td}$ on the event $\mathscr{A}'$.

Proof (Theorem 2). The proof follows from Theorem 6 and Lemma 9, which is stated below. Without loss of generality, assume $k_i = i, i \in [n']$, and $l_j = j, j \in [n'']$. Let $\mathscr{V}\Psi := Diag(\mathscr{V}\psi)\cdot \mathbf{Z}$, so $\tilde{n}'_a := \|\mathscr{V}\Psi_{[n'],a}\|_2^2, a \in [K]$. Let $\tilde{\mathbf{U}} \in \mathbb{R}^{n'\times K}$ consist of K leading eigenvectors of $\mathscr{E}\mathbf{A}$. Now we state Lemma 9. Its proof is in the Supplement.

Lemma 9. *There is an orthogonal matrix $\tilde{\mathbf{Q}} \in \mathbb{R}^{K\times K}$ such that if $\tilde{\mathbf{U}}^+$ consists of normalized rows of $\tilde{\mathbf{U}}\tilde{\mathbf{Q}}$ and if $\check{S}' := \{j \in [n''] : \|(\check{\mathbf{Z}}^+\check{\mathbf{X}} - \tilde{\mathbf{U}}^+)_{j,*}\|_2 < 1/4\}$, then all nodes in $\check{S}'$ are correctly classified, and there is a constant $C(\varepsilon)$ such that*
$n' - |\check{S}'| \leqslant C[K\tilde{n}'_{max}/(\psi_{min}\lambda\tilde{n}'_{min})^2 + \sqrt{K}(\sum_{k\in[K]}\tau_k)^{1/2}(n/\tilde{n}'_{min})(Td\lambda)^{-1}\|\mathbf{A} - \mathscr{E}\mathbf{A}\|]$.

On the event $\mathscr{A}'$ (defined before (Proof Theorem 1)), the number of misclassified nodes is $\leqslant n - |\check{S}'|$ which is at most the RHS of (9) by Lemma 9. Thus, (Lemma 9) proves the first assertion of Theorem 2. The second assertion holds because $\tau_k = O(m^2), n_k = O(n) = \tilde{n}_k$ for all k and $\psi_{\min} \gg n^{-1/2}$.

Proof (Theorem 6). We will need the following lemma whose proof is in the Supplement.

Lemma 10. *(A) For any $c \in (0,1)$ and $\delta > 0$, there are constants $C_1, C_2 > 0$ such that if $n \geqslant 3K, Td \geqslant C_2(K/\lambda)^{1+1/\delta}$ and $\mathscr{A}_1 := \{C_1^{-1}T\bar{d} \leqslant Td \leqslant (T\bar{d})^{1+\delta}\}$, then $\mathbb{P}(\mathscr{A}_1^c) \leqslant 2e^{-cTd\lambda}$.*
(B) Let $e(I,J) := \sum_{i\in I, j\in J}(A_0)_{i,j}, I, J \subset [n]$. If $\mathscr{A}_2 := \cap_{|I|\leqslant|J|\leqslant 4\sqrt{n}} \Big\{e(I,J)\log\frac{e(I,J)}{|I|\cdot|J|T(d/n)} \leqslant (Td)^{2\eta}|J|\log\frac{n}{|J|}\Big\} \cup \{e(I,J) \leqslant e^{4.4}|I|\cdot|J|T\frac{d}{n}\}$ for $\eta > 0$, then $\mathbb{P}(\mathscr{A}_2^c) \leqslant n^{6-(Td)^{2\eta}/4}$.

Given $c \in (0,1)$ and $\delta, \eta > 0$, let C_1, C_2 be the constants and $\mathscr{A}_1, \mathscr{A}_2$ be the events appearing in Lemma 10. We will take $\mathscr{A} := \mathscr{A}_1 \cap \mathscr{A}_2 \cap \mathscr{A}_3$, where $\mathscr{A}_3$ is defined in (10), and $C' = C_2$. We will write $\bar{\mathbf{A}}^{(t)}$ (resp. $\bar{\mathbf{A}}$) to denote $\mathbf{A}^{(t)} - \mathscr{E}\mathbf{A}^{(t)}$ (resp. $\mathbf{A} - \mathscr{E}\mathbf{A}$).

In order to bound $\|\bar{\mathbf{A}}\|$, we will use the fact (see *e.g.*, [17, Lemma B.1]) that if $S := \{\mathbf{x} = (x_1, x_2, \ldots, x_{n'}) : \|\mathbf{x}\|_2 \leqslant 1, 2\sqrt{n'}x_i \in \mathbb{Z}\ \forall\ i\}$, then $\|\mathbf{W}\| \leqslant 4\sup_{\mathbf{x},\mathbf{y}\in S}|\mathbf{x}^T\mathbf{W}\mathbf{y}|$ for any symmetric matrix $\mathbf{W} \in \mathbb{R}^{n'\times n'}$. Our argument for bounding $\sup_{\mathbf{x},\mathbf{y}\in S}|\mathbf{x}^T\bar{\mathbf{A}}\mathbf{y}|$ involves the following two main steps: bounding the contribution of (1) *light pairs* and (2) *heavy pairs*. For $\mathbf{x}, \mathbf{y} \in S$, we split the pairs (x_i, y_j) into *light pairs* L and *heavy pairs* $\bar{L}$:

$$L(\mathbf{x},\mathbf{y}) := \left\{(i,j) : |x_i y_j| \leqslant \frac{1}{n}(Td)^{1/2-\eta}\right\}, \bar{L}(x,y) := [n'] \times [n'] \setminus L(\mathbf{x},\mathbf{y}).$$

1. **Bounding the contribution of light pairs.** Here we show that if

$$\mathscr{A}_3 := \left\{ \sup_{\mathbf{x},\mathbf{y}\in S} \left| \sum_{(i,j)\in L(\mathbf{x},\mathbf{y})} x_i y_j \bar{A}_{ij} \right| \leqslant (Td)^{1/2+\eta} \right\}, \tag{10}$$

then $\mathbb{P}(\mathscr{A}_3) \geqslant 1 - \exp(-\frac{1}{8}(Td)^{2\eta} n)$, provided Td is large enough.
2. **Bounding the contribution of heavy pairs.** Here we show that

$$\sup_{\mathbf{x},\mathbf{y}\in S} \left| \sum_{(i,j)\in \bar{L}(\mathbf{x},\mathbf{y})} x_i y_j \bar{A}_{ij} \right| \leqslant C_4 (Td)^{1/2+\eta+\delta} \text{ on the event } \mathscr{A}_1 \cap \mathscr{A}_2. \tag{11}$$

for some constant $C_4 > 0$. (10) and (11) will complete the proof of Theorem 6. Proofs of (10) and (11) are presented in the Supplement.

References

1. Bhattacharyya, S., Bickel, P.J.: Community detection in networks using graph distance. arXiv preprint arXiv:1401.3915 (2014)
2. Chen, P.Y., Hero, A.O.: Multilayer spectral graph clustering via convex layer aggregation: theory and algorithms. IEEE Trans. Signal Inf. Process. Netw. **3**(3), 553–567 (2017)
3. Dong, X., Frossard, P., Vandergheynst, P., Nefedov, N.: Clustering with multi-layer graphs: a spectral perspective. IEEE Trans. Signal Process. **60**(11), 5820–5831 (2012)
4. Drineas, P., Kannan, R., Mahoney, M.W.: Fast Monte Carlo algorithms for matrices II: computing a low-rank approximation to a matrix. SIAM J. Comput. **36**(1), 158–183 (2006)
5. Feige, U., Ofek, E.: Spectral techniques applied to sparse random graphs. Random Struct. Algorithms **27**(2), 251–275 (2005)
6. Feldman, D., Monemizadeh, M., Sohler, C.: A PTAS for k-means clustering based on weak coresets. In: Proceedings of the Twenty-Third Annual Symposium on Computational Geometry, pp. 11–18. ACM (2007)
7. Fiedler, M.: Algebraic connectivity of graphs. Czechoslovak Math. J. **23**(98), 298–305 (1973)
8. Fortunato, S.: Community detection in graphs. Phys. Rep. **486**(3–5), 75–174 (2010)
9. Halko, N., Martinsson, P.G., Tropp, J.A.: Finding structure with randomness: probabilistic algorithms for constructing approximate matrix decompositions. SIAM Rev. **53**(2), 217–288 (2011)
10. Han, Q., Xu, K., Airoldi, E.: Consistent estimation of dynamic and multi-layer block models. In: International Conference on Machine Learning, pp. 1511–1520 (2015)
11. Holland, P.W., Laskey, K.B., Leinhardt, S.: Stochastic blockmodels: first steps. Soc. Netw. **5**(2), 109–137 (1983)
12. Jin, J., et al.: Fast community detection by score. Ann. Stat. **43**(1), 57–89 (2015)
13. Karrer, B., Newman, M.E.: Stochastic blockmodels and community structure in networks. Phys. Rev. E **83**(1), 016107 (2011)

14. Kumar, A., Rai, P., Daumé III, H.: Co-regularized spectral clustering with multiple kernels (2010)
15. Kumar, A., Sabharwal, Y., Sen, S.: A simple linear time (1+ ε)-approximation algorithm for k-means clustering in any dimensions. In: Annual Symposium on Foundations of Computer Science, pp. 454–462 (2004)
16. Lehoucq, R.B., Sorensen, D.C., Yang, C.: ARPACK Users' Guide: Solution of Large-Scale Eigenvalue Problems with Implicitly Restarted Arnoldi Methods, vol. 6. SIAM, Philadelphia (1998)
17. Lei, J., Rinaldo, A., et al.: Consistency of spectral clustering in stochastic block models. Ann. Stat. **43**(1), 215–237 (2015)
18. Lu, L., Peng, X.: Spectra of edge-independent random graphs. Electron. J. Comb. **20**(4), P27 (2013)
19. von Luxburg, U., Belkin, M., Bousquet, O.: Consistency of spectral clustering. Ann. Statist. **36**(2), 555–586 (2008). https://doi.org/10.1214/009053607000000640
20. Matias, C., Miele, V.: Statistical clustering of temporal networks through a dynamic stochastic block model. J. R. Stat. Soc. Ser. B (Stat. Methodol.) **79**(4), 1119–1141 (2017)
21. Ng, A.Y., Jordan, M.I., Weiss, Y., et al.: On spectral clustering: analysis and an algorithm. Adv. Neural Inf. Process. Syst. **2**, 849–856 (2002)
22. Paul, S., Chen, Y.: Spectral and matrix factorization methods for consistent community detection in multi-layer networks. arXiv preprint arXiv:1704.07353 (2017)
23. Paul, S., Chen, Y., et al.: Consistent community detection in multi-relational data through restricted multi-layer stochastic blockmodel. Electron. J. Stat. **10**(2), 3807–3870 (2016)
24. Pensky, M., Zhang, T., et al.: Spectral clustering in the dynamic stochastic block model. Electron. J. Stat. **13**(1), 678–709 (2019)
25. Rohe, K., Chatterjee, S., Yu, B.: Spectral clustering and the high-dimensional stochastic blockmodel. Ann. Statist. **39**(4), 1878–1915 (2011). https://doi.org/10.1214/11-AOS887
26. Shi, J., Malik, J.: Normalized cuts and image segmentation. IEEE Trans. Pattern Anal. Mach. Intell. **22**(8), 888–905 (2000)
27. Sussman, D.L., Tang, M., Fishkind, D.E., Priebe, C.E.: A consistent adjacency spectral embedding for stochastic blockmodel graphs. J. Am. Stat. Assoc. **107**(499), 1119–1128 (2012)
28. Tang, W., Lu, Z., Dhillon, I.S.: Clustering with multiple graphs. In: Ninth IEEE International Conference on Data Mining 2009. ICDM 2009, pp. 1016–1021. IEEE (2009)
29. Xu, K.S., Hero, A.O.: Dynamic stochastic blockmodels for time-evolving social networks. IEEE J. Sel. Top. Signal Process. **8**(4), 552–562 (2014)
30. Young, S.J., Scheinerman, E.R.: Random dot product graph models for social networks. In: International Workshop on Algorithms and Models for the Web-Graph, pp. 138–149. Springer, Heidelberg (2007)
31. Zhang, A.Y., Zhou, H.H., et al.: Minimax rates of community detection in stochastic block models. Ann. Stat. **44**(5), 2252–2280 (2016)
32. Zhang, X., Moore, C., Newman, M.E.: Random graph models for dynamic networks. Eur. Phys. J. B **90**(10), 200 (2017)

Processes

Zealotry and Influence Maximization in the Voter Model: When to Target Partial Zealots?

Guillermo Romero Moreno(✉), Edoardo Manino, Long Tran-Thanh, and Markus Brede

School of Electronics and Computer Science, University of Southampton, Southampton, UK
grm1g17@soton.ac.uk

Abstract. In this paper, we study influence maximization in the voter model in the presence of biased voters (partial zealots) on complex networks. Under what conditions should an external controller with finite budget who aims at maximizing its influence over the system target partial zealots? Our analysis, based on both analytical and numerical results, shows a rich diagram of preferences and degree-dependencies of allocations to partial zealots and normal agents varying with the budget. We find that when we have a large budget or for low levels of zealotry, optimal strategies should give larger allocations to partial zealots and allocations are positively correlated with node degree. In contrast, for low budgets or highly-biased zealots, optimal strategies give higher allocations to normal agents, with some residual allocations to partial zealots, and allocations to both types of agents decrease with node degree. Our results emphasize that heterogeneity in agent properties strongly affects strategies for influence maximization on heterogeneous networks.

Keywords: Influence maximization · Voter model · Zealots · Complex networks

1 Introduction

Perhaps motivated by the increasing prevalence of social media and their influence on public opinion, processes of opinion formation on social networks have found much attention in the recent literature [1,9]. Models in this domain have addressed general properties of opinion diffusion on static and co-evolving networks (refer to, e.g., [8,32] for reviews), but also basic mechanisms underlying phenomena such as radicalization [29] and the role of external influence [12,27].

The authors acknowledge support from the Alan Turing Institute (grant R-SOU-006) and the Royal Society (grant IES\R2\192206). M. Brede thanks the UNSW Canberra and DST Australia for support during this work.

H. Barbosa et al. (Eds.): *Complex Networks XI*, SPCOM, pp. 107–118, 2020.
https://doi.org/10.1007/978-3-030-40943-2_10

Whilst typical models of opinion formation often include the role of external influence, another relevant aspect is the question of its placing, a problem that has been termed *influence maximization (IM)*. Influence maximization has huge relevance for a variety of important applications that range from advertising [13], political and public information campaigns [33,36], to questions on how to optimally encourage a developing economy [2]. Starting with [18], the above problem has a huge pedigree in the computer science literature and has traditionally been studied in the context of the independent cascade (IC) model [15] or variants thereof such as threshold models [16]. In the typical problem setting for IC-like models, one seeks optimal seed nodes from which a one-off cascade can reach as large as possible a part of the social network.

More recent examples of strategically exerted interference on social media aiming at disrupting the democratic process [4,23] have sparked the interest of influence maximization on election events [34]. Here, we want a mechanism of opinion formation in which the agents can repeatedly change their opinion. In contrast to the typical IC literature, where flips of opinion only occur in one direction, these models are more suitable for volatile opinions or longer time frames, as these conditions fit better the political domain. In the following, we focus on the voter model (VM) [11,17] because of its simplicity and prominence in the literature. A key feature of the VM is its linearity, which makes it possible to establish many of its properties [30]. Previous work on the VM has also considered aspects of agent heterogeneity. For this purpose, existing literature has introduced so-called zealots, i.e. agents that have decreased chances of adopting particular opinions. The VM has been studied with perfect zealots whose opinions are completely unaffected by other individuals [19,22,24–26] (also referred to as "inflexible", "committed", "stubborn" or "frozen") and *partial zealots* [21] (also "partisans"), who can change their opinion albeit changes are biased towards one of the opinions. Dynamics including zealots from different opinions typically result in mixed equilibrium states.

Previous work on influence maximization in the VM initially sought the initial opinion distribution in the network that would lead to the preferred opinion in the consensus state with the highest probability [14]. However, in the presence of zealots equilibrium states of the VM are independent of initial opinion distributions, so later works shifted the goal to optimally transforming nodes into zealots to reach improved mixed-equilibrium states [19,35]. Another approach, more akin to problems in network control [20,28], regards perfect zealots as external controllers who optimally build unidirectional influence links (edges) to nodes in the network [6,7,22]. In this approach, external controllers can be seen as political parties, mass media or advertising companies that must choose the targets of their campaign policies.

As pointed out by Aral and Dhillon [3], past approaches to influence maximization often overlook aspects of heterogeneity in agent behavior. Specifically, in the context of the IC, it has been shown that results of IM are strongly affected by agents' susceptibility to adopt opinions [3]—or, translated into the context of the VM, *level of zealotry*. However, whereas demonstrating that zealotry influences optimal allocations in the IM, Aral and Dhillon [3] have not explored detailed mechanisms for such differences. Here, we extend previous work on IM for the voter model with external controllers by considering the effects of different levels of zealotry in the population. An optimal campaign manager with limited resources would never target perfect zealots whose opinions cannot be influenced. However, under what conditions would she target partial zealots and how does this relate to the zealot's topological position in the social network? Below, we shall explore these questions for different network topologies. As zealots represent partly radicalized agents, answers to the questions might help to find ways of reducing radicalism in social systems.

Our paper is organized as follows. Section 2 formulates the proposed IM problem and introduces solution methods. Section 3 first presents analytical and numerical solutions to the problem on simple graph topologies and later extends the analysis to scale-free networks. Section 4 discusses and summarizes our main conclusions.

2 Model and Methods

We introduce a scenario where an external controller wants to influence a group of N individuals connected via a social network. We assume individuals hold binary opinions $o_i \in \{A, B\}, i = 1, \ldots, N$ and define x_i as the probability of node i holding opinion A. Opinion diffusion occurs following the usual VM, where a random node copies the opinion of a random neighbor at each time-step. We assume nodes have a bias against opinion A, which we model by an intrinsic level of zealotry $q_i \in [0,1]$ that gives the probability of not copying opinion A from a neighbor j who holds $o_j = A$. Below, we are interested in optimal strategies for an external controller who wants to promote opinion A in the network. As in [6,7,22], a controller is modeled as a node with fixed opinion that exerts influence by establishing unidirectional links to nodes in the network with a specific link weight $a_i > 0$. Overall, the controller wants to optimize the allocation of the a_i to maximize her vote share, subject to a budget constraint $\mathscr{B}_a \geq \sum_i a_i$. Note that, unlike previous work, we relax the conditions on the a_i by allowing them to adopt any value within $\mathbb{R}$.

To proceed, we write rate equations for the probabilities of nodes holding opinion A, which evolve according to

$$\frac{dx_i}{dt} = (1 - q_i)(1 - x_i)\frac{\sum_j w_{ij}x_j + a_i}{s_i + a_i} - x_i\frac{\sum_j w_{ij}(1 - x_j)}{s_i + a_i}, \tag{1}$$

where $W = (w_{ij})$ is the weighted adjacency matrix and $s_i = \sum_j w_{ij}$ is the strength of node i. Note that the left term on the right-hand side of the equation accounts for the probability of a node holding opinion B and copying the opinion of a neighbor with A (which is reduced by the level of zealotry q_i), while the right term reflects the opposite situation.

We focus on influence maximization in the expected equilibrium state, which is unique and asymptotically reached irrespective of initial conditions [35].

The probabilities of adopting A in equilibrium, x_i^*, can be determined from $dx_i/dt = 0$, leading to a system of N second-order equations

$$0 = (1 - q_i) \sum_j w_{ij} x_j^* + q_i \, x_i^* \sum_j w_{ij} x_j^* - x_i^* \left(s_i + (1 - q_i) \, a_i\right) + (1 - q_i) \, a_i. \quad (2)$$

Since this system of equations is hard to solve numerically, vote shares in the equilibrium state can alternatively be found by performing numerical integration of (1).

The external controller aims to distribute her unidirectional connections in such a way as to maximize the expected vote share in equilibrium $X^* = \frac{1}{N} \sum_i x_i^*$, leading to the optimization problem

$$\max_{\boldsymbol{a}} X^*(W, \boldsymbol{a}, \boldsymbol{q}) \qquad s.t. \quad \mathscr{B}_a \geq \sum_i a_i, \quad a_i \geq 0. \quad (3)$$

As we lack an explicit expression for the total vote share X^* on general networks, we resort to numerical methods for the optimization process. The continuous definition of influence allocations allows gradient ascent techniques [31], for which we need to compute the gradient of the equilibrium vote share with respect to the allocations, $\nabla_{\boldsymbol{a}} X^*$. By applying partial derivatives on (2), we obtain

$$0 = (1 - q_i) \sum_j w_{ij} \frac{\partial x_j^*}{\partial a_l} + q_i \left(\frac{\partial x_i^*}{\partial a_l} \sum_j w_{ij} x_j^* + x_i^* \sum_j w_{ij} \frac{\partial x_j^*}{\partial a_l} \right) - \frac{\partial x_i^*}{\partial a_l} [s_i + (1 - q_i) \, a_i] + \delta_{i,l} (1 - q_i)(1 - x_i^*),$$

which is a system of linear equations whose solution gives us the final gradients

$$\nabla_{\boldsymbol{a}} X^* = \frac{1}{N} (\nabla_{\boldsymbol{a}} \boldsymbol{x}^*)^T \mathbf{1} = \frac{1}{N} \operatorname{diag}\left[(1 - q_i)(1 - x_i^*)\right] \left[\operatorname{diag}(s_i + (1 - q_i) \, a_i) - \operatorname{diag}(q_i) \operatorname{diag}(W \boldsymbol{x}^*) - W \operatorname{diag}(1 - q_i + q_i x_i^*)\right]^{-1} \mathbf{1}, \quad (4)$$

where $\delta_{i,l}$ is the Kronecker delta and the values of $\boldsymbol{x}^*$ are computed via numerical integration, as explained above.

3 Results

To gain intuition about the role of partial zealots in influence maximization in the VM and have some analytical reference, we first explore simple graph topologies in Sect. 3.1. We then extend the experiments to networks with heterogeneous degree distributions in Sect. 3.2, as their network structure is closer to the ones found in real social network data. For simplicity, we only consider undirected networks in all experiments below, with $w_{ij} \in \{0, 1\}$.

3.1 Simple Network Topologies

Complete Graph. We start with an infinite complete graph where a fraction ρ of nodes are partial zealots with equal level of zealotry q_z and the rest of the nodes are normal (unbiased) agents with $q_n = 0$. We assume that the external controller allocates the same link weight to agents of the same type, where a_z and a_n are the weights of link allocations to partial zealots and normal agents as a fraction of nodes in the network. Inserting the budget constraint, we find

$$\mathscr{B}_a/N^2 = \langle a\rangle = \rho\, a_z + (1-\rho)\, a_n = \alpha\langle a\rangle + (1-\alpha)\langle a\rangle,$$

where $\alpha \in [0,1]$ is the unique decision parameter for the distribution of the budget among both types of agents. The whole influence budget is focused on normal agents when $\alpha = 0$, on partial zealots when $\alpha = 1$, and link weights are equal for agents of both types when $\alpha = \rho$.

The total vote share in the equilibrium for these settings can easily be derived from the rate equation in (1), leading to

$$X^* = \rho\, x_z^* + (1-\rho)\, x_n^* = \frac{\langle a\rangle}{q_z \rho}\,\frac{(\alpha - \alpha\, q_z)\langle a\rangle(1-\alpha) + (1-\rho)\rho(1-\alpha\, q_z)}{\langle a\rangle(1-\alpha) + (1-\rho)\rho}, \tag{5}$$

with a boundary at $X^* = 1$. The vote share generally increases with budget availability $\langle a\rangle$ and decreases with the fraction of partial zealots ρ and their level of zealotry q_z. Optimal allocations α^* can be found by solving $\partial X^*/\partial\alpha = 0$, giving

$$\alpha^* = 1 - \frac{\rho(1-\rho)}{\langle a\rangle}\left(\frac{1}{\sqrt{1-q_z}} - 1\right), \tag{6}$$

which is bounded at $\alpha^* = 0$. This result shows that, for a given ρ, optimal allocations target partial zealots the more the bigger the budget $\langle a\rangle$ and the smaller the level of zealotry q_z. We can find the switching point of zealotry q_z^* at which normal agents start to be favored over zealots (when $\alpha^* = \rho$) as

$$q_z^* = 1 - \left(\frac{1}{\langle a\rangle/\rho + 1}\right)^2, \tag{7}$$

which increases with the budget $\langle a\rangle$ and decreases with the density of zealots ρ.

We use the analytical solutions developed above to evaluate the quality of numerical results obtained via gradient ascent as introduced in Sect. 2. For this purpose, we analyze how (re-scaled) optimal allocations to zealots $a_z^*/\langle a\rangle = \alpha^*/\rho$ vary with the level of zealotry q_z and budget size $\langle a\rangle$ for both methods (Fig. 1–left) and observe the resulting equilibrium vote shares X^* (Fig. 1–right). We note that for low values of zealotry q_z or high budget $\langle a\rangle$, the controller achieves full control $X^* = 1$ by both methods despite differences in their solutions of an optimal strategy. The gradient ascent algorithm for numerical results is initialized at targeting the two types of agents equally ($a_z = a_n$) and performs steps in the direction of the gradient the minimum required to achieve $X^* = 1$. Conversely, analytical optimizations do not take into account the boundary at

$X^* = 1$ and provide the most robust strategy. For higher values of q_z or lower budgets $\langle a \rangle$—where full control is not possible—allocation strategies from both methods are found to be in perfect agreement, as well as the equilibrium vote shares achieved by them.

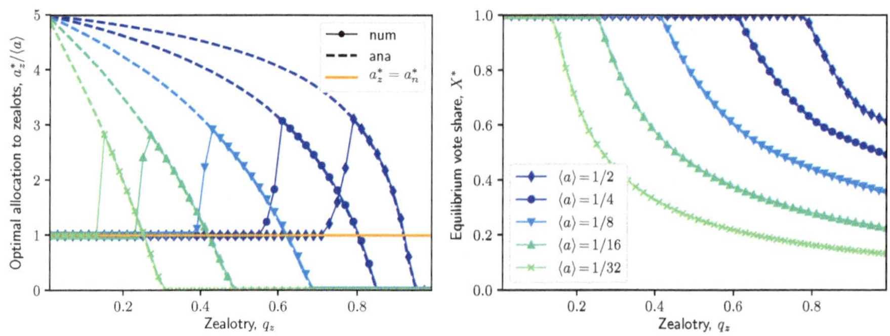

Fig. 1. Re-scaled link allocations given to partial zealots (left) and resulting equilibrium vote shares (right) obtained via analytical (*dashed*) and numerical (*symbols*) methods on a complete network for varying zealotry q_z and varying budget $\langle a \rangle = \mathscr{B}_a/N^2$. The network has size $N = 100$, with 20% of zealots with zealotry q_z and 80% of normal agents with $q_n = 0$. The horizontal line (*orange*) represents equal link weights given to both types of agents.

In general, optimal strategies favor allocations to partial zealots over allocations to normal agents for low values of q_z and high values of $\langle a \rangle$. This behavior gradually decreases with q_z, switching to favoring normal agents at some q_z^* (in the figure, crossing the horizontal orange line) and fully avoiding zealots ($a_z^* = 0$) at another critical level of zealotry q_z^{**}. Both critical points depend on the available budget $\langle a \rangle$. Equilibrium vote shares reach full control $X^* = 1$ for low values of q_z, experience a steep drop just after leaving the full control regime and gradually decelerate in their decrease as q_z increases.

Bipartite Graph. Next, as a way to explore the interplay of degree heterogeneity with zealotry, we explore complete bipartite graphs, i.e. graphs divided into two groups where all nodes from a group are connected to all nodes in the other group. We assume that nodes in the smaller group (*hubs*) comprise a fraction $\rho < 0.5$ of the total network and have zealotry q_h, while nodes from the larger group (*periphery*) have zealotry $q_p = 1 - q_h$. The equilibrium vote share X^* can also be obtained in close-form. Optimal allocations α^* can again be found by solving $\partial X^*/\partial \alpha = 0$, which results in a fourth-order polynomial equation that we evaluate numerically.

Figure 2 analyzes optimal allocations given to partial zealots on a bipartite network for the levels of zealotry defined above, obtained with both analytical and numerical methods. Figure 2–left gives (re-scaled) optimal link allocations to hubs $a_h^*/\langle a \rangle = \alpha^*/\rho$ for different levels of zealotry q_h and budget size $\langle a \rangle$. We note that, for a large budget (*diamond* symbols), there is a consistent preference in allocations toward hub nodes regardless of which group holds a higher level of zealotry—except in the limit of hubs with almost-perfect zealotry ($q_h \approx 1$). On the contrary, for a low budget (*crosses*), optimal allocations quickly switch from fully targeting hubs to fully targeting periphery nodes, even when the level of zealotry of hubs is larger than that of periphery nodes ($q_h < q_p$). Figure 2–right gives the resulting equilibrium vote shares X^* for optimal allocations. We note that, for low budgets (*crosses*), the vote shares obtained in optimal allocations are higher when stronger zealots are concentrated on hubs. For high budgets (*diamonds*), moderate levels of zealotry in both groups ($q_z \approx q_h \approx 0.5$) lead to higher vote shares from optimal allocations.

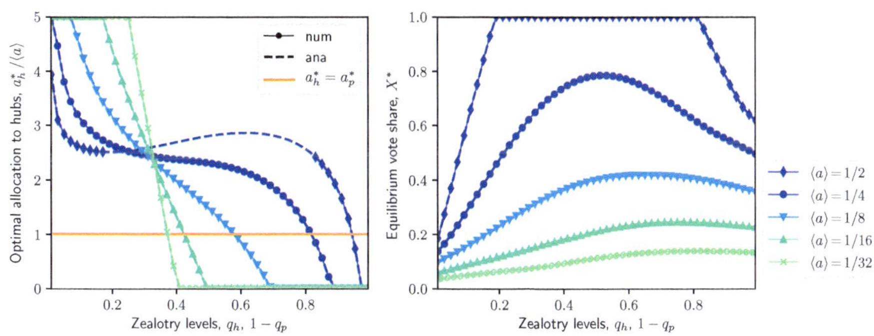

Fig. 2. Re-scaled influence allocations given to hubs (left) and resulting equilibrium vote shares (right) for different levels of zealotry and budget availability $\langle a \rangle$ obtained via analytical (*dashed*) and numerical (*symbols*) methods on a bipartite network of size $N = 100$. Hubs, with zealotry q_h, compose 20% of the network, while periphery nodes compose the remaining 80% and have zealotry $q_p = 1 - q_h$. Optimal influence allocations from numerical methods are omitted when $X^* = 1$ for improved clarity of the graph. Resulting equilibrium vote shares from numerical and analytical approaches match almost perfectly (see overlapping lines).

To summarize our findings on simple network topologies, we see that influence allocations to partial zealots are preferred when their level of zealotry is low or the available budget is high, while allocations to normal agents are preferred otherwise. When nodes of different degrees are present, allocations to hubs are preferred over periphery nodes when the budget availability is high, even when the level of zealotry of hubs is higher than that of periphery nodes. In contrast, for low budgets, higher influence allocations to periphery nodes can be preferable in cases where their level of zealotry is higher than that of hubs.

3.2 Scale-Free Networks. Zealotry and Node Degree

Next, we focus on Barabasi-Albert (BA) networks [5] to further explore how the interplay of zealotry and node degree affects optimal targeting on heterogeneous networks. For our experiments, we generate networks following preferential attachment rules with every new node linking to two existing ones (leading to networks with $\langle k \rangle \approx 4$) and then we allow a random sample of the population to become partial zealots with uniform zealotry q_z, while keeping the remaining nodes as normal (unbiased) agents.

We first explore general behaviors by looking at average optimal allocations given to partial zealots and the resulting equilibrium vote shares (Fig. 3). We note remarkable similarities between the results obtained here and those for the complete graph (Fig. 1) regarding both optimal allocations and equilibrium vote shares. Note that, while for the complete graph the per-node average allocation is expressed as fractions of nodes in the network ($\langle a \rangle = \mathscr{B}_a/N^2$), we take absolute values for the experiments on BA-networks ($\langle a \rangle = \mathscr{B}_a/N$). This results in a re-scaling of $\langle k \rangle/N$ for exerting a similar influence on BA-networks than on complete networks, as discussed in [10]. Again, zealot targeting is preferred when levels of zealotry are low or the available budgets are high. This preference slowly shifts to normal agents as q_z increases, eventually assigning them higher influence links (at q_z^*) and even focusing the whole budget on them (at q_z^{**}).

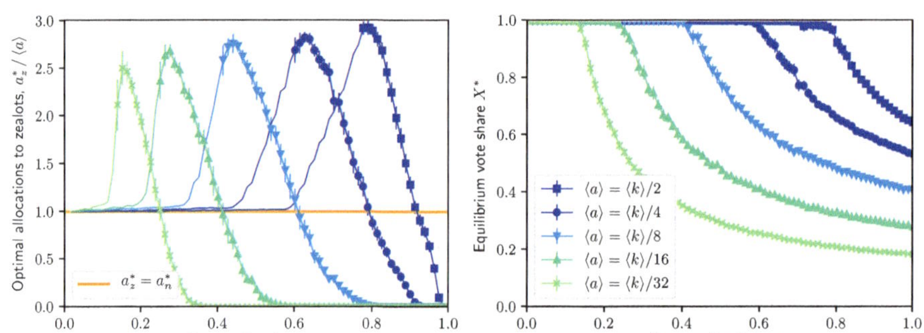

Fig. 3. Average optimal allocations given to partial zealots (left) and resulting equilibrium vote shares (right) on BA networks for varying zealotry q_z and varying per-node budget $\langle a \rangle = \mathscr{B}_a/N$. Networks are of size $N = 1000$, mean degree $\langle k \rangle \approx 4$, and with a random sub-sample of 20% of the network becoming partial zealots. Every point is the average over 50 realizations of the experiment, with error bars giving three standard deviations from the mean. Symbols are omitted for optimal allocations that lead to full control ($X^* = 1$). The horizontal line (*orange*) represents equal link weights given to both types of agents.

We next analyze the relationship between optimal allocations and node degree. More specifically, we want to find whether the link preferences to partial zealots or normal agents are uniformly held across different node degrees.

Figure 4 displays optimal allocations a_k^* grouped by node degree k for a given budget $\langle a \rangle = \langle k \rangle/16$ and three different zealotry parameters $q_z = 0.3, 0.5, 0.9$. We note clear correlations between optimal allocations and node degree in most cases. When the level of zealotry is relatively low ($q_z = 0.3$, *crosses*), the external controller exhibits a clear allocation preference to high-degree nodes, among both zealots and normal agents. For intermediate levels of zealotry ($q_z = 0.5$, *triangles*) optimal controls omit allocations to high-degree zealots, while mildly target zealots with the lowest degrees. This behavior relates to our findings for complete bipartite graphs above, where zealots at the periphery were preferred over hub zealots for low budgets and mild zealotry values. High-degree nodes are still preferred among normal agents in this scenario (right panel). Last, when zealots are highly stubborn ($q_z = 0.9$, *squares*), they remain untargeted, as well as normal agents on hubs, while the correlation with node degree generally decreases.

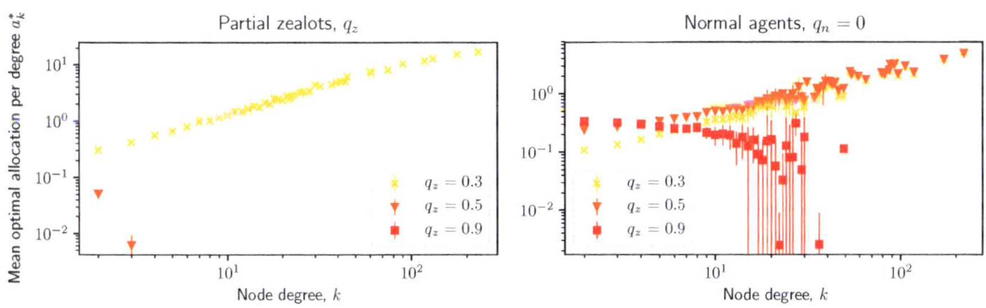

Fig. 4. Average per-degree allocations given to partial zealots (left) and normal agents (right) depending on the node degree k and for three values of q_z and budget $\langle a \rangle = \langle k \rangle/16$ on a BA-network size $N = 5000$. Each point is the mean allocation over all nodes of same degree, with error bars denoting three standard deviations from the mean. These results correspond to a single realization of the experiment.

We extend the correlation analysis in Fig. 4 to a wider range of zealotry values and available budgets. Figure 5 shows the Pearson correlations between a_k^* and k on BA-networks for different scenarios. We note again clear patterns that are in agreement with Fig. 4: hubs are preferred when levels of zealotry are low, with the preference decreasing with q_z and switching to periphery nodes at some q_z^*. The switching points are lower for partial zealots than for normal agents and increase with budget availability. Correlations on allocations to partial zealots eventually reach zero, marking the point where they are fully untargeted.

Combining the information from average allocations given to partial zealots (Fig. 3) and average per-degree allocations given to both groups (Fig. 5), we obtain the following general pattern. When levels of zealotry are low (but not enough for achieving full control of the network), targeting zealots is preferred over targeting normal agents and hubs receive more allocation within each group. Preferences for zealots and zealot hubs diminish as the level of zealotry increases, going through three different transition points. The first transition point q_z^* marks a shift of preference of allocations to normal agents over partial zealots and a shift of preference to zealot with low degree over zealot hubs. After the second transition point q_z^{**}, no allocation is given to zealots and the preference for hubs among normal agents starts to diminish. After the last transition point q_z^{***}, low-degree nodes are preferred over hubs among normal agents.

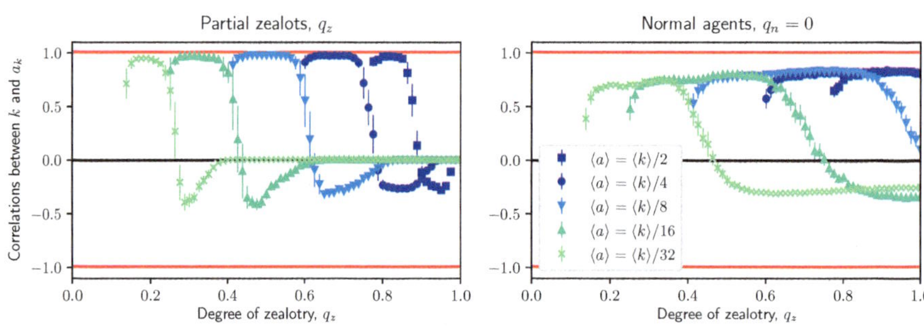

Fig. 5. Pearson correlations between per-degree optimal allocations a_k^* and node degree k for partial zealots (left) and normal agents (right) for different values of $\langle a \rangle$ and q_z on BA-networks of size $N = 1000$. Each point is the average correlation over 50 runs of the experiment with error bars accounting for three standard deviations from the mean, both computed in the Fisher transformation domain. Correlations are only shown for situations in which full control ($X^* = 1$) is not achieved. All p-values fall below 10^{-20}.

4 Conclusions

We have explored influence maximization on heterogeneous networks with some agents biased against the opinion of an external controller (partial zealots). Based on numerical experiments and analytical treatment, a general pattern in optimal influence allocations can be noted. We find that nodes that are harder to control (zealots) receive more influence allocation when the controller's budget is large, while nodes that are easier to control (unbiased) receive more allocation when the budget is small. The transition point between both regimes depends on the level of zealotry and fraction of zealots in the network. When networks with heterogeneous degree distributions are studied, a richer hierarchy emerges, with the following groups—sorted by their difficulty to be controlled in decreasing order—: zealot hubs, periphery zealots, hub normal agents, and periphery normal agents. Our findings fit in the general picture of previous literature which has found that optimal allocations tend to depend on a trade-off between budget availability and the difficulty to control nodes [6,7,31].

Although we have uncovered the essential effects that different levels of zealotry have on influence maximization, some additional research questions remain open. For instance, a natural extension to this work could include zealots biased towards different opinions and heterogeneous levels of zealotry in the same network. Similarly, the presence of two opposing external controllers would also be of high interest, as real social influence scenarios tend to include various actors that compete to spread their exclusive opinions. We plan to explore these questions in future work.

References

1. Acemoglu, D., Ozdaglar, A.: Opinion dynamics and learning in social networks. Dyn. Games Appl. **1**(1), 3–49 (2011)
2. Alshamsi, A., Pinheiro, F.L., Hidalgo, C.A.: Optimal diversification strategies in the networks of related products and of related research areas. Nat. Commun. **9**(1), 1328 (2018). https://doi.org/10.1038/s41467-018-03740-9
3. Aral, S., Dhillon, P.S.: Social influence maximization under empirical influence models. Nat. Hum. Behav. **2**, 375–382 (2018)
4. Badawy, A., Ferrara, E., Lerman, K.: Analyzing the digital traces of political manipulation: the 2016 Russian interference twitter campaign. In: 2018 IEEE/ACM International Conference on Advances in Social Networks Analysis and Mining (ASONAM), pp. 258–265, August 2018
5. Barabási, A.L., Albert, R.: Emergence of scaling in random networks. Science **286**(5439), 509–512 (1999)
6. Brede, M., Restocchi, V., Stein, S.: Resisting influence: how the strength of predispositions to resist control can change strategies for optimal opinion control in the voter model. Front. Robot. AI **5**, 34 (2018)
7. Brede, M., Restocchi, V., Stein, S.: Effects of time horizons on influence maximization in the voter dynamics. J. Complex Netw. **7**(3), 445–468 (2019)
8. Castellano, C., Fortunato, S., Loreto, V.: Statistical physics of social dynamics. Rev. Mod. Phys. **81**(2), 591–646 (2009)
9. Chen, W., Lakshmanan, L.V., Castillo, C.: Information and influence propagation in social networks. Synth. Lect. Data Manag. **5**(4), 1–177 (2013)
10. Chinellato, D.D., Epstein, I.R., Braha, D., Bar-Yam, Y., de Aguiar, M.A.M.: Dynamical response of networks under external perturbations: exact results. J. Stat. Phys. **159**(2), 221–230 (2015)
11. Clifford, P., Sudbury, A.: A model for spatial conflict. Biometrika **60**(3), 581–588 (1973). https://doi.org/10.1093/biomet/60.3.581
12. De, A., Bhattacharya, S., Ganguly, N.: Demarcating endogenous and exogenous opinion diffusion process on social networks. In: Proceedings of the 2018 World Wide Web Conference on World Wide Web - WWW 2018, pp. 549–558. ACM Press, New York, NY, USA (2018). https://doi.org/10.1145/3178876.3186121
13. Domingos, P., Richardson, M.: Mining the network value of customers. In: Proceedings of the Seventh ACM SIGKDD International Conference on Knowledge Discovery and Data mining - KDD 2001, pp. 57–66. ACM Press (2001)
14. Even-Dar, E., Shapira, A.: A note on maximizing the spread of influence in social networks. Inf. Process. Lett. **111**(4), 184–187 (2011)
15. Goldenberg, J., Libai, B., Muller, E.: Talk of the network: a complex systems look at the underlying process of word-of-mouth. Mark. Lett. **12**(3), 211–223 (2001). https://doi.org/10.1023/A:1011122126881
16. Granovetter, M.: Threshold models of collective behavior. Am. J. Sociol. **83**(6), 1420–1443 (1978). https://doi.org/10.1086/226707

17. Holley, R.A., Liggett, T.M.: Ergodic theorems for weakly interacting infinite systems and the voter model. Ann. Probab. **3**(4), 643–663 (1975)
18. Kempe, D., Kleinberg, J., Tardos, E.: Maximizing the spread of influence through a social network. In: Proceedings of the 9th ACM SIGKDD International Conference on Knowledge Discovery and Data Mining, KDD 2003, pp. 137–146. ACM (2003)
19. Kuhlman, C.J., Kumar, V.A., Ravi, S.: Controlling opinion propagation in online networks. Comput. Netw. **57**(10), 2121–2132 (2013)
20. Liu, Y.Y., Slotine, J.J., Barabási, A.L.: Controllability of complex networks. Nature **473**(7346), 167–173 (2011). https://doi.org/10.1038/nature10011
21. Masuda, N., Gilbert, N., Redner, S.: Heterogeneous voter models. Phys. Rev. E **82**, 010103 (2010)
22. Masuda, N.: Opinion control in complex networks. New J. Phys. **17**, 1–11 (2015). https://doi.org/10.1088/1367-2630/17/3/033031
23. McFaul, M., Kass, B.: Understanding Putins intentions and actions in the 2016 U.S. Presidential Election. Technical report, Standford University, June 2019
24. Mellor, A., Mobilia, M., Zia, R.K.P.: Characterization of the nonequilibrium steady state of a heterogeneous nonlinear q-voter model with zealotry. EPL (Europhys. Lett.) **113**(4), 48001 (2016). https://doi.org/10.1209/0295-5075/113/48001
25. Mobilia, M.: Does a single zealot affect an infinite group of voters? Phys. Rev. Lett. **91**, 028701 (2003)
26. Mobilia, M., Petersen, A., Redner, S.: On the role of zealotry in the voter model. J. Stat. Mech: Theory Exp. **2007**(08), P08029–P08029 (2007). https://doi.org/10.1088/1742-5468/2007/08/P08029
27. Palombi, F., Ferriani, S., Toti, S.: Influence of periodic external fields in multiagent models with language dynamics. Phys. Rev. E **96**(6), 062311 (2017)
28. Porfiri, M., di Bernardo, M.: Criteria for global pinning-controllability of complex networks. Automatica **44**(12), 3100–3106 (2008)
29. Ramos, M., Shao, J., Reis, S.D.S., Anteneodo, C., Andrade, J.S., Havlin, S., Makse, H.A.: How does public opinion become extreme? Sci. Rep. **5**, 10032 (2015)
30. Redner, S.: Reality-inspired voter models: a mini-review. C.R. Phys. **20**(4), 275–292 (2019). https://doi.org/10.1016/j.crhy.2019.05.004
31. Romero Moreno, G., Tran-Thanh, L., Brede, M.: Continuous influence maximisation for the voter dynamics: is targeting high-degree nodes a good strategy? Manuscript Submitted for Publication (2019)
32. Sîrbu, A., Loreto, V., Servedio, V.D.P., Tria, F.: Opinion dynamics: models, extensions and external effects, pp. 363–401. Springer, Heidelberg (2017)
33. Wilder, B., Ou, H.C., de la Haye, K., Tambe, M.: Optimizing network structure for preventative health. In: Proceedings of the 17th International Conference on Autonomous Agents and MultiAgent Systems, AAMAS 2018, pp. 841–849. International Foundation for Autonomous Agents and Multiagent Systems (2018)
34. Wilder, B., Vorobeychik, Y.: Controlling elections through social influence. In: Proceedings of the 17th International Conference on Autonomous Agents and MultiAgent Systems, AAMAS 2018, pp. 265–273. International Foundation for Autonomous Agents and Multiagent Systems, Richland, SC (2018)
35. Yildiz, E., Ozdaglar, A., Acemoglu, D., Saberi, A., Scaglione, A.: Binary opinion dynamics with stubborn agents. ACM Trans. Econ. Comput. **1**(4), 1–30 (2013). https://doi.org/10.1145/2538508
36. Zhang, H., Vorobeychik, Y., Letchford, J., Lakkaraju, K.: Data-driven agent-based modeling, with application to rooftop solar adoption. Auton. Agents Multi-Agent Syst. **30**(6), 1023–1049 (2016)

Collective Decision-Making on Triadic Graphs

Ilja Rausch(✉), Yara Khaluf, and Pieter Simoens

IDLab - Department of Information Technology, Ghent University - imec, Technologiepark 126, 9052 Ghent, Belgium
ilja.rausch@ugent.be

Abstract. Many real-world networks exhibit community structures and non-trivial clustering associated with the occurrence of a considerable number of triangular subgraphs known as triadic motifs. Triads are a set of distinct triangles that do not share an edge with any other triangle in the network. Network motifs are subgraphs that occur significantly more often compared to random topologies. Two prominent examples, the feedforward loop and the feedback loop, occur in various real-world networks such as gene-regulatory networks, food webs or neuronal networks. However, as triangular connections are also prevalent in communication topologies of complex collective systems, it is worthwhile investigating the influence of triadic motifs on the collective decision-making dynamics. To this end, we generate networks called Triadic Graphs (TGs) exclusively from distinct triadic motifs. We then apply TGs as underlying topologies of systems with collective dynamics inspired from locust marching bands. We demonstrate that the motif type constituting the networks can have a paramount influence on group decision-making that cannot be explained solely in terms of the degree distribution. We find that, in contrast to the feedback loop, when the feedforward loop is the dominant subgraph, the resulting network is hierarchical and inhibits coherent behavior.

Keywords: Complex networks · Triadic motifs · Collective decision-making · Group coherence · Feedforward loop · Hierarchality

1 Introduction

The network topology defining agent interactions plays a crucial role in swarm-inspired collective systems [4,8,11,14,17]. Researchers are beginning to unravel the influence of the topology on collective decision-making and propose engineering approaches that increase the performance of (artificial) collective systems [11,14,16]. To gain insights into how interaction networks impact decision-making, a popular approach is to generate well defined topologies that are inspired from real-world observations. Often these topologies are designed to reduce complexity and focus on the impact of specific network properties.

One of the properties worth investigating is the presence of triangular connections. Recently, an abundance of triangular subgraphs called *triadic motifs* has been discovered in numerous real-world networks [13,15,21]. Particularly, the feedforward loop

H. Barbosa et al. (Eds.): *Complex Networks XI*, SPCOM, pp. 119–130, 2020.
https://doi.org/10.1007/978-3-030-40943-2_11

was found, among others, in gene-regulatory networks, food webs or neuronal networks [1,15,21].

In light of these discoveries, a reliable method was developed for generating graphs with an abundance of such motifs, called *Triadic Random Graphs* (TRGs). A TRG is generated using exclusively triadic motifs $\mathscr{M}_i$ as building blocks [20]. This approach is based on conditionally independent sampling of triples from *Steiner Triple Systems* and the assignment of the motif-specific topology to each triple. As a result, the homogeneous TRG $\mathscr{T}_i$ is a directed graph that consists purely of T distinct triadic subgraphs of type $\mathscr{M}_i$. The benefits of TRGs are twofold: (i) the dynamics emerging on the local level of each triadic motif are directly influenced through no more than three nodes, i.e. higher-order correlations are reduced, (ii) the global network topology can be purely described in terms of motif type $\mathscr{M}_i$ and T (no higher order motif analysis is required). Network properties such as average degree, clustering coefficient or shortest path can be controlled, at least to some extent, by the choice of $\mathscr{M}_i$ and T. Thus, TRGs can provide a simple but powerful fundamental design for studying the impact of triadic connections on complex dynamic processes.

An increasingly popular example of complex dynamics is the collective decision-making observed in locust marching bands [2–4,8,16]. In essence, it represents a symmetry-breaking scenario in which a large group faces the choice between two equal-quality options. In dense groups, the corresponding communication network displays a high number of triangular connections whose impact on the group behavior is not yet fully understood [8,11]. In this paper we aim at enhancing this understanding by studying the influence of triadic motifs on the locust-inspired symmetry-breaking model.

A common approach for modeling swarming systems such as locust marching bands, is to allow the individuals influence each other's behavior based on physical proximity (i.e. euclidean distance) [2,6,10,18]. However, to focus the analysis solely on the contribution of the motif topology, the approach taken in the current study is different in that we only consider the non-euclidean, static group communication network, disregarding correlations in physical space. This approach is necessary as it allows applying pre-constructed well-defined triadic graphs and, thus, comparing the contribution of the different motif types in a controlled way. The findings obtained this way can then serve as a baseline for more sophisticated methods such as adaptive-network models [4,8,9].

Nevertheless, as we will show below, our network-driven approach is sufficient to reproduce the main characteristics of state trajectories known from the swarming model [2,3]. Moreover, the network approach offers the possibility to confirm the validity of the previously analytically derived upper limit to the coherence degree $\overline{|\phi|}$ (a measure of the group alignment) [16]. In particular, we will demonstrate that for some motif types, the group may reach this upper bound of $\overline{|\phi|}$ while other motifs have a comparably inhibitory impact. As we will show, a shared characteristic of the latter motif types is the presence of nodes with zero out-degree. Decreasing their number resulted in an increase of $\overline{|\phi|}$. Finally, we observed that the use of the feedforward loop led to a hierarchical network structure that was particularly detrimental to the group alignment. Therefore, our results suggest that the motif type can have important consequences for collective decision-making.

We introduce the TRGs in Sect. 2 and the locust-inspired collective decision-making model in Sect. 3. Subsequently, we report and discuss our results in Sect. 4, and conclude in Sect. 5.

2 Triadic Graphs

2.1 Triadic Motifs

In the current study we focus on the directed closed *triadic motifs* $\mathcal{M}_i$ illustrated in Fig. 1. Among these, are the *feedforward loop* $\mathcal{M}_1$, the *feedback loop* $\mathcal{M}_3$ and the *bi-directional loop* $\mathcal{M}_7$. To investigate the influence of triadic motifs on the collective decision-making we generated seven types of TGs, $\mathcal{T}_i$, following a procedure described below. For each $\mathcal{T}_i$ only one type of triadic motifs, $\mathcal{M}_i$, was used as building block, respectively. In each $\mathcal{T}_i$ every node is an element of at least one triadic motif. The total number of distinct triadic motifs was verified using the MFINDER software (version 1.2)[1].

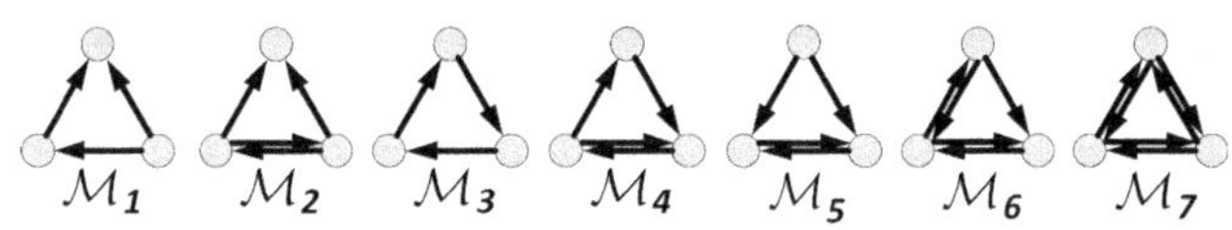

Fig. 1. The seven types of triadic motifs used as building blocks of TGs $\mathcal{T}_i$.

2.2 Steiner Triple Systems

A recent mechanism proposed by [20] specializes in network construction using triadic motifs as primary building blocks. This mechanism is based on the concept of *Steiner Triple Systems* (STS), a mathematical design of a structure consisting of distinct three element subsets (i.e. triples). The most profound feature of the STS is that any pair of elements can be connected through only one unique link that belongs to exactly one triad. Therefore, a network based on Steiner Triples contains exclusively distinct triadic subgraphs. To realize an STS with N elements, two necessary and sufficient conditions need to be satisfied [12]:

$$N \bmod 2 = 1, \; N(N-1) \bmod 3 = 0 \tag{1}$$

leading to an upper bound for the number of distinct triads T [20]:

$$T \leq \frac{1}{3}\frac{N(N-1)}{2}. \tag{2}$$

To satisfy Eqs. (1)–(2), all TRGs generated in the current study have a size of $N = 7^3 = 343$ and $T = 343n$ with $n \in \mathbb{N} < 8$.

[1] https://www.weizmann.ac.il/mcb/urialon/.

2.3 Triadic Graphs

Similar to the Erdős-Rényi model, where the assignment of edges is not conditioned on the current state of the network, in the model proposed by [20] the triad assignment is conditionally independent and the resulting graph is the TRG. A key aspect of this approach is that T triads are sampled from a pre-constructed STS G_{STS}. Each sampled triad is assigned the motif-specific topology (such as those shown in Fig. 1). Therefore, triadic motifs are the fundamental parts constituting (directed) TRGs. Note that this model allows the generation of random network ensembles that have the same number of edges and nodes.

However, to ensure that every generated network has only one connected component, we slightly deviated from the above-mentioned procedure. For each instance, before independently assigning the triads, we first created a seed network G_s (an example is shown in Fig. 2 (left)) by iterating through all nodes of G_{STS} and assigning a triad $\theta \in G_{STS}$ to G_s only if θ satisfies the following two conditions: (i) at least one node $(v \in \theta) \wedge (v \notin G_s)$, (ii) at least one $(v \in \theta) \wedge (v \in G_s)$. As a result, G_s has only one connected component with the same predefined number of nodes N as the G_{STS} from which the triads are sampled. Subsequently, to reach the predefined number T, G_s is assigned new triads by means of randomly drawing new θ from G_{STS} without replacement. Figure 2 (right) shows an example of a final TRG $\mathscr{T}_1$. Due to this procedure the graphs are not as strictly random as the original [20], thus we will henceforth refer to them as Triadic Graphs (TGs).

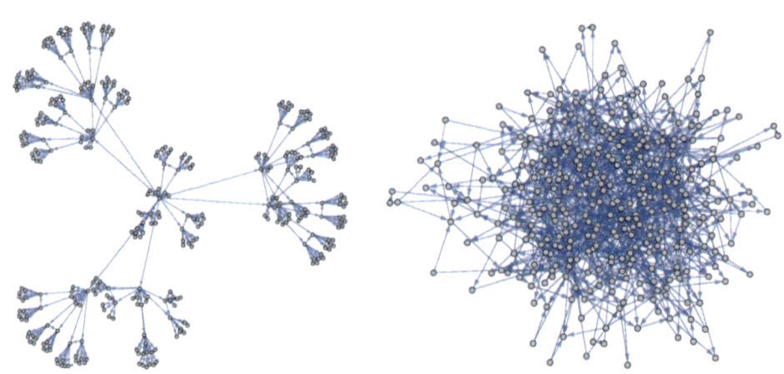

Fig. 2. Examples of **Left:** a seed network G_s, **Right:** a final TG. Both graphs were constructed from motifs of type $\mathscr{M}_1$.

We distinguish between homogeneous and heterogeneous TGs. In the former case the total number of triads is $T = T_i$, with T_i being the number of motifs of type $\mathscr{M}_i$. Whereas, in a heterogeneous TG, $T = \sum_{i=1}^{7} T_i$. The choice of the particular type of $\mathscr{M}_i$ for TG construction strongly determines various network properties, particularly the degree distribution.

The total degree of a node v is given by $k_{tot,v} = k_{in,v} + k_{out,v}$, where $k_{in,v}$ and $k_{out,v}$ are the node's in- and out-degree, respectively. For instance, consider the right-most node of $\mathscr{M}_1$ in Fig. 1. Its degree is given by $k_{in,v} = 0$, $k_{out,v} = 2$ and $k_{tot,v} = 2$. In contrast, the degree of the analogous node of $\mathscr{M}_2$ is given by $k_{in,v} = 1$, $k_{out,v} = 2$ and

$k_{tot,v} = 3$. As the triadic motifs are assigned conditionally independent of each other, the number of triadic motifs around v is binomially distributed for small TGs and Poisson distributed for large TGs. Moreover, the global average of the total degree is correlated with the type of $\mathscr{T}_i$. For instance, consider again the motifs $\mathscr{M}_1$ and $\mathscr{M}_7$. The total in-degree, summed over all three nodes, is $k_{in,\mathscr{M}_1} = 3$ and $k_{in,\mathscr{M}_7} = 6$, respectively (and similarly for the out-degree). Thus, the total degree distribution of a network is a function of the type and number of triadic motifs and can be deduced from the TG model [20]. Consequently, the average total degree can be calculated as a function of $\mathscr{T}_i$: $\langle k_{tot} \rangle = \frac{T_i}{N}(k_{in,\mathscr{M}_i} + k_{out,\mathscr{M}_i})$.

Finally, it can be useful to isolate the influence of the degree distribution on the dynamical processes from that of the triadic motifs. For this purpose, we compare the simulation outcomes of TGs to *null-models* which are the respective randomizations. The randomized version of each TG is generated using the a Markov Chain Monte Carlo rewiring algorithm [19]. In essence, the algorithm runs a predefined number of mutually independent, degree-preserving rewirings between pairs of nodes. As a consequence, the initial abundance of triadic motifs vanishes while the network size, the number of connections and the distribution of in- and out-degrees are preserved.

3 Decision Making Model

We examine the influence of triadic motifs on the decentralised decision-making dynamics in the canonical model of locust marching [3, 8, 11, 16]. This biologically inspired model represents a binary decision problem in which the agents need to collectively decide whether to go left or right [2, 6, 22].

Consider a one-dimensional opinion space o_i of an individual i, in which a commitment to option A (B) corresponds to $o_i < 0$ ($o_i > 0$), respectively. Then, similar to the locust velocity [2, 16, 22], the opinion of i is updated at each time step to

$$o_i(t+1) = \delta_s \left[G(\langle o_i(t) \rangle) + \zeta_i(t) \right], \tag{3}$$

where $\zeta_i(t) \in [-1.0, 1.0]$ is a real random number sampled from a uniform distribution (representing noise). Additionally, the average opinion of the individual's neighborhood $\langle o_i(t) \rangle$ is given by $\langle o_i(t) \rangle = \frac{1}{k_{in,i}} \sum_{j=0}^{N} o_j(t) A_{ji}$, where A_{ji} is an element of the network's $N \times N$ adjacency matrix, with $A_{ji} = 1$ if there is a direct link leading from j to i and $A_{ji} = 0$ otherwise. As a common simplification, self-loops are excluded, i.e. $A_{ii} = 0$ for all i; $k_{in,i} = \sum_{j=0}^{N} A_{ji}$ is the i's total in-degree. Moreover, the contribution of the individual's neighbors is maintained close to ± 1.0 by the piece-wise continuous function

$$G(\langle o_i(t) \rangle) = \frac{1}{2} \left[\langle o_i(t) \rangle + sgn(\langle o_i(t) \rangle) \right], \tag{4}$$

with $sgn()$ being the sign-function. Note that for $G(\langle o_i(t) \rangle) = \langle o_i(t) \rangle$, Eq. (3) would be similar to the majority vote model. Finally, δ_s in Eq. (3) is a binary digit that is $\delta_s = -1$ with probability p_s and $\delta_s = 1$ otherwise. While the Czirók model is obtained for $p_s = 0$, setting $p_s > 0$ extends this model and represents the ability of each individual to spontaneously change its opinion [4, 8, 11]. This spontaneous opinion switch does not necessarily represent noise but can also be attributed to the individual's drive to explore alternatives and perpetuate self-organization. Alternatively, it could be a malicious behavior originating from external attacks or a similar underlying mechanism unknown to the observer [16].

Similar to [11,16], the collective opinion of the N individuals, i.e. the collective state of the system is given by:

$$\phi(t) = \frac{1}{N}\sum_{i=0}^{N} sgn\left(o_i(t)\right). \tag{5}$$

The collective *coherence degree* $|\phi(t)|$ is defined as the absolute value of the collective opinion. When all individuals agree on an option the system reaches consensus with $|\phi(t)| = 1$.

Beside the spontaneous switching, there are two further significant differences between the standard locust marching model and the decision-making model considered here. First, unlike the previous studies, we do not include the spatial information of the individuals. Instead, we only focus on the opinion dynamics from an abstract network-driven perspective. While previously the neighbors were selected from the spatial proximity of an individual, i.e. within a certain range Δ around the individual [2,3,6,16,22], or from an adaptive network [4,8], here the neighbors are assigned by the adjacency matrix of a predefined static network. Thus, we intentionally isolate the problem from spatial correlation and dynamic link rewiring to focus exclusively on the impact of the subtle differences between the triadic motifs. Second, due to to the directed nature of the triadic motifs, the considered networks are directed which is in contrast to most previous works that focused mainly on bidirectional communication.

4 Results and Discussion

First, we demonstrate that Eqs. (3)–(5) can qualitatively reproduce the main characteristics of the empirically observed locust alignment trajectories [2,6,22]. For this, we simulated the collective decision making on TGs in various configurations. The initial number of individuals with $o_i > 0$ and $o_i < 0$ was always $\lceil \frac{N}{2} \rceil$ and $N - \lceil \frac{N}{2} \rceil$, respectively. The example shown in Fig. 3 is the simulation outcome for three instances of $\mathscr{T}_1$—which differ only in T—that illustrates the qualitative similarities to three major empirical findings: (i) for high enough communication degree, $\phi(t)$ fluctuates around a non-trivial stable state with a residence time $\tau \geq \tau_{tot}$ (dark blue data with $\langle k_{tot} \rangle = 24$); the residence time τ represents the time interval between two state transitions while τ_{tot} is the total experiment duration, with $\tau_{tot} = 5000ts$ for all our experiments; (ii) collective state may repeatedly undergo rapid transitions between temporary stable states with $1 \ll \tau < \tau_{tot}$ (blue data with $\langle k_{tot} \rangle = 12$); (iii) lower communication degree leads to lower group alignment and $\tau \rightarrow 1$ (light blue data with $\langle k_{tot} \rangle = 6$) [11,16].

Moreover, the maximum value of $\overline{|\phi(t)|}$ obtained from simulation agrees well with previous theoretical findings [16] suggesting an upper bound of:

$$|\phi_m| = \frac{|0.5 - p_s|}{1 - |0.5 - p_s|}. \tag{6}$$

Figure 3 (right) showcases this agreement for complete and regular networks. Each $\overline{|\phi|}$ trajectory was averaged over 50 simulation runs. As illustrated in the inset

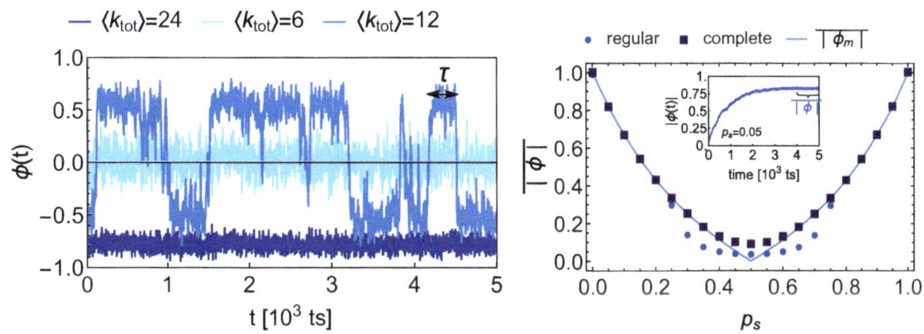

Fig. 3. Left: Collective state trajectories for $p_s = 0.05$ and three $\mathscr{T}_1$ instances with different T leading to different values of $\langle k_{tot} \rangle$. Each trajectory represents one sample, i.e. here the simulations were run with the same random number generator seed. An example of the residence time τ is illustrated for one of the temporary states. **Right:** Coherence degree $\overline{|\phi(t)|} = \overline{|\phi|}$ over p_s for a complete and a regular network with $\langle k_{tot} \rangle = 48$, both networks with $N = 343$. The inset shows the period over which the time average $\overline{|\phi|}$ of the collective state was taken.

of Fig. 3 (right), each data point represents the time average of $\overline{|\phi|}$ obtained over the last 1000 time steps to account for a transient period occurring in the early stages of the simulations. Similar to [16], the global coherence degree does not exceed the limit given by Eq. (6) for the regular network of $\langle k_{tot} \rangle = 48$ (this $\langle k_{tot} \rangle$ value is comparable to the high-degree simulations in [16]). Additionally, as expected for the complete network, $\overline{|\phi|}$ exceeds $\overline{|\phi_m|}$ only for $p_s \approx 0.5$, i.e. where Eq. (6) is not well defined. Note that for $p_s > 0.5$, $\overline{|\phi|}$ increases. For these p_s the agent is likely to switch its opinion within the same time step as its neighbours, leading to exceedingly synchronous switching and higher $\overline{|\phi|}$ [16].

To investigate the influence of the motifs, it is worthwhile examining the collective ability to achieve $\overline{|\phi_m|}$ for different $\mathscr{T}_i$. Figure 4 (left) shows the simulation outcome for different values of the spontaneous switch probability p_s. As Fig. 4 suggests, $\overline{|\phi|}$ can be influenced by the choice of $\mathscr{M}_i$ although the results are comparably similar for most motifs. The differences between the TGs with motifs $\mathscr{M}_{1-2}$ and $\mathscr{M}_{3-7}$ are particularly apparent with $\mathscr{M}_1$ and $\mathscr{M}_2$ having the largest inhibitory impact. This is reflected in the deviation of $\overline{|\phi|}$ from $\overline{|\phi_m|}$. On the one hand, for some values of p_s, this deviation more than doubles between the TGs $\mathscr{T}_1$ and $\mathscr{T}_{3-7}$ (see inset of Fig. 4 (left)). On the other hand, the coherence degree results are remarkably similar from $\mathscr{T}_3$ to $\mathscr{T}_7$.

To obtain a more detailed view on the impact of the motif topology, we generated null-models (i.e. degree-preserving TG randomizations as described in Sect. 2.3). The decision-making results for the null-models are shown in Fig. 4 (right). Strikingly, the simulation outcomes for $\mathscr{T}_{3-7}$ are very close to their randomized counterparts and the only noticeable difference is observed for $\mathscr{T}_1$, i.e. the TG generated from feedforward loops $\mathscr{M}_1$. This observation indicates that the k_{in} and k_{out} distributions have a more critical impact on $\overline{|\phi|}$ than $\mathscr{M}_i$, with the exception of $\mathscr{M}_1$. In Fig. 4 (right) one can see that despite the absence of motifs, the null-models of $\mathscr{T}_1$ and $\mathscr{T}_2$ still have an inhibitory impact on $\overline{|\phi|}$ in comparison to the other graphs. A shared characteristic of $\mathscr{T}_1$ and $\mathscr{T}_2$

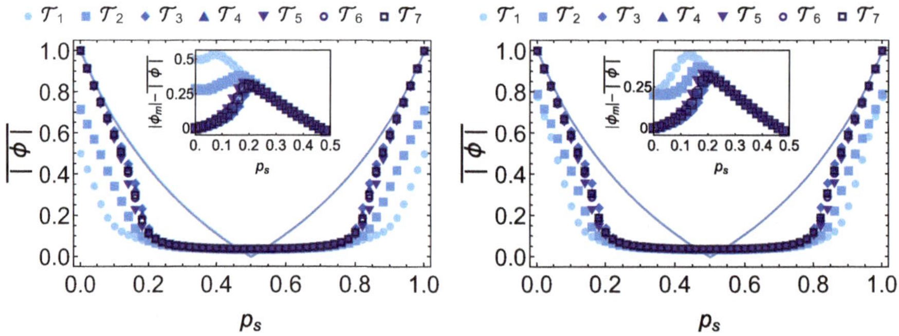

Fig. 4. Coherence degree as a function of $p_s \in [0, 1]$ and the network topology. For all networks, $N = 343$ and $\langle k_{tot} \rangle = 24$; The curve represents the theoretical value of maximum coherence degree $|\phi_m|$. **Left:** TGs $\mathcal{T}_i$, **Right:** Null-models of $\mathcal{T}_i$. The insets show the respective differences between the theoretical value $|\phi_m|$ and the measured time averaged coherence degree $\overline{|\phi|}$.

is the presence of nodes with $k_{out} = 0$ that are unable to communicate their opinion. Interestingly, the inhibitory effects of nodes with $k_{in} = 0$ are lower as evidenced by the comparison with $\mathcal{T}_5$ where all nodes have $k_{out} > 0$ but some nodes may have $k_{in} = 0$. Nodes with $k_{in} = 0$ are not affected by their neighborhood and are thus similar to 'stubborn' individuals commonly referred to as *zealots* [5].

Moreover, as was shown previously, the coherence degree is strongly correlated with $\langle k_{tot} \rangle$ [16]. Higher $\langle k_{tot} \rangle$ leads to higher $\overline{|\phi|}$, up to $\overline{|\phi_m|}$. This behaviour can also be observed for TGs by increasing T, as shown in Fig. 5 for $\mathcal{T}_1$ (left) and $\mathcal{T}_3$ (right). One can see that for a range of p_s values even after increasing $\langle k_{tot} \rangle$ by a factor of eight (i.e. for $T = 343 \times 8 = 2744$), $\mathcal{T}_1$ is associated with lower $\overline{|\phi|}$ than $\mathcal{T}_3$.

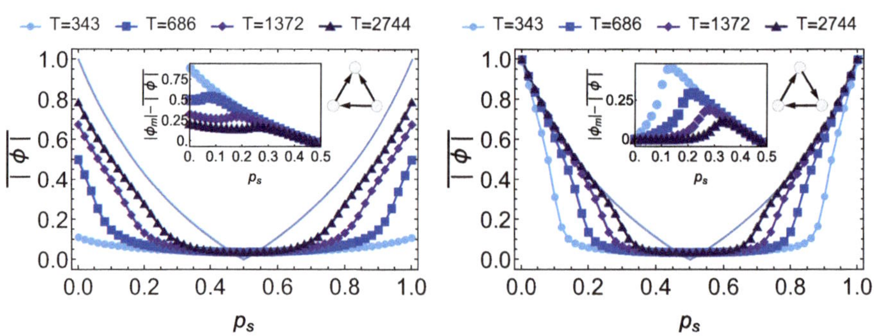

Fig. 5. Left: Simulation outcome for $\mathcal{T}_1$. **Right:** Simulation outcome for $\mathcal{T}_3$. In both plots, the lines between the data points are guides to the eye. The V-shaped curve indicates the theoretical $\overline{|\phi_m|}$. In each panel the insets show the corresponding difference $\overline{|\phi_m|} - \overline{|\phi|}$ as well as an illustration of the building block motif $\mathcal{M}_i$.

Transitioning from $\mathscr{T}_1$ to $\mathscr{T}_3$ by means of gradually replacing quantities of $\mathscr{M}_1$ with $\mathscr{M}_3$ leads to a decrease of nodes with $k_{in} = 0$ and $k_{out} = 0$ together with an increase of $\overline{|\phi|}$. This observation is demonstrated in Fig. 6 (left) where the data was collected for a set of heterogeneous TGs with T_3 motifs of type $\mathscr{M}_3$ and $T_1 = T - T_3$ motifs of type $\mathscr{M}_1$ (and for $p_s = 0.05$). Each heterogeneous TG had on average k^0_{out} nodes with $k_{out} = 0$ that decreased linearly with T_3 and reached zero for $T_3 \approx 231$ (with $\overline{|\phi|} \approx 0.64$). In contrast, the coherence degree increased with T_3 non-linearly. Thus, the inhibitory impact of $\mathscr{M}_1$ and $\mathscr{M}_2$ can only partially be explained by the presence of nodes with $k_{out} = 0$.

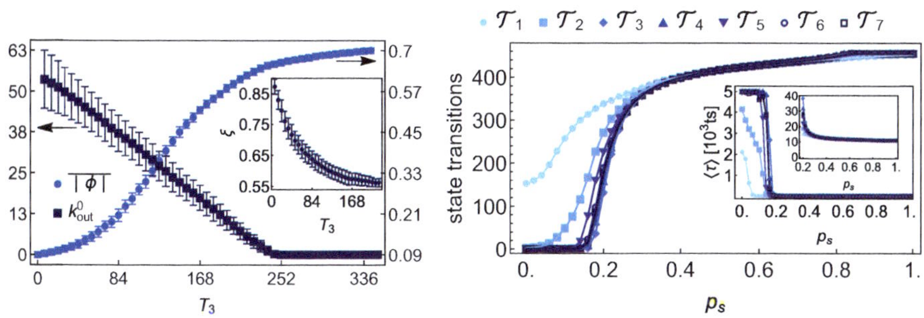

Fig. 6. Left: Number of nodes with $k_{out} = 0$ and $\overline{|\phi|}$ for the transition from $\mathscr{T}_1$ to $\mathscr{T}_3$ by replacing T_3 motifs of type $\mathscr{M}_1$ with $\mathscr{M}_3$ for $p_s = 0.05$. Inset: Current parameter ξ as a function of T_3. **Right:** Number of state transitions within 5000 time steps of simulation. The inset shows the average residence time $\langle\tau\rangle$ and a related close-up view for $p_s \geq 0.2$. The lines are guides to the eye.

Another relevant feature that changes with the transition from $\mathscr{T}_1$ to $\mathscr{T}_3$ is the *current parameter* ξ. It is a measure for the hierarchality or the inherent directionality of the system [7]. In the presence of nodes with $k_{out} = 0$ it can be easily obtained and it essentially represents the fraction of links that point upwards the hierarchical node ordering. Consequently, $\xi = 1$ in a perfectly hierarchical structure with very strong inherent directionality and $\xi \approx 0.5$ in a random graph with a vanishing level of inherent directionality and hierarchy [7]. For the set of homogeneous TGs $\mathscr{T}_i$ (as those used in Fig. 4), our measurements yielded $\xi = \{0.85 \pm 0.03, 0.50 \pm 0.01, 0.54 \pm 0.01, 0.51 \pm 0.01, 0.63 \pm 0.07, 0.50 \pm 0.01, 0.50 \pm 0.00\}$, where the first element, ξ_1, corresponds to $\mathscr{T}_1$, the second, ξ_2, to $\mathscr{T}_2$, etc. Note that while $\xi_1 \approx 0.85$, indicating strong hierarchality, $\xi_2 \approx 0.5$ indicating that the hierarchical property of $\mathscr{T}_2$ is almost non-existent. This suggests that the inhibitory impact of $\mathscr{M}_2$ on group coherence is mainly due to the presence of nodes with $k_{out} = 0$ (of which there are $k^0_{out} = 34 \pm 11$). This is in line with the observations that in contrast to $\mathscr{T}_1$, degree-preserved randomization of $\mathscr{T}_2$ does not significantly improve $\overline{|\phi|}$ (see Fig. 4).

Finally, Fig. 6 (right) shows that $\mathscr{M}_1$ and $\mathscr{M}_2$ increase the number of state transitions and, consequently, decrease the average residence time $\langle\tau\rangle$. The differences between the $\mathscr{T}_i$ rapidly vanish for $p_s > 0.2$ (see inset of Fig. 6 (right)). However, for $p_s \leq 0.2$ the higher number of state transitions of $\mathscr{T}_1$ suggest a potentially higher level of group adaptivity. Changing the state enables the group to explore the properties of this state and reassess its quality. Conversely, maintaining a stable state enables exploitative behavior. Motifs $\mathcal{M}_3 - \mathcal{M}_7$ appear to be beneficial for the latter while $\mathcal{M}_1$ and $\mathcal{M}_2$ for the former behavior.

5 Conclusion

In a binary collective decision-making task, noise coupled with social feedback mechanisms allow an initially symmetric system to self-organize and converge towards one of the two options. Social feedback mechanisms are defined by the interactions between the individuals, i.e. their underlying communication network. Therefore, it is paramount to understand the role of the network topology in promoting or inhibiting collective decision-making and coordination.

To investigate the impact of particular network types that include an abundance of triadic motifs, we generated random graphs that consist exclusively from such motifs. These graphs were then applied in simulations of binary collective decision-making scenarios. The results have shown that two specific types of motifs, in particular the feedforward loop, have a strong inhibitory impact on the coherence of collective behavior $\overline{|\phi|}$. In contrast, with motifs such as the feedback loop or the bidirectional loop (the motif in which all links are bidirectional), the system was able approach maximum coherence. Moreover, through comparison to null-models we have shown that the latter motif types achieve similar $\overline{|\phi|}$ to their degree-preserved randomizations, indicating that the in- and out-degree distributions may be the more critically influential properties.

More importantly, these results demonstrate that a number of motifs influence collective decision-making similarly to the bi-directional motif that can also be interpreted as undirected. On the one hand, this suggests that, depending on the purpose of the study, increasing the system complexity by including directionality may not be necessary. On the other hand, it appears that a group can reach maximum coherence even when local relationships are considerably asymmetric (compare the feedback loop to the bi-directional loop). This observation essentially lifts the constraint of symmetric relationships in which both nodes along an edge need to communicate to each other.

However, not all motifs lead to similar results and, particularly, the feedforward loop stands out with having a comparably inhibiting impact on the group coherence. We identified two possible characteristics, resulting from the motif topology, that contribute to such influence on collective decision-making. First, it is the presence of nodes with zero out-degree, i.e. nodes that do not communicate their opinion to their neighborhood. Surprisingly, their counterparts, nodes with zero in-degree, appear to not have similar inhibitory impact on collective decision-making. In fact, the motif where one node has zero in-degree but all three nodes have above-zero out-degree leads to similar group coherence as the feedback loop or the bi-directional loop but not as the feedforward loop. Second, the abundance of feedforward loops leads to hierarchical structures that are not beneficial to opinion alignment in group decision-making. Moreover, it leads to comparably unstable group commitment and a higher number of transitions between the options. Therefore, our results suggest that in certain cases the motif topology can have important consequences on collective decision-making.

In future research, it is worthwhile investigating in more detail the precise reasons behind the inhibitory influence of the prominent feedforward loop on collective decision-making, particularly with focus on the role of nodes with zero out-degree and hierarchality. Other network properties such as the path length or centrality may be relevant and should be included in the analysis. Moreover, the ability to lift the constraints of symmetrical relationships allows to examine scenarios of heterogeneous societies in which nodes are assigned particular roles based on the motif topology. Finally, the study can be further extended to include stochastic models as well as other motif types such as open triadic or quadratic motifs.

References

1. Alon, U.: Network motifs: theory and experimental approaches. Nat. Rev. Genet. **8**(6), 450–461 (2007)
2. Ariel, G., Ayali, A.: Locust collective motion and its modeling. PLoS Comput. Biol. **11**(12), e1004522 (2015)
3. Buhl, J., Sumpter, D.J., Couzin, I.D., Hale, J.J., Despland, E., Miller, E.R., Simpson, S.J.: From disorder to order in marching locusts. Science **312**(5778), 1402–1406 (2006)
4. Chen, L., Huepe, C., Gross, T.: Adaptive network models of collective decision making in swarming systems. Phys. Rev. E **94**(2), 022415 (2016)
5. Colaiori, F., Castellano, C.: Consensus versus persistence of disagreement in opinion formation: the role of zealots. J. Stat. Mech.: Theory E **2016**(3), 033401 (2016)
6. Czirók, A., Barabási, A.L., Vicsek, T.: Collective motion of self-propelled particles: Kinetic phase transition in one dimension. Phys. Rev. Lett. **82**, 209–212 (1999)
7. Domínguez-García, V., Pigolotti, S., Muñoz, M.A.: Inherent directionality explains the lack of feedback loops in empirical networks. Sci. Rep. **4**, 7497 (2014)
8. Huepe, C., Zschaler, G., Do, A.L., Gross, T.: Adaptive-network models of swarm dynamics. New J. Phys. **13**(7), 073022 (2011)
9. Khaluf, Y., Hamann, H.: Modulating interaction times in an artificial society of robots. In: The 2018 Conference on Artificial Life (ALIFE), pp. 372–379. MIT Press (2019)
10. Khaluf, Y., Pinciroli, C., Valentini, G., Hamann, H.: The impact of agent density on scalability in collective systems: noise-induced versus majority-based bistability. Swarm Intell. **11**(2), 155–179 (2017)
11. Khaluf, Y., Rausch, I., Simoens, P.: The impact of interaction models on the coherence of collective decision-making: a case study with simulated locusts. In: Dorigo, M., Birattari, M., Blum, C., Christensen, A.L., Reina, A., Trianni, V. (eds.) Swarm Intelligence: 11th International conference, ANTS 2018. LNCS, vol. 11172, pp. 252–263. Springer, Cham (2018)
12. Kirkman, T.P.: On a problem in combinations. Camb. Dublin Math. J **2**(191–204), 1847 (1847)
13. Klaise, J., Johnson, S.: The origin of motif families in food webs. Sci. Rep. **7**(1), 16197 (2017)
14. Mateo, D., Horsevad, N., Hassani, V., Chamanbaz, M., Bouffanais, R.: Optimal network topology for responsive collective behavior. Sci. Adv. **5**(4), eaau0999 (2019)
15. Milo, R., Shen-Orr, S., Itzkovitz, S., Kashtan, N., Chklovskii, D., Alon, U.: Network motifs: simple building blocks of complex networks. Science **298**(5594), 824–827 (2002)
16. Rausch, I., Reina, A., Simoens, P., Khaluf, Y.: Coherent collective behaviour emerging from decentralised balancing of social feedback and noise. Swarm Intell. **13**(3), 321–345 (2019)
17. Shang, Y., Bouffanais, R.: Influence of the number of topologically interacting neighbors on swarm dynamics. Sci. Rep. **4**, 4184 (2014)
18. Vicsek, T., Czirók, A., Ben-Jacob, E., Cohen, I., Shochet, O.: Novel type of phase transition in a system of self-driven particles. Phys. Rev. Lett. **75**, 1226–1229 (1995)

19. Winkler, M., Reichardt, J.: Node-specific triad pattern mining for complex-network analysis. In: 2014 IEEE International Conference on Data Mining Workshop, pp. 605–612. IEEE Press, New York (2014)
20. Winkler, M., Reichardt, J.: Motifs in triadic random graphs based on steiner triple systems. Phys. Rev. E **88**, 022805 (2013)
21. Yao, Y., Carretero-Paulet, L., Van de Peer, Y.: Using digital organisms to study the evolutionary consequences of whole genome duplication and polyploidy. PLOS One **14**(7), 1–21 (2019)
22. Yates, C.A., Erban, R., Escudero, C., Couzin, I.D., Buhl, J., Kevrekidis, I.G., Maini, P.K., Sumpter, D.J.T.: Inherent noise can facilitate coherence in collective swarm motion. P. Natl. Acad. Sci. **106**(14), 5464–5469 (2009)

Reconstruction of Demand Shocks in Input-Output Networks

Chengyuan Han[1,2](✉), Johannes Többen[3,4], Wilhelm Kuckshinrichs[1], Malte Schröder[5], and Dirk Witthaut[1,2]

[1] Institute for Energy and Climate Research (IEK-STE), Forschungszentrum Jülich, 52428 Jülich, Germany
ch.han@fz-juelich.de
[2] Institute for Theoretical Physics, University of Cologne, 50937 Köln, Germany
[3] Gesellschaft für Wirtschaftliche Strukturforschung, 49080 Osnabrück, Germany
[4] Potsdam Institute for Climate Impact Research, Social Metabolism and Impacts, 14412 Potsdam, Germany
[5] Chair for Network Dynamics, Center for Advancing Electronics Dresden (cfaed) and Institute of Theoretical Physics, Technical University of Dresden, 01062 Dresden, Germany

Abstract. Input-Output analysis describes the dependence of production, demand and trade between sectors and regions and allows to understand the propagation of economic shocks through economic networks. A central challenge in practical applications is the availability of data. Observations may be limited to the impact of the shocks in few sectors, but a complete picture of the origin and impacts would be highly desirable to guide political countermeasures. In this article we demonstrate that a shock in the final demand in few sectors can be fully reconstructed from limited observations of production changes. We adapt three algorithms from sparse signal recovery and evaluate their performance and their robustness to observation uncertainties.

Keywords: Economic networks · Input-output analysis · Compressed sensing

1 Introduction

Input-Output (IO) analysis, developed by Wassily W. Leontief, enables the quantitative analysis of the dependence of production, resource requirements and trade between sectors and regions in economic networks [6]. The central quantity in IO analysis is the Input-Output matrix, describing the underlying interdependencies among the sectors or regions (required inputs from one sector or region to another) in terms of a linear mapping [6]. One central application of Input-Output analysis is to understand and predict the influence of economic shocks. For instance, it predicts how the production in different sectors or regions

H. Barbosa et al. (Eds.): *Complex Networks XI*, SPCOM, pp. 131–140, 2020.
https://doi.org/10.1007/978-3-030-40943-2_12

reacts to a sudden change of the final demand. But what happens if a shock is unknown and we only observe its consequences?

The availability and quality of economic input data is a central challenge for the assessment of economic shocks and for Input-Output analysis in general. Natural and man-made disasters, for example storms, floods or terrorist attacks, cause shocks to both the demand and supply sides in certain sectors or regions. In many cases, it is only possible to measure the impacts of a disaster on a few sectors directly, but knowing the indirect impacts on other sectors is of great importance for designing policies that aim at enhancing the economy's resilience [4,8,9]. Hence, the application of powerful techniques for data analysis and reconstruction is central in IO analysis.

In this contribution, we analyze to which extend demand shocks can be reconstructed from limited and potentially noisy observations of production changes. This problem cannot be exactly solved in general, unless additional structural information is available. We demonstrate that shocks remain reconstructable if they are sparse, i.e. if the initial shock is limited to only few sectors or regions. This case is of high relevance, as political decisions, strikes or man-made disasters often affect only few sectors directly, while natural disaster often take place in a limited geographical region. We adapt different methods from the theory of sparse signal recovery [13] and evaluate their performance and robustness to observation noise for IO network data from the World Input-Output Database [12].

2 Fundamentals and Notations

We briefly review the fundamentals of Input-Output analysis analysis following [6]. In IO analysis, the economic system is divided into S sectors. The total output (production) of the sector i is denoted as x_i and is either sold to other sectors as input for their production or sold to final consumers (households, governments, capital formation and exports), resulting in the balance equation [6]:

$$x_i = \sum_{j=1}^{S} Z_{ij} + f_i. \tag{1}$$

Here, f_i is the final demand for output of sector i, and Z_{ij} denotes the inter-industry sales from sector i to sector j, where all quantities as measured in equivalent monetary units. To describe the interdependence between sectors i and j, Input-Output Analysis assumes fixed production processes and a constant amount A_{ij} of input from sector j required for a unit of output in sector i, such that $A_{ij} = Z_{ij}/x_j$. We can then rewrite Eq. (1) as

$$x_i = \sum_{j=1}^{S} (A_{ij} x_j) + f_i. \tag{2}$$

and recast this equation in vector form

$$\vec{x} = \mathbf{A}\vec{x} + \vec{f} \qquad \Leftrightarrow \qquad (\mathbf{I} - \mathbf{A})\vec{x} = \vec{f}, \tag{3}$$

where $\mathbf{I}$ is the $S \times S$ identity matrix. The column sums of A satisfy $\sum_{j=1}^{S} A_{ij} < 1$ for all i, such that one unit of output in sector i requires less than one unit of total input. Under this assumption $\mathbf{I} - \mathbf{A}$ is diagonally dominant and thus invertable. This allows us to reverse Eq. (3) to express $\vec{x}$ in term of $\vec{f}$ as

$$\vec{x} = (\mathbf{I} - \mathbf{A})^{-1}\vec{f} = \mathbf{L}\vec{f}, \tag{4}$$

where $\mathbf{L} = (\mathbf{I} - \mathbf{A})^{-1}$, known as the **Leontief inverse** or the **total requirements matrix**.

3 Reconstruction of Shocks

3.1 Sparse Shocks and Limited Observability

Input-Output Analysis describes how the demand drives the production across economic sectors. In particular it allows to assess the impact of changes of the demand: Assume that an external shock affects the final demand, $\vec{f} \to \vec{f} + \Delta\vec{f}$. On short timescales, we assume that the matrix $\mathbf{A}$ remains unaffected, as it mainly depends on production technology. Still, due to the dependence of the sectors on each other, the exogenous shock will affect the output of all sectors such that $\vec{x} \to \vec{x} + \Delta\vec{x}$. Due to the linearity of Eq. (4) the changes of production and demand are also related by $\Delta\vec{x} = \vec{L}\,\Delta\vec{f}$. Even if we cannot obtain direct information on the shock, we can in principle reconstruct the shock as $\Delta\vec{f} = \vec{L}^{-1}\,\Delta\vec{x}$ from observations of the total output changes.

However, situations arise, were we may not be able to obtain full information on the output changes $\Delta\vec{x}$. In this article we consider what happens if our knowledge of $\Delta\vec{x}$ is limited: let K_x denote the set of sectors where information about the production changes Δx_i is available. Instead of the full system, we then only have the following linear system of equations relating the initial shock $\Delta\vec{f}$ to the observations:

$$\sum_{j=1}^{S} L_{ij}\Delta f_j = \Delta x_i, \quad \text{for} \quad i \in K_x. \tag{5}$$

If the set K_x does not include all sectors, this equation is underdetermined and thus not uniquely solvable. Hence, an exact reconstruction of the initial shock $\Delta\vec{f}$ becomes impossible in general. However, we can overcome this problem if we have additional *structural information* about $\Delta\vec{f}$ that we can exploit. If we know that only a few sectors were disturbed initially, then just a few entries of $\Delta\vec{f}$ are non-zero. We can thus assume that the true solution of the underdetermined equation is obtained by minimizing the number of non-zero entries:

$$\min \|\Delta\vec{f}\|_0 \quad \text{such that} \quad \sum_{j=1}^{S} L_{ij}\;\Delta f_j = \Delta x_i \quad \text{for} \quad i \in K_x, \tag{6}$$

where $\|\cdot\|_0$ denotes the number of non-zero elements in a vector. This problem is NP-hard to solve in general such that a direct solution is not achievable. However, the last decade has witnessed significant progress in the development of efficient methods for an indirect solution applicable to many cases of practical importance [5,11]. A review of different approaches can be found in [13].

3.2 Sparse Reconstruction Methods

In this article we test the capability of three algorithms to reconstruct sparse shocks of the demand in an IO network:

1. Convex relaxation (ℓ_1 minimization): The optimization problem (6) is hard to solve since $\|\cdot\|_0$ is non-convex. Fortunately, in many cases one can still get the correct solution by replacing the non-convex pseudo-norm $\|\cdot\|_0$ by the convex norm $\|\cdot\|_1$, greatly simplifying the optimization problem [1,3].
2. The Orthogonal Matching Pursuit (OMP) approach is based on the following observation [2,10]. If the vector $\Delta\vec{f}$ is sparse, then only few columns of the matrix $\vec{L}$ enter the product $L\Delta\vec{f}$. Hence OMP tries to approximate the signal Δx by a superposition of only few columns of $\vec{L}$, which are chosen one-by-one to reduce the error of the approximation as much as possible in each step.
3. The Compressive sampling matching pursuit (CoSaMP) extends and improves OMP in several ways [7]. For instance, several new columns of $\vec{L}$ can be chosen in each step. In particular, instead of choosing new columns one-by-one, the algorithm adds several columns in each step and subsequently removes the ones that contribute the least to a correct approximation.

3.3 First Example

We illustrate the problem of reconstructing an initial shock using an elementary example from [6], Sect. 2.3.4, describing the US economy on coarse scales with only $S = 7$ sectors. The IO matrix $\mathbf{A} \in \mathbb{R}^{7\times 7}$ is given by

$$\mathbf{A} = \begin{bmatrix} 0.2008 & 0.0000 & 0.0011 & 0.0338 & 0.0001 & 0.0018 & 0.0009 \\ 0.0010 & 0.0658 & 0.0035 & 0.0219 & 0.0151 & 0.0001 & 0.0026 \\ 0.0034 & 0.0002 & 0.0012 & 0.0021 & 0.0035 & 0.0071 & 0.0214 \\ 0.1247 & 0.0684 & 0.1801 & 0.2319 & 0.0339 & 0.0414 & 0.0726 \\ 0.0855 & 0.0529 & 0.0914 & 0.0952 & 0.0645 & 0.0315 & 0.0528 \\ 0.0897 & 0.1668 & 0.1332 & 0.1255 & 0.1647 & 0.2712 & 0.1873 \\ 0.0093 & 0.0129 & 0.0095 & 0.0197 & 0.0190 & 0.0184 & 0.0228 \end{bmatrix}, \tag{7}$$

and the Leontief inverse matrix can be derived from Eq. (4). In this example, there is a change in the final demand in sector 1 (agricultural items) and sector 4 (manufactured items) due to foreign demand. In particular,

$$\Delta\vec{f} = \left[1.2, 0, 0, 6.8, 0, 0, 0\right]^\top \tag{8}$$

(in million dollars), using the symbol $\top$ to denote the transpose of a matrix or vector. This causes a change of output

$$\Delta\vec{x} = \mathbf{L}\Delta\vec{f} = \left[1.9114, 0.2444, 0.0526, 9.1249, 1.2421, 2.2709, 0.2788\right]^{\top}. \quad (9)$$

Now suppose we measure all entries of $\Delta\vec{x}$, then we can simply recover the cause of disturbance $\Delta\vec{f}$ by Eq. (3). But what happens if we have incomplete information? Say we only have the information

$$\Delta x_1 = 1.9114, \quad \Delta x_3 = 0.0526, \quad \text{and} \quad \Delta x_6 = 2.2709. \quad (10)$$

Assuming $\Delta\vec{f}$ is sparse we can answer this question as discussed above by appling the algorithms introduced in Sect. 3.2 to approximately solve the optimization problem

$$\Delta\vec{f}_R = \arg\min_{\Delta\vec{f}} \|\Delta\vec{f}\|_0, \text{ s.t. } \Delta x_1 = 1.9114,\ \Delta x_3 = 0.0526, \text{ and } \Delta x_6 = 2.2709. \quad (11)$$

In this example, we find that both OMP and CoSaMP yield the correct solution for $\Delta\vec{f}$ as shown in Eq. (8). But is reconstruction possible in general and which algorithm is most appropriate? To answer this question, we consider a larger, more realistic representation of an IO network and try different algorithms in the next section.

4 Result

4.1 Impact Recovery for the WIOD Dataset

To test whether sparse signal recovery is possible in a real-world setting, we analyze the performance of the reconstruction methods mentioned in Sect. 3.2 for the World Input-Output Database (WIOD) [12].

The WIOD provides multi-regional input-output tables from different years to represent the trade between any two sectors in the world. The latest 2014 table consists of 28 EU countries, 15 other major countries, and the "rest of the world" entries to complete the data set. Each country's economy is divided into 56 sectors to portray the different industries. Here we consider only the information of the individual sectors, aggregating the input-output-dependency over all countries such that the IO Matrix $\vec{A} \in \mathbb{R}^{56\times 56}$ with entries corresponding to average inter-industry sales $\langle A_{ij}\rangle \approx 10^{-2}$.

To test the accuracy of the three algorithms mentioned in Sect. 3.2, we evaluate the success rates of the reconstruction, varying both the sparsity of the initial shock $\Delta\vec{f}$, measured in term of the number s of non-zero entries, and the number of observations n_o of the output changes $\Delta\vec{x}$. For each combination of values (s, n_o), we synthetically generate a large ensemble of $R = 10^2$ test cases as follows. We uniformly randomly select s sectors and choose the entries of $\Delta\vec{f}$ in these sectors as random values sampled uniformly from the interval $(0, 10]$.

The remaining entries are set to zero. We then compute $\Delta\vec{x} = \mathbf{L}\,\Delta\vec{f}$ and uniformly randomly select n_o sectors to be observed. That is, we randomly choose the set K_x such that $|K_x| = n_o$. For these choices we exclude sector 56 from both the shocks and the observations. This sector summarizes the "activities of extraterritorial organizations and bodies" and only contributes to consumption, not production.

For each of these test cases we attempt to reconstruct the initial demand shock via the three algorithms listed in Sect. 3.2. That is, we compute a vector Δf_R which satisfies the linear constraints (5) and which shall minimize the sparsity $\|\Delta f_R\|_0$. We compare this reconstructed signal $\Delta\vec{f}_R$ with the original shock $\Delta\vec{f}$. If $\|\Delta\vec{f}_R - \Delta\vec{f}\|_2 \leq 10^{-5}$, where $\|\cdot\|_2$ is the Euclidean norm, then the reconstruction is considered successful. This procedure is repeated for each of the $R = 10^2$ realizations for each combination of values (s, n_o) to obtain the average success rate.

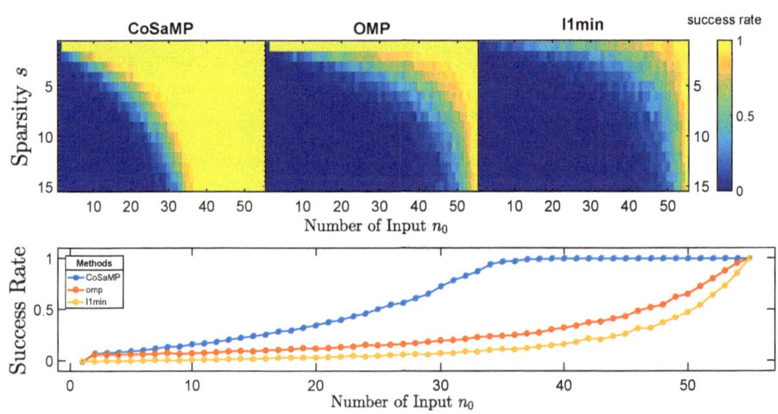

Fig. 1. Performance of different algorithms for the reconstruction of demand shocks in IO analysis. Top: the arrays show the success rate of (left to right) CoSaMP, OMP and ℓ_1 minimization in a color scale plot as a function of the sparsity s of the desired output of the reconstruction $\Delta\vec{f}$ and the number of observations n_o used as input for the reconstruction. Bottom: Success rate as a function of the number of observations n_o, averaged over all values of the sparsity $s \in [1, 15]$.

The numerical results of these tests demonstrate that sparse demand shocks can be reconstructed from limited observations for real-world IO systems (cf. Fig. 1). However, the success rates of the three methods differ vastly. In particular, convex relaxation (ℓ_1 minimization) performs poorly for the current task. A success rate above 90% can be achieved only if we have almost complete information about $\Delta\vec{x}$, i.e. if n_0 is close to the total number of sectors $S = 56$. In contrast, CoSaMP shows a very promising performance and outperforms the other algorithms for all values of s and n_o. For strongly sparse signals, $s \leq 4$, the algorithm has a success rate of 100% even if we measure Δx for less than half of the sectors. Even for values as high as $s = 15$, we find a perfect success rate of 100% with a number of inputs n_o well below 40. Because of its superior performance, we focus on the CoSaMP algorithm in the following and analyze its robustness to noisy inputs.

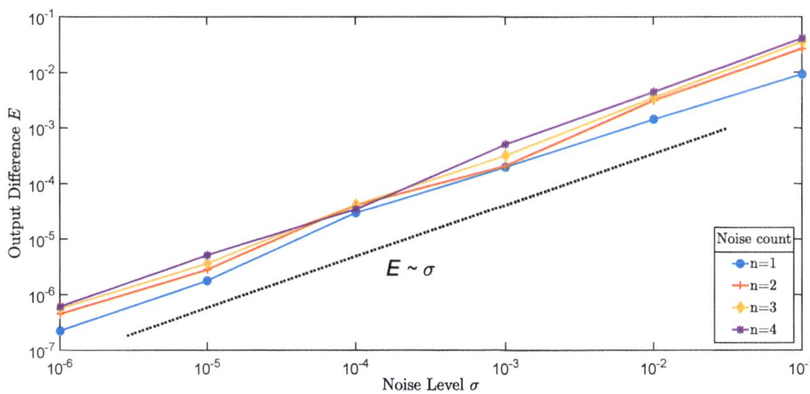

Fig. 2. Propagation of observation uncertainty ('noise') through the sparse reconstruction. We observe that the error of the reconstructed signal $E = \|\Delta\vec{f}_R - \Delta\vec{f}\|_2$ increases approximately linearly with the noise level. Points correspond to the average over $R = 10^2$ repetitions for different values of n_n, lines are drawn to guide the eye. The remaining parameters are $s = 5$ and $n_o = 50$. The dotted line shows a linear scaling $E \sim \sigma$ for comparison.

4.2 Robustness to Measurement Noise

Trade data can be subject to various forms of inaccuracies. Hence any reconstruction algorithm must be robust to inaccurate or noisy input signals to be useful in practice. To test the robustness of the reconstruction method, we add noise to the observed output changes $\Delta\vec{x}$. Test data is generated as follows. We create an initial shock of sparsity s and select n_o outputs as above. We then select n_n of the n_o observations and add noise to then, drawn independently and uniformly at random from the interval $[0, \sigma]$. The parameter $\sigma \in [0, 10^{-1}]$ quantifies the noise strength. The key question is then how this observation noise propagates via the reconstruction algorithm and whether a reconstruction remains possible in principle. To address these questions, we evaluate the Euclidean norm of the difference between the original signal $\Delta\vec{f}$ and the reconstructed signal $\Delta\vec{f}_R$:

$$E = \|\Delta\vec{f}_R - \Delta\vec{f}\|_2. \tag{12}$$

As above, we define reconstruction be successful if $E < 10^{-5}$ to compute the success rate. We repeat this process for $R = 10^2$ realizations to compute the success rate and the average error.

Propagation of Uncertainty. Figure 2 illustrates the propagation of noise in the reconstruction process. We find that the reconstruction error E increases approximately linearly with the noise level σ, a finding that is largely independent of the number of noisy observations n_n. We conclude that the reconstruction process is robust in the sense that a small amount of noise in the observations causes only a small error in the reconstruction Δf_R. In particular, a limited amount of noise does *not* render the results of the reconstruction algorithm unfeasible.

Success Rate. The reconstruction method does not 'amplify' the uncertainty of the observation, but still the performance will be degraded if the noise becomes too strong. To assess these limitations in more detail, we evaluate the success rate as a function of the noise level σ and the number of noisy observations n_n in detail in Figs. 3 and 4. Results are shown for three values $s = 1, 7, 15$, representing cases of low, medium, and high sparsity. In each case, the number of observations n_o was then chosen such that the points are on the boundary of 100% success rate in the noiseless case (compare Fig. 1). That is, a success rate of 100% is found in the noiseless case for the given value of n_o, but not for smaller values of n_o. This procedure yields the values $(s, n_o) \in \{(1, 5), (7, 34), (15, 40)\}$.

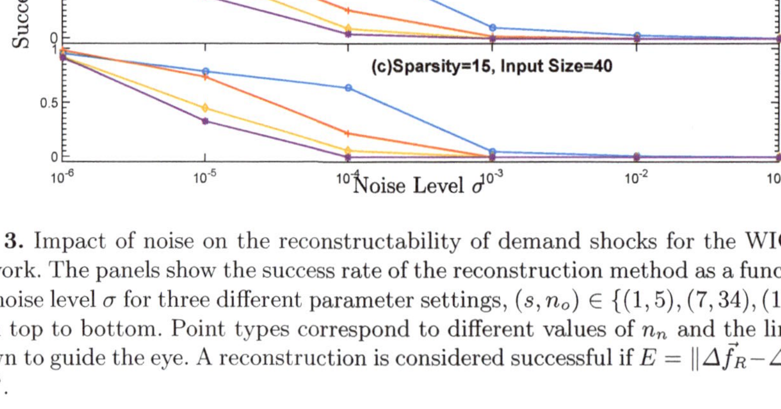

Fig. 3. Impact of noise on the reconstructability of demand shocks for the WIOD IO network. The panels show the success rate of the reconstruction method as a function of the noise level σ for three different parameter settings, $(s, n_o) \in \{(1, 5), (7, 34), (15, 40)\}$ from top to bottom. Point types correspond to different values of n_n and the lines are drawn to guide the eye. A reconstruction is considered successful if $E = \|\Delta \vec{f}_R - \Delta \vec{f}\|_2 < 10^{-5}$.

The presence of observation uncertainty can indeed significantly reduce the success rate as shown in Fig. 3 - but this depends strongly on the remaining parameters. A limited amount of noise σ has only a limited influence on the reconstruction $\Delta \vec{f}$, but can be enough to increase the error rate E above the

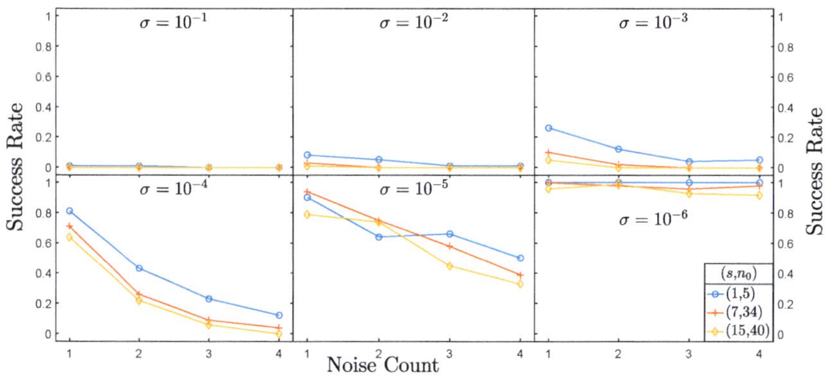

Fig. 4. Success rate of reconstruction of the signal $\Delta\vec{f}$ versus the number n_n of measurement subject to noise (the "noise count"). Each plot represents a level of noise σ. The three lines in each plot represent a different parameter choice of sparsity s and number of observation n_o. The success rate of the reconstruction is approximately inversely proportional to the count of the noise n added to the input signal when the noise level σ is close to the threshold of reconstructability. Lower sparsity s (blue circles) is more robust than the other two cases for almost all situations.

threshold value 10^{-5}defining a successful reconstruction. In particular for the given parameter values (s, n_o) at the boundary of the 100% success region, already weak noise can exceed this threshold. As a consequence, the success rate shown in Fig. 3 decreases with σ and drops to zero for values between 10^{-4} and 10^{-1} depending on the remaining parameters. The higher the number of noisy observations n_n the faster the success rate decreases.

The number of uncertain observations n_n can have a strong influence on the success rate if the parameters (s, n_o, σ) are such that the E is of the order of the threshold value 10^{-5} as shown in Fig. 4. For $\sigma = 10^{-4}$ the success rate drops by more than 60% when n_n is increased from 1 to 4. In other cases, when we are not close to the threshold, the success rate is largely independent of n_n

In conclusion, we have demonstrated the robustness of the reconstruction process and we mapped out the consequences of imperfect observation. Uncertainty of the observations ('noise') propagates but does not render the algorithm unfeasible in principle. In practice, it crucially depends on the tolerable error of the reconstruction whether noise may be problematic or not.

5 Discussion

In this contribution we have analyzed the inference of economic shocks from limited observations in IO networks. We have demonstrated that it is possible to reconstruct a change of the final demand Δf_i in few sectors from limited observations of production changes Δx_j. The key step is to utilize structural information about the initial demand shocks. If shocks emerge from few sectors, the vector of demand changes Δf is sparse, enabling the use of advanced methods for sparse signal reconstruction. The best performance was obtained using the Compressive sampling matching pursuit (CoSaMP) algorithm and it was demonstrated that this approach is robust against small uncertainties in the observations. Nevertheless, large uncertainties can be crucial depending on the required accuracy of the reconstructed signals. Even in cases where the reconstruction is not quantitatively successful, this approach may still help to identify which sectors were the source of the initial demand shock.

We note that a successful reconstruction of Δf also allows to reconstruct the missing information about the production changes via the relation $\Delta\vec{x} = \mathbf{L}\,\Delta\vec{f}$. Hence, a reconstruction yields both the origin and the impacts of shocks, which is of great importance to design policies to enhance the resilience of economic networks [4,8,9].

Acknowledgments. We gratefully acknowledge support from the Helmholtz association (grant no. VH-NG-1025), the German Ministry for Education and Research (BMBF grant no. 03SF0472) and the German Research Foundation (DFG) through the Cluster of Excellence *Center for Advancing Electronics Dresden* (cfaed) and the project 'Bilinear Compressed Sensing'.

References

1. Candès, E.J., Romberg, J.K., Tao, T.: Stable signal recovery from incomplete and inaccurate measurements. Commun. Pure Appl. Math. **59**, 1207–1223 (2006)
2. Davis, G., Mallat, S., Avellaneda, M.: Adaptive greedy approximations. Constr. Approx. **13**, 57–98 (1997)
3. Donoho, D.L.: Compressed sensing. IEEE Trans. Inf. Theory **52**, 1289–1306 (2006)
4. Hallegatte, S.: An adaptive regional input-output model and its application to the assessment of the economic cost of Katrina. Risk Anal.: Int. J. **28**, 779–799 (2008)
5. Marques, E.C., Maciel, N., Naviner, L., Cai, H., Yang, J.: A review of sparse recovery algorithms. IEEE Access **7**, 1300–1322 (2018)
6. Miller, R.E., Blair, P.D.: Input-Output Analysis - Foundations and Extensions. Cambridge University Press, Cambridge (2009)
7. Needell, D., Tropp, J.: CoSaMP: iterative signal recovery from incomplete and inaccurate samples. Appl. Comput. Harmon. Anal. **26**, 301–321 (2009)
8. Okuyama, Y., Santos, J.R.: Disaster impact and input-output analysis. Econ. Syst. Res. **26**, 1–12 (2014)
9. Oosterhaven, J., Többen, J.: Wider economic impacts of heavy flooding in germany: a non-linear programming approach. Spat. Econ. Anal. **12**, 404–428 (2017)
10. Pati, Y.C., Rezaiifar, R., Krishnaprasad, P.S.: Orthogonal matching pursuit: recursive function approximation with applications to wavelet decomposition. In: Proceedings of 27th Asilomar Conference on Signals, Systems and Computers, pp. 40–44 (1993)
11. Timme, M., Casadiego, J.: Revealing networks from dynamics: an introduction. J. Phys. A: Math. Theor. **47**(34), 343001 (2014)
12. Timmer, M.P., Dietzenbacher, E., Los, B., Stehrer, R., Vries, G.J.: An illustrated user guide to the world input-output database: the case of global automotive production. Rev. Int. Econ. **23**, 575–605 (2015)
13. Tropp, J.A., Wright, S.J.: Computational methods for sparse solution of linear inverse problems. Proc. IEEE **98**, 948–958 (2010)

Biomedical Applications

Boolean Threshold Networks as Models of Genotype-Phenotype Maps

Chico Q. Camargo[1](✉) and Ard A. Louis[2]

[1] Oxford Internet Institute, University of Oxford, Oxford, UK
chico.camargo@oii.ox.ac.uk

[2] Rudolf Peierls Centre for Theoretical Physics, University of Oxford, Oxford, UK

Abstract. Boolean threshold networks (BTNs) are a class of mathematical models used to describe complex dynamics on networks. They have been used to study gene regulation, but also to model the brain, and are similar to artificial neural networks used in machine learning applications. In this paper we study BTNs from the perspective of genotype-phenotype maps, by treating the network's set of nodes and connections as its genotype, and dynamic behaviour of the model as its phenotype. We show that these systems exhibit (1) Redundancy, that is many genotypes map to the same phenotypes; (2) Bias, the number of genotypes per phenotypes varies over many orders of magnitude; (3) Simplicity bias, simpler phenotypes are exponentially more likely to occur than complex ones; (4) Large robustness, many phenotypes are surprisingly robust to random perturbations in the parameters, and (5) this robustness correlates positively with the evolvability, the ability of the system to find other phenotypes by point mutations of the parameters. These properties should be relevant for the wide range of systems that can be modelled by BTNs.

Keywords: Boolean networks · Gene regulatory networks · Genotype-phenotype maps · Input-output maps

1 Introduction

Boolean networks (BN) were first introduced by Stuart Kauffman as models of gene regulatory networks (GRNs) [28]. Each gene is abstracted as a binary (Boolean) variable that can be either on or off. A temporal dynamics is imposed, where each gene interacts with others through a Boolean function. For example, if $S_i(t)$ is the state of a node i at time t, then

$$S_1(t+1) = (\, S_1(t) \texttt{ AND } S_2(t)\,) \texttt{ OR NOT } (\, S_3(t) \texttt{ AND } S_4(t)\,) \tag{1}$$

is the state of the node S_1 at time $t+1$, and this is influenced by the states of nodes S_2, S_3, and S_4. The state of every node in the network is then updated synchronously with all other nodes in the network with rules similar to the example above. The requirement of synchronicity is sometimes dropped, in what

H. Barbosa et al. (Eds.): *Complex Networks XI*, SPCOM, pp. 143–155, 2020.
https://doi.org/10.1007/978-3-030-40943-2_13

is called asynchronous updating. In Kauffman's original formulation [28], he defined an NK network of N nodes with K connections. Figure 1 shows an example of this kind of network, for $N = 3$ and $K = 3$, showing how different initial configurations of the network can lead to different attractors in its state space. Here attractors refer to steady state behaviour that the system falls into.

Boolean function

(t) A B C	(t+1) A B C
0 0 0	0 1 1
0 0 1	1 0 1
0 1 0	1 1 1
0 1 1	0 1 1
1 0 0	1 1 1
1 0 1	1 1 1
1 1 0	1 0 1
1 1 1	1 1 0

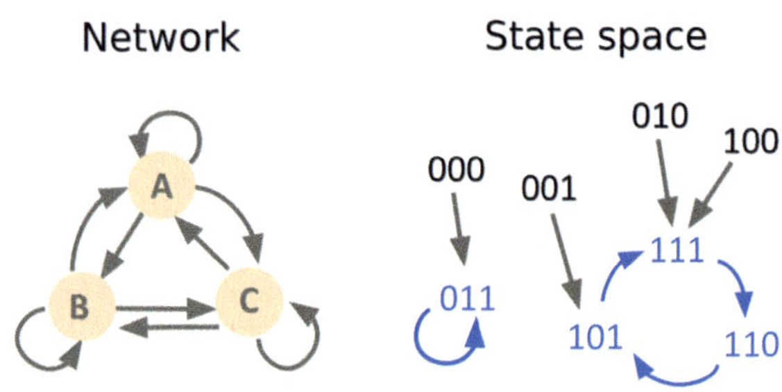

Fig. 1. Example of Boolean network with $N = 3$, $K = 3$. In this network, all nodes are regulated by the whole network ($N = K$). The table on the left defines the Boolean functions that describe how the state of nodes A, B and C will change according to their states in the previous time step. The $2^3 = 8$ states in state space are organised in two attractors, one fixed point and one 3-cycle, both shown in blue.

Even though they are highly simplified models, BNs are known to be extremely versatile. They have been used to represent complex systems at multiple scales, ranging from gene regulatory networks to ecosystems [29].

In the context of gene regulation, one popular application has been to model pluripotent cells as a gene network with many attractors [29]. Each cell type then corresponds to a different attractor. Another famous application of the BN model was to segment polarity in *Drosophila* [2], where it was shown that simplifying a highly complex differential equation model of the GRN to a much simpler BN model nevertheless led to good agreement with experiments. Similar successful applications have since proliferated, including BN models for the GRNs regulating flower development [18], signal transduction in human fibroblasts [25], plant cell signalling [33], and mammalian cortical development [20]. BNs have also been used to study more general properties of GRNs, such as robustness [5,11,12,41] and evolvability [6,11,12,41].

In the context of neural networks, different attractors have been interpreted as different memories stored in the network [26,27,35]. Attractors can also represent different heartbeat rhythms for cardiac systems [21], or alternative species distributions when modelling ecosystems [34].

The versatility of BN models comes at the price of having a large parameter space. As each node in a NK network is assigned a Boolean function on the 2^K possible states of those K nodes, there are 2^{2^K} possible functions per node, meaning that for $N = 3$ and $K = 3$, as in Fig. 1, there are over 3×10^9 possible networks, whereas for a network with $N = 5$ and $K = 3$ this number grows to approximately 10^{17}. One way to overcome the impractical size of the parameter spaces of BNs is by focussing on a subset of all BNs. Boolean threshold networks (BTNs) correspond to one of these subsets, and are characterised by simple interaction rule:

$$S_i(t+1) = \begin{cases} 1, & \text{if } \sum_{j=0}^{N} w_{ij}\, S_j(t) > 0 \\ 0, & \text{if } \sum_{j=0}^{N} w_{ij}\, S_j(t) < 0 \end{cases} \tag{2}$$

where the weights w_{ij} indicate the strength and sign of the regulation of node i by node j. The value of $S_i(t+1)$ when $\sum_{j=0}^{N} w_{ij}\, S_j(t) = 0$ varies across studies, having values such as $S_i(t+1) = 1$, 0, or $S_i(t)$ itself. This choice has been shown not to have a big impact on results [14, 31, 36]. In this work, we will model GRNs as defined in Eq. (2), with $S_i(t+1)$ when $\sum_{j=0}^{N} w_{ij}\, S_j(t) = 0$, and $w_{ij} \in \{1, 0, 1\}$, respectively indicating upregulation, downregulation and lack of interaction between genes. This choice reflects the strong assumption that the effect of activatory and inhibitory interaction between genes is essentially additive, and that all gene-gene interactions are equally strong [48].

The threshold function in Eq. (2) inevitably reduces the range of behaviours of a GRN [48], but it also reduces its parameter space by orders of magnitude. Instead of requiring N Boolean functions to be defined, as would be the case for a Boolean network, a BTN can be completely specified by its $N \times N$ adjacency matrix w_{ij}. Each GRN or genotype can also be treated as a node in a genotype network where two genotypes are neighbours if they differ in only one term in their w_{ij} matrix [12]. Since every connection can be either -1, 0 or 1 the whole genotype space is composed of $3^{N \times N}$ gene networks, each one connected to $2N^2$ neighbours.

BTNs have also been used to successfully model gene expression, in GRNs such as the ones regulating lymphocyte differentiation [38], signal transduction in human fibroblasts [25], and the mammalian cell cycle [19]. One of the most successful applications of BTNs is in modelling the yeast cell cycle [14, 15, 31], which led to studies predicting knockout mutant phenotypes [7, 15], as well as multiple studies providing explanations for the designability and robustness of the wild-type phenotype [3, 4, 8, 10].

2 Genotype-Phenotype Maps

When used to study the evolutionary properties of GRNs, the mapping from a Boolean network's wiring to its dynamic properties can be studied as a genotype-phenotype (GP) map. GP maps have proven to be a useful lens in understanding the origins of phenotypic variation, and how this variation can steer evolution even before natural selection comes into play [39, 46]. Much of this understanding comes from a growing body of research structural properties shared by multiple GP maps [1, 24, 46]. These properties include: (1) Redundancy, meaning that many genotypes often map to the same phenotypes, (2) Bias, referring to how the majority of genotypes map only to a handful of phenotypes [1,23], (3) Simplicity bias, i.e. how simple phenotypes correspond to larger fractions of genotype space [16], (4) a degree of mutational robustness which is much larger than what would be expected for a random uncorrelated GP map [1,22,24,24], and (5) a positive correlation between robustness and evolvability [23,45,46].

Provided that the genotype of a GRN can be described by its topology and represented by its adjacency matrix w_{ij}, one still needs to choose how to represent its phenotype. Different definitions of phenotype might be more or less appropriate depending on what is being measured. For example, when modelling the regulation of circadian rhythm by a GRN that produces oscillating behaviour for multiple initial conditions, it might be convenient to define the phenotype of that GRN as its cyclic attractor in state space. If one is interested in the possible cell fates of a pluripotent cell, it might be better to look at the whole list of attractors. Alternatively, one might define a phenotype concerning a specific set of initial conditions, or the transitions between certain states. With that in mind, we explore two different phenotype definitions, that capture aspects of different kinds of biologically relevant phenotypes:

Phenotype 1: we define the phenotype of a GRN as the attractor occupying the largest fraction of its state space.

Phenotype 2: we define the phenotype as a list of the attractors corresponding to all possible initial states of the network.

The two phenotype definitions described above are very general, and might not be suited for all GRNs. For instance, it might be the case that not all initial states of the network are biologically relevant, or that BTNs that produce very similar behaviour from a dynamical systems point of view might represent very different biological behaviours. In spite of these limitations, we find that all phenotype definitions studied here result in GP maps which show many of the structural properties of GP maps listed above.

3 Results for Phenotype 1: Dominant Attractor

To identify the attractor with the largest basin for each genotype, we use the same method as Nochomovitz and Li [36]: first, we use the update function from Eq. (2) to produce a list of 2^N "next states", which we then connect as a directed graph from state to state, as illustrated in Fig. 1. We then use Tarjan's algorithm to calculate the strongly connected components of this directed graph, that is, the sections of state space where every state can reach every other state [42]. As the threshold update function is deterministic, every state will only lead to one other state, and the strongly connected components of the state graph will be the attractors of the BTN state space. Finally, we also take into account how multiple attractors might be equivalent under symmetry operations. These operations include permuting gene order, shifting the cyclic attractors by any number of steps, or swapping 1s for 0s for a given gene. This last form of symmetry has been described as a "gauge symmetry" [11,12]. We calculate the dominant attractor for every GRN in the genotype space of all BTNs with N genes. In the sections below, we present the properties observed for this GP map, for the full enumeration of all $3^{16} \approx 43$ million $N = 4$ networks.

3.1 Phenotype Frequencies Vary over Orders of Magnitude

As there are many more genotypes than phenotypes in this GP map, it is natural that the map will show some redundancy – genotypes which map to the same phenotypes. In the GP map literature, the concept of redundancy is measured in terms of *neutral networks*, i.e. regions of a discrete genotype space that map to the same phenotype. The number of genotypes in a neutral network has been given different names in the literature, such as the phenotype's degeneracy level [9], abundance [13], designability [32,36], genotype set size [37], neutral set size [39] or neutral network size (NNS) [40,45]. In this work, we will also use the word frequency, meaning the NSS of a phenotype divided by the size of the whole genotype space. The frequency of a phenotype can also be understood as the probability that a randomly chosen genotype will map to that phenotype.

We performed a full enumeration of all $3^{16} \approx 4.3 \times 10^7$ GRNs with $N = 4$ genes. As can be seen in Fig. 2a, the frequencies of 2759 phenotypes observed for $N = 4$ networks range over many orders of magnitude. The distribution of phenotype frequencies is very skewed, with the most common attractor type, which is a single fixed point, covering 67% of genotype space. This uneven distribution of phenotype frequency is also observed within cyclic outputs of the

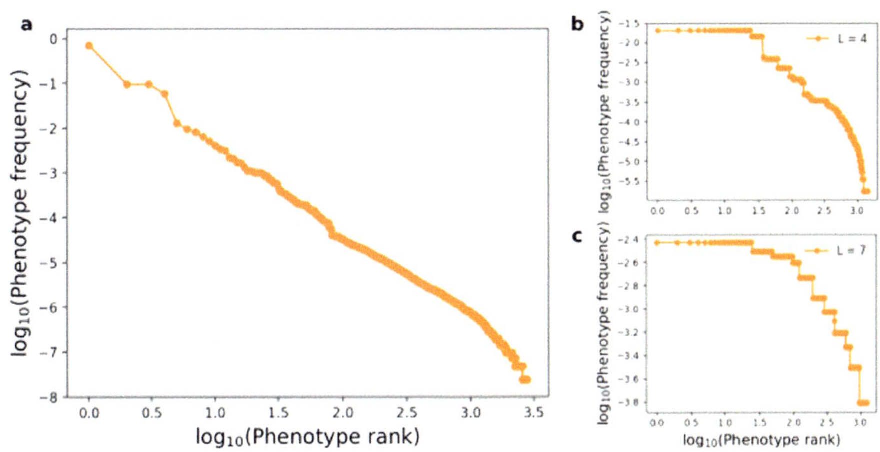

Fig. 2. This GP map shows redundancy and bias. **(a)** Rank plot for the number of gene networks that produce each one of all the 2759 dominant attractors found from a full enumeration of of $N = 4$ gene networks. It is a very uneven distribution, with 99% of all genotypes mapping to 0.38% of all phenotypes, and the most designable phenotype, a single fixed point, corresponding to 67% of the genotype space. The skewed distribution of neutral network sizes is also observed within cyclic outputs of the same length, shown in **(b)** and **(c)** for lengths $L = 4$ to 7 respectively.

same length, as already noted for $N = 4$ BTNs by Nochomovitz and Li [36]. Figures 2b and c illustrate this pattern for cyclic attractors of lengths $L = 4$ and $L = 7$ respectively.

3.2 Low-Complexity Attractors Correspond to More Genotypes

Not only do Figs. 2b and c show that there is a wide distribution of phenotype frequencies even among cyclic attractors of the same length, but they also show that the frequency of $L = 7$ cyclic attractors is orders of magnitude lower than that of $L = 4$ attractors. This is shown in Fig. 3c for other attractor lengths. Overall, the larger the cycle, i.e. the more states in the cycle, the smaller the number of GRNs which will produce that phenotype.

This negative relation between phenotype frequency and cycle length is part of a wider pattern, where low-complexity phenotypes take over larger fractions of genotype space. This behaviour, which Dingle et al. call *simplicity bias* [16], is found in numerous input-output maps, including but not limited to RNA sequence-to-structure maps [17], stochastic models in financial mathematics [16] and parameter-to-behaviour maps in deep neural networks [44]. This pattern has also been reported for genotype-phenotype maps for protein quaternary structures and macromolecular self-assembly and generative models of tree geometries [16]. And even though the specific definition of a phenotype depends on

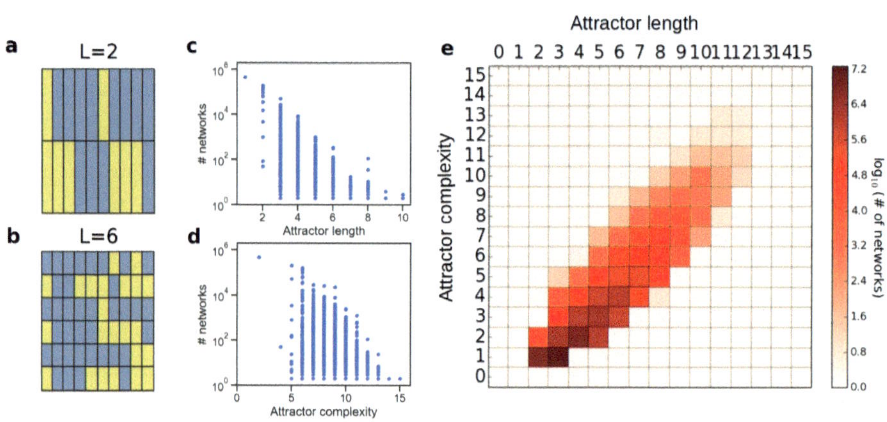

Fig. 3. This GP map shows simplicity bias. **(a)** and **(b)** show two cyclic attractors, of length $L = 2$ states and $L = 6$ states respectively, for a GRN with $N = 10$ genes. In both panels, every column represents the binary state of a node, with yellow for 1 and blue for 0. **(c)** and **(d)** show this GP map is biased towards low-complexity phenotypes. Both measures show the number of GRNs mapping to a phenotype is bounded by an exponential decay in its complexity, as measured by their attractor length and Kolmogorov complexity respectively. Finally, **(e)** shows a heat map for all $N = 4$ GRNs, comparing attractor cycle length to attractor complexity, indicating that more GRNs produce phenotypes of lower complexity, even when comparing phenotypes of the same cycle length. Darker colours represent higher phenotype frequency.

the details of every GP map or input-output map, as long as a phenotype can be represented a binary string, its Kolmogorov complexity can be approximated using the Lempel-Ziv algorithm [30], as described by Dingle et al. [16]

Figures 3a and b respectively show examples of simple and complex phenotypes, for a GRN with $N = 10$ genes. Both panels represent cyclic attractors, with every column representing the binary state of a node and every row indicating a state in the cyclic attractor, with yellow for `1` and blue for `0`; for example, Fig. 3a indicates the 2-state cycle `1000110001` → `1110001110`, while Fig. 3b indicates a 6-state cycle starting with `0000001010` → `1001110111`.

Even though the length of a cyclic attractor in a BTN can be used as a proxy for its complexity [36], cycle length alone does not explain the variation of phenotype frequency that also happens among the cycles of the same equal length, as shown above in Figs. 2b and c.

Since two cycles with the same length can represent two different patterns of gene activation, an ideal measure of phenotype complexity should take these differences into account. Here we quantify these differences in complexity between different attractors by "stacking" all states of the cycle in a single binary string, such that a sequence such as `0110` → `1111` cycle would be represented as `01101111`. We then estimate the Kolmogorov complexity of the resulting string using the canonical Lempel-Ziv method [16,30]. The result of this analysis, presented in Fig. 2d, shows the same trend revealed by looking at cycle length in Fig. 2c: the number of genotypes mapping to a phenotype is bounded by a function that decreases exponentially with complexity – a hallmark of simplicity bias, as discussed by Dingle et al. [16].

Given that both cycle length and Kolmogorov complexity point towards the presence of simplicity bias in this GP map, as shown by Figs. 2c and d respectively, it could be the case that both measures are simply revealing the bias towards shorter attractors. This is, however, not the case. Figure 2e shows a heatmap where every column represents attractors of the same cycle length, such as the ones grouped in the same subplots in Fig. 2e. In the figure, the darker shades of red are always in the lower, low-complexity side of the red-shaded part of every column. Since darker colours represent more genotypes, this plot indicates that more genotypes map to phenotypes of lower complexity, even when comparing phenotypes of the same attractor length.

Even though defining the phenotype of a network as its main attractor in its state space is a practical way to model the behaviour of a GRN, this phenotype definition discards information about which initial conditions lead to that main attractor, while also ignoring other attractors present in the GRN state space. In a case where all biologically relevant initial conditions lead to the same attractors, such as the yeast cell cycle studied by Li et al. [31], these limitations are not a problem, but this definition of phenotype is not suited to represent a bistable system, or a cell that, provided the right initial conditions, might differentiate into multiple cell types [29].

With that in mind, in the next section we study a fine-grained GP map for GRNs, defining a phenotype taking into account the attractor corresponding to every possible initial condition in state space. Or, in biological terms, which gene expression patterns lead to which cellular behaviours.

4 Results for Phenotype 2: Multiple Attractors

For this analysis, we define the phenotype of a GRN with N genes as a 2^N-long string of the attractors corresponding to each one of its 2^N possible states. For example, for $N = 4$, a phenotype would be a string such as AABACBDBCCCDDCCD, indicating that that states 0000 and 0001 map to attractor A, state 0010 maps to attractor B, and so on.

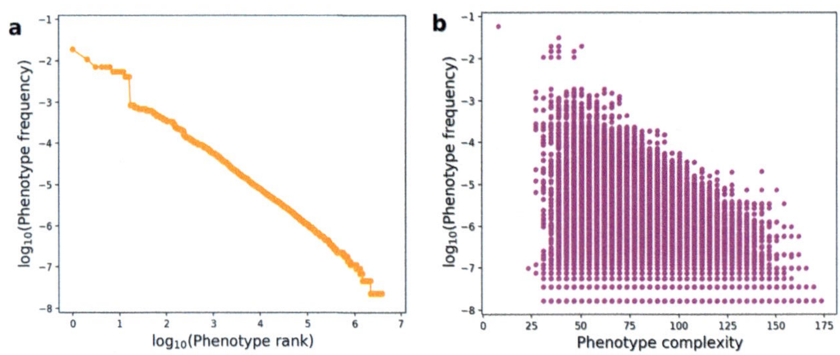

Fig. 4. **(a)** Rank plot for $N = 4$ GRNs showing a very biased distribution of phenotype frequency, for the phenotype defined as the list of attractors. **(b)** Scatter plot of phenotype frequency versus phenotype complexity, measured using the Lempel-Ziv algorithms, showing a strong bias towards simple phenotypes.

Having the phenotype written down as a string also allows us to calculate its Kolmogorov complexity using the Lempel-Ziv method. We do this by first iterating over all $3^{N\times N}$ possible genotypes for a given number of genes, and producing their corresponding strings of attractors. We then translate those phenotype strings into the shortest binary strings that can accommodate the range of attractors presented by this GP map. In the case of $N = 4$, the whole genotype space produces a total of 8172 different attractors, meaning that the index corresponding to each attractor can be represented in $\lceil \log_2(8172) \rceil = 13$ bits, from 0000000000001 to 1111111101100, and the whole phenotype string made of the $2^N = 16$ attractors corresponding to every initial state can be represented as a string with $16 \times 13 = 208$ bits, and its complexity can then be measured using the Lempel-Ziv method.

4.1 Redundancy and Simplicity Bias

As this GP map uses a finer-grained definition of phenotype, it produces a larger number of phenotypes than the more coarse-grained GP map presented in the previous section. Figure 4a shows a rank plot of the frequency of the 3.8×10^6 phenotypes produced by this GP map for $N = 4$ GRNs. This GP map also shows simplicity bias, as indicated in Fig. 4b. The scatter plot exhibits the characteristic triangular shape shown for the previous GP map in Figs. 3c and d, with the most frequent phenotype being when all initial conditions lead to 0000 as a fixed point, representing when all genes turn inactive.

5 Robustness and Evolvability

In addition to having a highly biased distribution of neutral network sizes and a strong bias towards simple outputs, both GP maps studied in this paper show evidence of a very correlated neutral network landscape. This can be measured by comparing the frequency f_P of a given phenotype P, i.e. the fraction of the genotype space that maps to P, with the phenotype's mutational robustness ρ_P, which is defined as the chance that a mutation to one of those genotypes which map to P produces a phenotype which still maps to P. In other words, it is the chance that a mutation to one of those phenotypes is a *neutral mutation*, that is, a mutation that leads to the same phenotype [1,5,11,12,24,41].

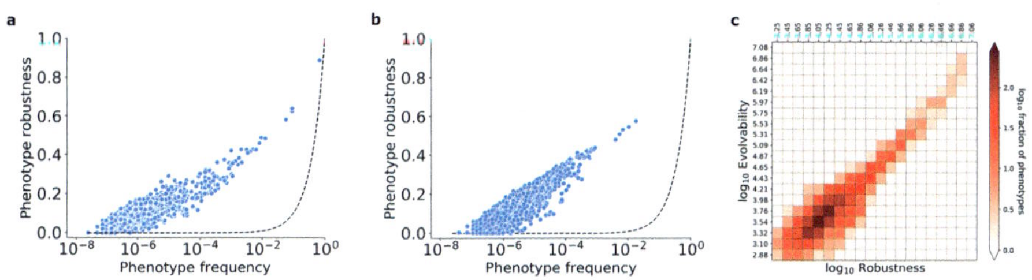

Fig. 5. Robustness and evolvability. Panels **(a)** and **(b)** show phenotype frequency versus phenotype robustness for phenotypes 1 and 2 respectively. In both cases, phenotype robustness grows proportionally to the logarithm of phenotype frequency. The dashed black line represents the random GP map expectation where $\rho_p = f_p$. Panel **(c)** shows a heatmap comparing phenotype robustness and evolvability for phenotype 2, showing that the most robust phenotypes are also the most evolvable ones.

For a random GP map where there is no correlation between neighbour, the chance of finding P in the neighbourhood of P's neutral network should be equal to the chance of finding P anywhere else in genotype space. Put simply, $\rho_P = f_P$. This is indicated by the black dashed lines in Figs. 5a and b. The GP maps studied here do not show this pattern: instead, phenotype robustness to mutations scales linearly with the logarithm of phenotype frequency, for both phenotype definitions. Finally, these GP maps also have a high correlation between phenotype robustness and phenotype evolvability, defined as the number of different phenotypes within one point mutation of a given phenotype [45]. This is illustrated for phenotype 2 in Fig. 5c. This behaviour echoes what has been observed for other GP maps in the literature, which often show correlated landscapes, where robust phenotypes are also the most evolvable [1,24,39].

6 Conclusion

In this paper, we have studied the map from a Boolean threshold network to its dynamic behaviour as a GP map, and showed that this map presents a series of properties observed in the GP maps literature, such as high levels of redundancy, a wide distribution of neutral network sizes, high robustness and evolvability, and most notably a bias towards low-complexity phenotypes.

Although these results are interesting, they only concern the space of GRN topologies. One could ask if we would obtain the same results if we had instead fixed the network topology and varied the strength of gene-gene interactions, or varied topology and interactions – both interesting research directions.

More broadly, the presence of a strong bias towards simplicity raises many questions about the systems modelled by Boolean threshold networks. GRNs are behind cell signalling, differentiation, embryonic development, and other complex biological patterns, but BTNs are also the building block of neural network models in machine learning [35]. Following the analogy between both types of networks, the evolutionary search for a gene network with a given set of biological properties can be compared to the machine learning task of finding a set of parameters that minimises a cost function. The parallels between evolution and computation, in particular between evolution and (machine) learning, are discussed in a large body of literature [43,44,47]. We believe that the study of GP maps might contribute to this literature. Based on the work presented in this paper, one could expect that phenomena such as simplicity bias, and even more biologically relevant properties such as robustness and evolvability, might have a computational or learning equivalent. Needless to say, the translation of concepts from one field to another leads to more questions than answers, and plenty of interesting work ahead.

References

1. Ahnert, S.E.: Structural properties of genotype–phenotype maps. J. R. Soc. Interface **14**(132), 20170275 (2017). https://doi.org/10.1098/rsif.2017.0275
2. Albert, R., Othmer, H.G.: The topology of the regulatory interactions predicts the expression pattern of the segment polarity genes in Drosophila melanogaster. J. Theor. Biol. **223**(1), 1–18 (2003)
3. Aral, N., Kabakçıoğlu, A.: Coherent regulation in yeast's cell-cycle network. Phys. Biol. **12**(3), 036002 (2015)
4. Aral, N., Kabakçıoğlu, A.: Coherent organization in gene regulation: a study on six networks. Phys. Biol. **13**(2), 026006 (2016)
5. Azevedo, R.B., Lohaus, R., Srinivasan, S., Dang, K.K., Burch, C.L.: Sexual reproduction selects for robustness and negative epistasis in artificial gene networks. Nature **440**(7080), 87 (2006)
6. Bergman, A., Siegal, M.L.: Evolutionary capacitance as a general feature of complex gene networks. Nature **424**(6948), 549 (2003)
7. Boldhaus, G., Bertschinger, N., Rauh, J., Olbrich, E., Klemm, K.: Robustness of Boolean dynamics under knockouts. Phys. Rev. E **82**(2), 021916 (2010)

8. Boldhaus, G., Klemm, K.: Regulatory networks and connected components of the neutral space. Eur. Phys. J. B-Condens. Matter Complex Syst. **77**(2), 233–237 (2010)
9. Borenstein, E., Krakauer, D.C.: An end to endless forms: epistasis, phenotype distribution bias, and nonuniform evolution. PLoS Comput. Biol. **4**(10), e1000202 (2008)
10. Chen, H., Wang, G., Simha, R., Du, C., Zeng, C.: Boolean models of biological processes explain cascade-like behavior. Sci. Rep. **7** (2016). Article number 20067. https://www.nature.com/articles/srep20067
11. Ciliberti, S., Martin, O.C., Wagner, A.: Innovation and robustness in complex regulatory gene networks. Proc. Natl. Acad. Sci. **104**(34), 13591–13596 (2007)
12. Ciliberti, S., Martin, O.C., Wagner, A.: Robustness can evolve gradually in complex regulatory gene networks with varying topology. PLoS Comput. Biol. **3**(2), e15 (2007)
13. Cowperthwaite, M.C., Economo, E.P., Harcombe, W.R., Miller, E.L., Meyers, L.A.: The ascent of the abundant: how mutational networks constrain evolution. PLoS Comput. Biol. **4**(7), e1000110 (2008)
14. Davidich, M.I., Bornholdt, S.: Boolean network model predicts cell cycle sequence of fission yeast. PLoS One **3**(2), e1672 (2008)
15. Davidich, M.I., Bornholdt, S.: Boolean network model predicts knockout mutant phenotypes of fission yeast. PLoS One **8**(9), e71786 (2013)
16. Dingle, K., Camargo, C.Q., Louis, A.A.: Input-output maps are strongly biased towards simple outputs. Nat. Commun. **9**(1), 761 (2018)
17. Dingle, K., Schaper, S., Louis, A.A.: The structure of the genotype-phenotype map strongly constrains the evolution of non-coding RNA. Interface Focus **5**(6), 20150053 (2015)
18. Espinosa-Soto, C., Padilla-Longoria, P., Alvarez-Buylla, E.R.: A gene regulatory network model for cell-fate determination during Arabidopsis thaliana flower development that is robust and recovers experimental gene expression profiles. Plant Cell **16**(11), 2923–2939 (2004)
19. Fauré, A., Naldi, A., Chaouiya, C., Thieffry, D.: Dynamical analysis of a generic Boolean model for the control of the mammalian cell cycle. Bioinformatics **22**(14), e124–e131 (2006)
20. Giacomantonio, C.E., Goodhill, G.J.: A Boolean model of the gene regulatory network underlying mammalian cortical area development. PLoS Comput. Biol. **6**(9), e1000936 (2010)
21. Glass, L., Mackey, M.C.: From Clocks to Chaos: The Rhythms of Life. Princeton University Press, Princeton (1988)
22. Greenbury, S.F., Ahnert, S.E.: The organization of biological sequences into constrained and unconstrained parts determines fundamental properties of genotype-phenotype maps. J. R. Soc. Interface **12**(113), 20150724 (2015)
23. Greenbury, S.F., Johnston, I.G., Louis, A.A., Ahnert, S.E.: A tractable genotype-phenotype map modelling the self-assembly of protein quaternary structure. J. R. Soc. Interface **11**(95), 20140249 (2014)
24. Greenbury, S.F., Schaper, S., Ahnert, S.E., Louis, A.A.: Genetic correlations greatly increase mutational robustness and can both reduce and enhance evolvability. PLoS Comput. Biol. **12**(3), e1004773 (2016)
25. Helikar, T., Konvalina, J., Heidel, J., Rogers, J.A.: Emergent decision-making in biological signal transduction networks. Proc. Natl. Acad. Sci. **105**(6), 1913–1918 (2008)

26. Hopfield, J.J., Tank, D.W.: Collective computation with continuous variables. In: Disordered Systems and Biological Organization, pp. 155–170. Springer (1986)
27. Hopfield, J.J., Tank, D.W.: Computing with neural circuits - a model. Science **233**(4764), 625–633 (1986)
28. Kauffman, S.A.: Homeostasis and differentiation in random genetic control networks. Nature **224**(5215), 177–178 (1969)
29. Kauffman, S.A.: The Origins of Order: Self-Organization and Selection in Evolution. Oxford University Press, Oxford (1993)
30. Lempel, A., Ziv, J.: On the complexity of finite sequences. IEEE Trans. Inf. Theory **22**(1), 75–81 (1976)
31. Li, F., Long, T., Lu, Y., Ouyang, Q., Tang, C.: The yeast cell-cycle network is robustly designed. Proc. Natl. Acad. Sci. U.S.A. **101**(14), 4781–4786 (2004)
32. Li, H., Helling, R., Tang, C., Wingreen, N.: Emergence of preferred structures in a simple model of protein folding. Science **273**(5275), 666 (1996)
33. Li, S., Assmann, S.M., Albert, R.: Predicting essential components of signal transduction networks: a dynamic model of guard cell abscisic acid signaling. PLoS Biol. **4**(10), e312 (2006)
34. May, R.M.: Models for two interacting populations. In: Theoretical Ecology: Principles and Applications, pp. 49–70 (1976)
35. McCulloch, W.S., Pitts, W.: A logical calculus of the ideas immanent in nervous activity. Bull. Math. Biophys. **5**(4), 115–133 (1943)
36. Nochomovitz, Y.D., Li, H.: Highly designable phenotypes and mutational buffers emerge from a systematic mapping between network topology and dynamic output. Proc. Natl. Acad. Sci. U.S.A. **103**(11), 4180–4185 (2006)
37. Raman, K., Wagner, A.: The evolvability of programmable hardware. J. Roy. Soc. Interface **8**(55), rsif20100212 (2010). https://royalsocietypublishing.org/doi/full/10.1098/rsif.2010.0212#ref-list-1
38. Remy, E., Ruet, P., Mendoza, L., Thieffry, D., Chaouiya, C.: From logical regulatory graphs to standard petri nets: dynamical roles and functionality of feedback circuits. In: Transactions on Computational Systems Biology VII, pp. 56–72. Springer (2006)
39. Schaper, S., Louis, A.A.: The arrival of the frequent: how bias in genotype-phenotype maps can steer populations to local optima. PLoS One **9**(2), e86635 (2014)
40. Schuster, P., Fontana, W., Stadler, P.F., Hofacker, I.L.: From sequences to shapes and back: a case study in RNA secondary structures. Proc. Roy. Soc. London B: Biol. Sci. **255**(1344), 279–284 (1994)
41. Steiner, C.F.: Environmental noise, genetic diversity and the evolution of evolvability and robustness in model gene networks. PLoS One **7**(12), e52204 (2012)
42. Tarjan, R.: Depth-first search and linear graph algorithms. SIAM J. Comput. **1**(2), 146–160 (1972)
43. Valiant, L.: Probably Approximately Correct: Nature's Algorithms for Learning and Prospering in a Complex World. Basic Books, New York (2013)
44. Valle-Pérez, G., Camargo, C.Q., Louis, A.A.: Deep learning generalizes because the parameter-function map is biased towards simple functions. In: International Conference on Learning Representations (ICLR) (2019)
45. Wagner, A.: Robustness and evolvability: a paradox resolved. Proc. Roy. Soc. London B: Biol. Sci. **275**(1630), 91–100 (2008)
46. Wagner, A.: Robustness and Evolvability in Living Systems. Princeton University Press, Princeton (2013)

47. Watson, R.A., Szathmáry, E.: How can evolution learn? Trends Ecol. Evol. **31**(2), 147–157 (2016)
48. Zañudo, J.G., Aldana, M., Martínez-Mekler, G.: Boolean threshold networks: virtues and limitations for biological modeling. In: Information Processing and Biological Systems, pp. 113–151 (2011)

Subsystem Cooperation in Complex Networks - Case Brain Network

Vesa Kuikka(✉)

Finnish Defence Research Agency,
Tykkikentäntie 1, PO BOX 10, 11311 Riihimäki, Finland
vesa.kuikka@mil.fi

Abstract. Modelling processes and interactions on complex networks can provide insight into the analysis of networks. Appropriate models should be developed to describe processes under interest. On the other hand, social network processes are various and individuals interact in many social networks with different intensities. Many biological networks, such as brain networks, are only partially understood. We discuss modelling approaches and use a recent brain network study as a basis. In order to model subsystem organisation and cooperation, we use a detailed model of the network topology. Two different network models are used to illustrate the ideas: classical network connectivity and influence spreading models represent connectivity based and spreading processes. The use of the influence spreading model is illustrated with calculations of centrality and betweenness measures for discovering and analysing hubs in brain networks. In this paper, the subsystem detection approach in the brain is not based on commonly applied hierarchical clustering methods but instead on a general community detection method. The proposed method enables discovering subsystems and their cooperation not restricted by hierarchical organisation structure. The two example network models show that modelling decisions can lead to different results at least on detailed levels.

Keywords: Complex networks · Brain network · Subsystem cooperation in networks · Processes on networks

1 Introduction

1.1 Background and Motivation

Data mining and data analysis involve a number of methods that can be used in many different applications domains. Examples of the methods are cluster analysis, genetic algorithms and text mining, and examples of application areas are big data, bioinformatics, and domain driven data mining. It is obvious that methods should be selected for the application area that are most useful in studying the specific research question.

H. Barbosa et al. (Eds.): *Complex Networks XI*, SPCOM, pp. 156–169, 2020.
https://doi.org/10.1007/978-3-030-40943-2_14

We discuss how modelling interactions and processes on networks can provide more insight into understanding phenomena and organisation of complex networks. We compare results from two different network models: the standard network connectivity model [2] and an influence spreading model [9–11]. In this paper, we don't provide an exact definition of a process because it means different things in different contexts. For example, in the area of communication networks, one description is that a process can exchange information or synchronise their operation through several methods of inter-process communication. Many of the contexts have both synchronous and asynchronous methods of communication.

We use as a case study the brain network structure published in [14]. This network is selected because interactions on brain subsystems and their relationship with diverse human or animal functions are only partially understood. This is why the case of brain networks is particularly interesting as we don't know which network models realistically describe complex processes and structure in brain networks. Nevertheless, we can compare the results from different models and make some conclusions based on common outcomes of the models, and explore probable application domains of the methods. Here, our aim is not to perform or repeat the analysis of the previous work [14] but instead to discuss and highlight aspects of complex network research and illustrate the ideas with examples.

Subsystem (module) detection in brain networks is often based on hierarchical clustering methods. We use a general community detection method that enables discovering subsystems and their cooperation without assuming a hierarchical subsystem structure. The method of this paper has two phases: network modelling and community detection [11]. The two alternative network models are based on different assumptions: connectivity between node pairs or influence spreading. The community detection method is explained briefly in Sect. 3.

1.2 Processes on Complex Networks

In the research of complex systems, general data analysis and statistical methods can often be supplemented with specialised methods to take into account complex network characteristics. We expect that the main modelling principles are common for a wide domain of applications. However, interactions and processes on social, biological and technological networks can play an important role. Especially, it would be interesting to investigate whether similar methods and models are sufficient for describing these complex network systems, or do these network classes require dedicated modelling approaches. Humans can form communities and substructures in different roles associated with their work, hobbies and free time. One individual is a part of numerous social groups in the society and is networked around beliefs and opinions on many levels.

Organisation and processes on biological complex networks are also an interesting research question. Connections in the brain and networks describing protein interactions are clearly different. The term 'process' refers to dynamic and time dependent changes on network structure. These can be spreading processes or physical changes of the network topology. We don't know exactly how the brain's subsystems function, while the research methods themselves should not have biased assumptions, which can lead to erroneous outcomes. For instance, assuming characteristics of the network processes which are not valid may result in misleading conclusions.

Many of the current network analysis methods are static or do not consider complex network topology in full extent. Network topology is an important aspect of complex networks. Networks are usually modelled with nodes, and directed connections between nodes. Nodes and links have attributes like strength, state of spreading and different static information about network elements.

In this study, we use two different network models: the classical connectivity model [2] and an influence spreading model [10]. The spreading model is less restrictive than the network connectivity model, and this is the reason why it may better describe activities in human social networks. In the spreading model, we also allow loops in interactions between nodes in the network. This feature is often believed to characterise social network processes.

In both models we assume that the probability of connectivity or the spreading probability do not depend on the states of the mediating nodes or links between the two communicating nodes. In human social networks this assumptions is not always exactly true. In cases, where nodes have no memory, it does not make a difference whether the node has received the same information already or not. In social interactions, the information content and the way it is presented changes during the process. This kind of opinion formation may, in some degree, be described by state independent modelling.

2 Related Work

Several review articles cover methods of community detection [6,13]. Hierarchical clustering tree (see for example [7]), spectral graph partitioning and modularity maximisation are three examples in the wide context of community detection methods. Solutions to the community detection problem are typically solutions to the optimisation problem of a quality function. The Louvain algorithm and Infomap [3,6] are two examples that are scalable for very large networks. Recently, a community detection method based on influence spreading processes has been proposed in [11]. Modularity measures the strength of division of a network into modules. One description of a community is a locally dense connected subgraph in a network. The definition of modularity is based on the adjacency matrix and the fraction of links within the group minus the expected fraction if links were distributed at random [3].

Multiresolution consensus clustering method (MRCC) is a method for identifying community structure at different scales based on multiresolution modularity and consensus clustering. MRCC includes a strategy for sampling the entire range of possible solutions for a multiresolution modularity quality function. A hierarchical consensus clustering procedure is used to construct a hierarchical consensus structure given a set of input partitions. The method allows one to obtain a solution among many different community structures often as the result of runs of a stochastic algorithm [12].

In MRCC an iterated variant of the Louvain [4] algorithm is used for modularity optimisation. The algorithm has two-phases. First, the algorithm moves single nodes between communities. When there are no more single moves that increase modularity, the identified communities are aggregated to form new combinations. These phases are repeated until no increase in modularity is identified. In order to further improve modularity, the algorithm is then restarted using the new communities as the initial partition until no increase in modularity is identified [12].

Brain networks have been studied extensively in the literature. References can be found, for example, in review article [1]. In this paper, we use the organisation of a brain network as the case study. Subsystem organisation of axonal connections within and between the right and left cerebral cortex and cerebral nuclei (endbrain) have been studied in [14]. The organisation of cortical association and commissural connections in the rat were investigated. The endbrain is primarily responsible for supporting cognition and affect. It consists of right and left cerebral hemispheres. One deals with auditory and visual information, and the other corresponds mainly to the default mode network [14].

The network of 244 gray matter regions is connected by 10,002 macroconnections producing a complete endbrain subconnectome. Organisation of the network structure was analysed with a multiresolution consensus clustering (MRCC) method [12]. The result was a hierarchy of four modules at the top level and 60 components at the bottom level [14].

The authors in [14] concluded that the size and coverage of anatomic neighbourhood have effects on the status of a region as a connectivity hub. Hub is a node occupying a central position in the overall organisation of a network. Hubs are important for enabling efficient neuronal signalling and communication. Hubs support diverse functional roles across cognitive tasks and dynamic coupling within and across functional networks [8].

3 Complex Network Modelling

In this paper we optimise a basic quality function of the sums of rows and columns of an interaction probability matrix C.

$$P = \sum_{s,t \in V} C_{s,t} + \sum_{s,t \in (G-V)} C_{s,t} \tag{1}$$

The original network G is separated into two divisions V and G-V. In the network connection model and in the influence spreading model different definitions of the probability matrix are used. The elements of the matrix are connection probabilities or influence spreading probabilities at time T between node pairs in the network.

The optimal value of Eq. (1) can be used for ranking the communities according to the strength of their internal cohesion. The highest value of Eq. (1) does not guarantee that the community is spontaneously formed with a high probability starting from a random initial division. This probability is an alternative measure for ranking divisions of the network. In fact, it depends on the initial state of the original network which one of the possible outcomes will realise. However, in many cases, these two measures provide similar highest rankings for communities. Later in this study we have decided to use the probability of forming a division of the network as the measure for ranking solutions. We will refer to this measure as 'statistical community measure' [11].

The influence spreading model used in computing the influence spreading matrix has been presented in [10]. Features of the spreading model are: exact network topology of a network, directed links, weighted nodes and links, and time dependency of the spreading process. Any probabilistic distribution function can be used for describing the time dependency of the spreading over links in the network. In our numerical examples a limiting value of time approaching infinity describes the equilibrium state of developing community structures. The model allows to set a maximum path length in order to investigate local interactions. Also, the maximum number of visits on one node in a path can be limited. This is useful for studying self-avoiding paths. We assume that loops (but no self-loops) are allowed in social interactions. In this paper, we also assume that loops are allowed for interactions on the brain network.

Specifically, the influence spreading model [10] takes into account all paths, not just the shortest paths, between all node pairs of the network (limited by the maximum path length L_max because of computational reasons or for describing local interactions). This kind of modelling can provide more accurate results for analysing processes on networks. Equation (1) does not include cross terms in the summands. This is a common practice in many other models: the standard definition of modularity [3] has a sum over node pairs within the community. Cross terms may have an interpretation in the context of social networks. Social pressure exists between communities to convince others of their beliefs and opinions. In the cases of technological and biological networks, it is more difficult to find an interpretation for non-zero cross terms.

4 Results

4.1 Centrality and Betweenness Measures

Centrality and betweenness measures are useful for discovering and analysing hubs in brain networks. In the influence spreading model centrality measures can be defined for output and input interactions. There is only one version of the betweenness measure. These measures characterize different roles of hubs in the network.

The influence spreading model is used to compute node centrality measures for in-centrality and out-centrality as defined in [10]. In Fig. 1 in-centrality measures are indicated with blue and out-centrality measures with orange bars. The out-centrality measure is designed for describing the influence spreading power of nodes in the network. The in-centrality measure describes the inbound influence from other nodes of the network. These measures are different from the classical centrality measures in the literature [3]. Figure 1 shows the numerical values of in-centrality and out-centrality for the 122 first regions (nodes) of the brain network of [14].

Names of the regions are indicated as labels to allow comparison with the original study [14]. Two digits in the labels indicate the two cerebral hemispheres and the four modules detected by MRCC algorithm [12]. One observation is that the values of in-centrality and out-centrality are unbalanced in many cases: in-centrality values are high for {11 NDP, 21 NDP, 13 PERI, 13 PL, 13 BLAa, 13 ORBm, 13 ORBv, ...} and out-centrality values are high for {11 ACAd, 13 CP, 13 ACB, ...}. In-centrality and out-centrality measures of the influence spreading model provide more information than the standard node degree results presented in [14]. The remaining results of centrality measures for regions 123–244 are documented in Appendix.

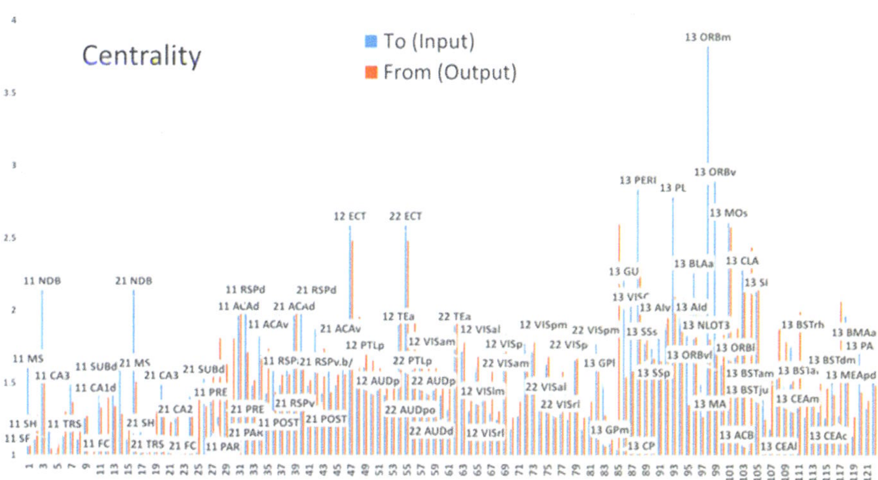

Fig. 1. Values of node centrality measures as defined in [10] for the first 122 nodes (regions in [14]) of the brain network. Out- and in-centrality values are indicated by orange and blue bars.

Figure 2 shows the values of the betweenness measure as defined in [10]. The centrality measures and the betweenness measure are correlated but interesting differences can be observed in the details. The betweenness measure is not defined for in- or out-measures, instead it is a compromise between the two centrality measures. Betweenness measures the role of a node as a mediator [5] of influence, and the relative impact on the influence spreading process if a node is removed from the network. The remaining results of the betweenness measure for regions 123–144 are documented in Appendix.

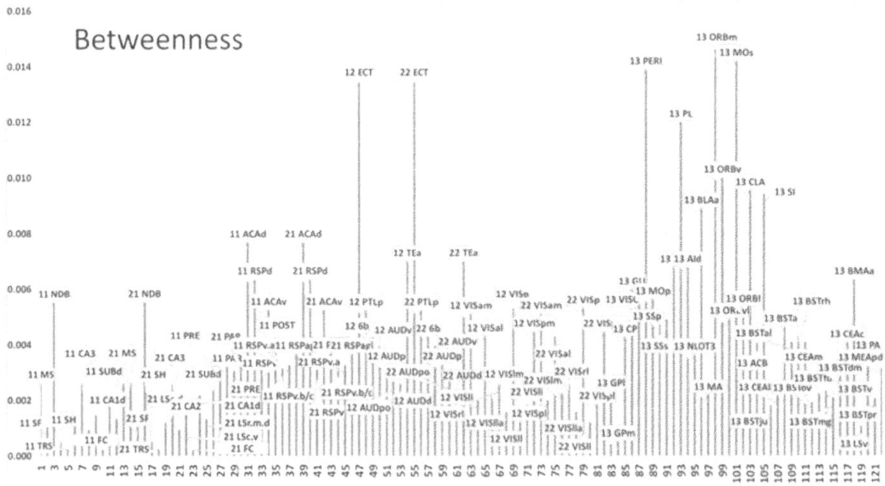

Fig. 2. Values of node betweenness measure as defined in [10] for the first 122 nodes.

4.2 Detected Subsystems

Next, we discuss briefly some representative results of the community detection method. The algorithm optimising the quality function of Eq. (1) discovers seven different divisions of the entire brain network of [14]. In Sect. 4.2 we refer to both partitions of a division as separate subsystems. The pattern of subsystem organisation is shown in Table 1. The divisions into two halves are indicated by 0s and 1s in the table. The regions (nodes) inside both of these two factions are connected or cooperating together in some way. In the following we assume that this interpretation holds. The first line indicates exactly the strongest division into two hemispheres of the brain network as in [14]. Table 2 shows the same information with node numbers (Figs. 1, 2 and 3 use the same numbering). The first line in Tables 1 and 2 corresponds to the highest hierarchy level detected by the multiresolution consensus clustering (MRCC) method in [14]. Names of the regions corresponding to the numbering in Tables 1, 2, 3, 4 and Fig. 3 are listed in Table 5 in Appendix. Table 1 displays subsystem cooperation across the hierarchical levels detected in [14]. These kind of interactions between subsystems can be important in analysing subsystem organisation. Different views can exist simultaneously and they can be useful in understanding different aspects of complex systems such as brain networks. Table 1 visualises regions of the brain network subsystems that cooperate as predicted by optimising Eq. (1). The last column in Table 2 shows the numerical values of the statistical community measure describing the probability of subsystem formation given that the initial division is random. The first line dominates all the other alternatives. The four modules of regions 1–46, 47–82, 83–163, 164–244 in [14] are not directly detected by optimising Eq. (1). Lines 2–7 show combinations of regions with several configurations where individual regions cooperate across the module and hemisphere boundaries. Table 1 visualises a hierarchical structure with a fine structure of some individual regions that are differently associated in lines 2–7. In fact, these regions can be in important roles in coordinating functions in the network. Regions can be identified in more detail with the help of Table 2 and Table 5 in Appendix. We conclude from Table 2 of line 2 that the following combinations of regions probably cooperate: {11 FC, 11 CA1d, 11 SUBd, 21 FC, 21 CA1d, 21 SUBd, 11 PRE, ..., 21 RSPagl}, {12 ECT, ..., 22 VISpm}, {13 GPI, 13 Ald,13 BLAa, 13 ORBm, ..., 13 CLA, 13 LA, 13 Alp, 13 ENTI, 13 IG, ..., 24 Ald, 24 BLAa, 24 ORBm, ..., 24 CLA, 24 LA, 24 Alp, 24 ENTI, 24 IG, 24 ENTm}. All the regions in module 2 are included in this configuration. On the other hand, we expect that the remaining regions {11 MS, ..., 11 CA2, 11 LSr.m.d, ...} is also a set of regions coordinating their actions. This kind of analysis can reveal subsystem cooperation between different modules and the two hemispheres, not just cooperation inside modules.

Table 1. Subsystems in the brain network [14] discovered by optimising the community detection measure of Eq. (1). The first line indicates the strongest division into two hemispheres of the brain network.

Table 2. Nodes included in divisions of the network (the other half consists of the remaining nodes of the original network of 244 nodes). The lines correspond to the lines in Table 1. The values of the statistical community measure are shown in the last column. (0.003 times the numerical binned values in [14] are used as link weights in the influence spreading model)

1	14-26,29-30,39-46,55-62,73-82,164-244	0.6399 %
2	10-11,13,23-24,26-94,96,98-103,145,156,159,162-175,177,179-184,226,237,240,243-244	0.0074 %
3	13,24,26-94,98-103,156,163-175,177,179-184,226,237,240-244	0.0036 %
4	10-11,13,23-24,26-96,98-103,145,156,159-160,162-177,179-184,226,237,240-241,243-244	0.0034 %
5	11,13,26-94,96,98-103,145,156,159,163-175,179-184,237-244	0.0017 %
6	10-11,13,23-24,26-94,96,98-103,145,156,159,161-175,177,179-184,226,237,240,242-244	0.0012 %
7	10-11,13,23-24,26-96,98-103,145,156,159-177,179-184,226,237,240-244	0.0005 %

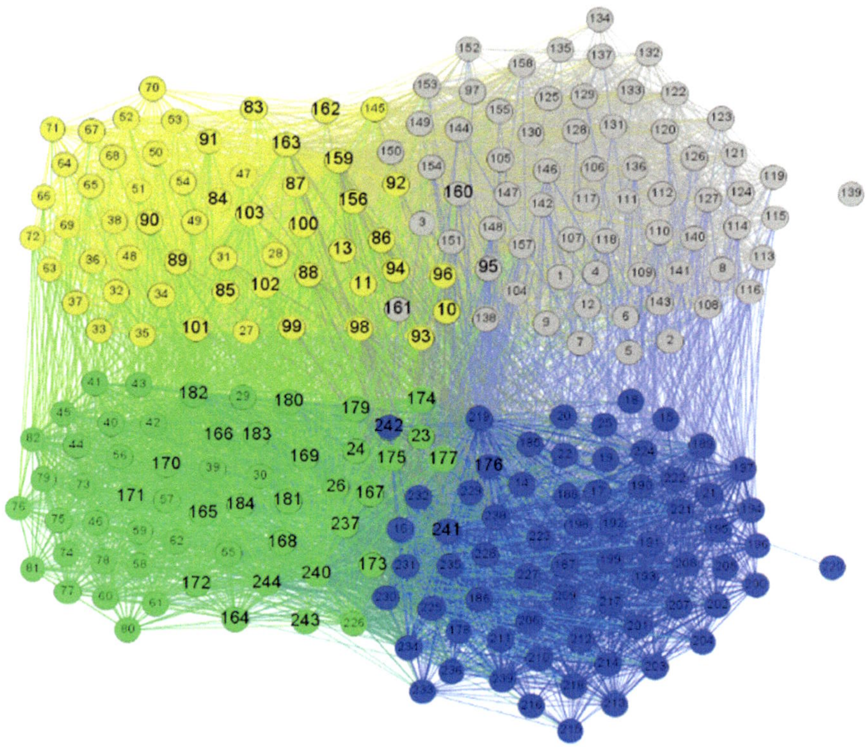

Fig. 3. The main division into two hemispheres (line 1 in Tables 1 and 2) is clearly seen in the upper and lower parts of the figure. The second order division (line 2 in Tables 1 and 2) is indicated by colours. Regions in detected subsystems that cooperate across the main division are indicated by larger font numbers (figure is created by Force Atlas layout of Gephi software).

Table 3. Results from the network connection model that can be compared with Table 2. The seven strongest divisions are shown. (0.1 times the numerical binned values in [14] are used as connection probabilities in the network connectivity model)

1	14-26,29-30,39-46,55-62,73-82,164-244	2.7558 %
2	10-11,13,23-24,26-94,96,98-103,145,156,159,162-175,177,179-184,226,237,240,243-244	0.0549 %
3	10-11,13,23-24,26-94,96,98-103,145,156,159,161-175,177,179-184,226,237,240,242-244	0.0363 %
4	2,5,8,15,21,95,97,104-144,146-155,157-158,160,176,178,185-225,227-236,238-239,241	0.0223 %
5	10-11,13,23-24,26-96,98-103,145,156,159,162-177,179-184,226,230-232,234-235,237,240-241,243-244	0.0103 %
6	10-11,13,23-24,26-96,98-103,145,149-151,153-154,156,159-160,162-177,179,184,226,237,240,243-244	0.0035 %
7	10-11,13,23-24,26-94,98-103,156,159,161-175,179-184,237,240,242-244	0.0018 %

Figure 3 shows the main division (line 1 in Tables 1 and 2) into two hemispheres. In addition the second order division (line 2) is indicated with colours. In Fig. 3 individual regions cooperating across the boundaries of the main division are collected (from lines 2–7) and highlighted with large font size numbers. Tables 1 and 2 can be used to list subsystems that cooperate with each of these special regions. Note that the high-level layout of Fig. 3 generated by Gephi software coincides very well with the community detection algorithm of Eq. (1) and also with the results in [14]. For comparison, the results calculated from the network connectivity model [2] are shown in Table 3. The seven strongest divisions of the network are shown. The first two lines agree with the influence spreading model results in Table 2. The results in Table 3 are close to lines in Table 2, but differences exist in details. Both models reveal one special configuration, but they are different: line 3 in Table 2 and line 4 in Table 3. Other differences in lines 1–4 are nodes 162 and 243 (13 IG and 24 IG) in the connection model and nodes 95, 160, 176, and 241 (13 NLOT3, 13 BLAp, 24 NLOT3, 24 BLAp) in the spreading model. From Fig. 3 we can see that these nodes are examples of nodes near boundaries (predicted by Gephi software) of the left and right partitions (line 2 in Tables 2 and 3). The conclusion is that the main results are similar but interaction characteristics can have an impact on detailed results. Note that the results in these section are calculated with link weight values appropriate for demonstrating purposes. More granular results and more detailed structure is revealed with lower link weight values. Tables 1, 2 and 3 would contain more lines for analysing.

4.3 Intersections and Subtractions

As the last tool for analysing subsystems in networks we present a method of computing intersections and subtractions of the detected subsystems. Detailed analysis of these results can provide useful information about the cooperation of subsystems in the network structure. In the brain network the most important sets of mediating regions (nodes) can be discovered by taking intersections between the main division and weaker divisions of the network structure. Subtractions reveal regions cooperating possibly in a different process with another subsystem. Intersections and subtractions of subsystems may be or may not be optimal solutions of Eq. (1) depending on their mutual strength of interactions. The results can be presented in the same form as Tables 1 and 2. Table 4 shows intersections and subtractions of the subsystems in Table 2. Intersections of subsystems can be ranked by using the product of the values of the two related subsystem measures. Lines in Table 4 are ordered according to their rankings by the statistical community measure.

Table 4. Lines 1–12 are intersections of all two subsystem pairs in Table 2. Lines 13–23 are the strongest subtractions of subsystem pairs in Table 2.

1	23-24,26,29-30,39-46,55-62,73-82,164-175,177,179-184,226,237,240,243-244
2	24,26,29-30,39-46,55-62,73-82,164-175,177,179-184,226,237,240-244
3	23-24,26,29-30,39-46,55-62,73-82,164-177,179-184,226,237,240-241,243-244
4	26,29-30,39-46,55-62,73-82,164-175,179-184,237-244
5	23-24,26,29-30,39-46,55-62,73-82,164-175,177,179-184,226,237,240,242-244
6	23-24,26,29-30,39-46,55-62,73-82,164-177,179-184,226,237,240-244
7	10-11,13,23-24,26-94,96,98-103,145,156,159,162-175,177,179-184,226,237,240,243-244
8	13,24,26-94,98-103,156,163-175,177,179-184,226,237,240-244
9	13,26-94,98-103,156,163-175,179-184,237-244
10	11,13,26-94,96,98-103,145,156,159,163-175,179-184,237-244
11	10-11,13,23-24,26-96,98-103,145,156,159-160,162-177,179-184,226,237,240-241,243-244
12	10-11,13,23-24,26-94,96,98-103,145,156,159,161-175,177,179-184,226,237,240,242-244
13	14-22,25,176,178,185-225,227-236,238-239,241-242
14	10-11,13,27-28,31-38,47-54,63-72,83-94,96,98-103,145,156,159,162-163
15	14-23,25,176,178,185-225,227-236,238-239,241-243
16	13,27-28,31-38,47-54,63-72,83-94,98-103,156,163
17	14-22,25,178,185-225,227-236,238-239,242
18	10-11,13,27-28,31-38,47-54,63-72,83-96,98-103,145,156,159-160,162-163
19	14-25,176-178,185-236,238-243
20	11,13,27-28,31-38,47-54,63-72,83-94,96,98-103,145,156,159,163
21	14-22,25,176,178,185-225,227-236,238-239,241
22	10-11,13,27-28,31-38,47-54,63-72,83-94,96,98-103,145,156,159,161-16314-22,25,178,185-225, 227-236,238-239
23	10-11,13,27-28,31-38,47-54,63-72,83-96,98-103,145,156,159-163

5 Conclusions

We discuss how interactions and processes on complex networks can be studied. The use of the influence spreading model is illustrated with calculations of centrality and betweenness measures for analysing brain networks and discovering important nodes. These measures reveal different characteristics of hubs in brain networks. Next, we study whether two different network models lead to different subsystem structures. Subsystems are sets of regions (nodes) in the brain network that optimise a quality function. The influence spreading model and the standard network connectivity model provide mainly the same results but essential differences exist. Assuming different interactions on networks and different modelling decisions can have effects on the results of discovered subsystems. Finally, our aim is to discuss supplementary or alternative community detection methods to hierarchical clustering methods previously used to investigate brain network organisation. The community detection method used here is not based on hierarchical organisation. These kind of methods can reveal subsystem cooperation across organisation hierarchies. Examples presented in this paper illustrate how cooperation among subsystems across module and hemisphere boundaries can be studied. In this paper, we have discussed modelling methods and possible new properties of complex network subsystem cooperation and organisation. Assuming that particular types of network processes describe cooperation among subsystems, such as a connectivity or a spreading model, can lead to different results and conclusions. If mechanisms and network processes are not completely understood, alternative models can provide new insights. Different models are useful in confirming or rejecting alternative presumptions about systems' behaviours.

Appendix

(See Table 5, Figs. 4 and 5)

Table 5. Names of regions 1–244. Two digits before names indicate the two hemispheres and the four modules documented in [14].

1	11 MS	21	21 LSc.v	41	21 RSPv	61	22 AUDd	81	22 VISpl	101	13 MOs	121	13 BSTpr	141	13 LSr.dl	161	13 DG	181	24 ORBvl	201	24 PA	221	24 LSr.vl	241	24 BLAp
2	11 SF	22	21 CA2	42	21 ACAv	62	22 TEa	82	22 VISpm	102	13 ORBl	122	13 MEApd	142	13 SUBv	162	13 IG	182	24 MOs	202	24 BSTpr	222	24 LSr.dl	242	24 DG
3	11 NDB	23	21 FC	43	21 POST	63	12 VISam	83	13 GPl	103	13 CLA	123	13 BSTd	143	13 LSr.m.v	163	13 ENTm	183	24 ORBl	203	24 MEApd	223	24 SUBv	243	24 IG
4	11 SH	24	21 CA1d	44	21 RSPv.b/c	64	12 VISli	84	13 GPm	104	13 ACB	124	13 BSTse	144	13 CA1v	164	24 GPl	184	24 CLA	204	24 BSTd	224	24 LSr.m.v	244	24 ENTm
5	11 TRS	25	21 LSr.m.d	45	21 RSPv.a	65	12 VISal	85	13 CP	105	13 SI	125	13 AAA	145	13 LA	165	24 GPm	185	24 ACB	205	24 BSTse	225	24 CA1v		
6	11 LSc.d	26	21 SUBd	46	21 RSPagl	66	12 VISim	86	13 GU	106	13 BSTju	126	13 BSTtr	146	13 BMAp	166	24 CP	186	24 SI	206	24 AAA	226	24 LA		
7	11 CA3	27	11 PRE	47	12 ECT	67	12 VISrl	87	13 VISC	107	13 CEAl	127	13 BSTif	147	13 OT	167	24 GU	187	24 BSTju	207	24 BSTtr	227	24 BMAp		
8	11 LSc.v	28	11 PAR	48	12 PTLp	68	12 VISlla	88	13 PERI	108	13 BSTam	128	13 COApl	148	13 NLOT	168	24 VISC	188	24 CEAl	208	24 BSTif	228	24 OT		
9	11 CA2	29	21 PRE	49	12 6b	69	12 VISp	89	13 MOp	109	13 BSTal	129	13 PAA	149	13 TTd	169	24 PERI	189	24 BSTam	209	24 COApl	229	24 NLOT		
10	11 FC	30	21 PAR	50	12 AUDv	70	12 VISll	90	13 SSp	110	13 BSTrh	130	13 IA	150	13 TTv	170	24 MOp	190	24 BSTal	210	24 PAA	230	24 TTd		
11	11 CA1d	31	11 ACAd	51	12 AUDp	71	12 VISpl	91	13 SSs	111	13 CEAm	131	13 MEAad	151	13 AOA	171	24 SSp	191	24 BSTrh	211	24 IA	231	24 TTv		
12	11 LSr.m.d	32	11 RSPd	52	12 AUDpo	72	12 VISpm	92	13 AIv	112	13 BSTov	132	13 MEAav	152	13 MOB	172	24 SSs	192	24 CEAm	212	24 MEAad	232	24 AOA		
13	11 SUBd	33	11 RSPv	53	12 AUDd	73	22 VISam	93	13 PL	113	13 BSTfu	133	13 MEApv	153	13 EPd	173	24 AIv	193	24 BSTov	213	24 MEAav	233	24 MOB		
14	21 MS	34	11 ACAv	54	12 TEa	74	22 VISli	94	13 AId	114	13 BSTv	134	13 BA	154	13 PIR	174	24 PL	194	24 BSTfu	214	24 MEApv	234	24 EPd		
15	21 SF	35	11 POST	55	22 ECT	75	22 VISal	95	13 NLOT3	115	13 BSTmg	135	13 AOB	155	13 EPv	175	24 AId	195	24 BSTv	215	24 BA	235	24 PIR		
16	21 NDB	36	11 RSPv.b/c	56	22 PTLp	76	22 VISim	96	13 BLAa	116	13 BSTdm	136	13 COApm	156	13 AIp	176	24 NLOT3	196	24 BSTmg	216	24 AOB	236	24 EPv		
17	21 SH	37	11 RSPv.a	57	22 6b	77	22 VISrl	97	13 MA	117	13 CEAc	137	13 COAa	157	13 FS	177	24 BLAa	197	24 BSTdm	217	24 COApm	237	24 AIp		
18	21 TRS	38	11 RSPagl	58	22 AUDv	78	22 VISlla	98	13 ORBm	118	13 BMAa	138	13 ILA	158	13 TR	178	24 MA	198	24 CEAc	218	24 COAa	238	24 FS		
19	21 LSc.d	39	21 ACAd	59	22 AUDp	79	22 VISp	99	13 ORBv	119	13 LSv	139	13 BAC	159	13 ENTl	179	24 ORBm	199	24 BMAa	219	24 ILA	239	24 TR		
20	21 CA3	40	21 RSPd	60	22 AUDpo	80	22 VISll	100	13 ORBvl	120	13 PA	140	13 LSr.vl	160	13 BLAp	180	24 ORBv	200	24 LSv	220	24 BAC	240	24 ENTl		

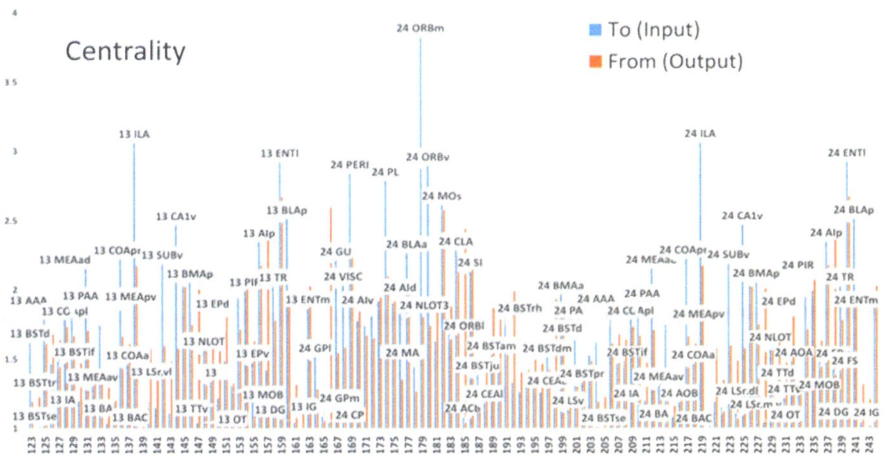

Fig. 4. Node centrality measures as defined in [10] for nodes 123–244 of the brain network [14]. Out-centrality is indicated by "From" and in-centrality by "To".

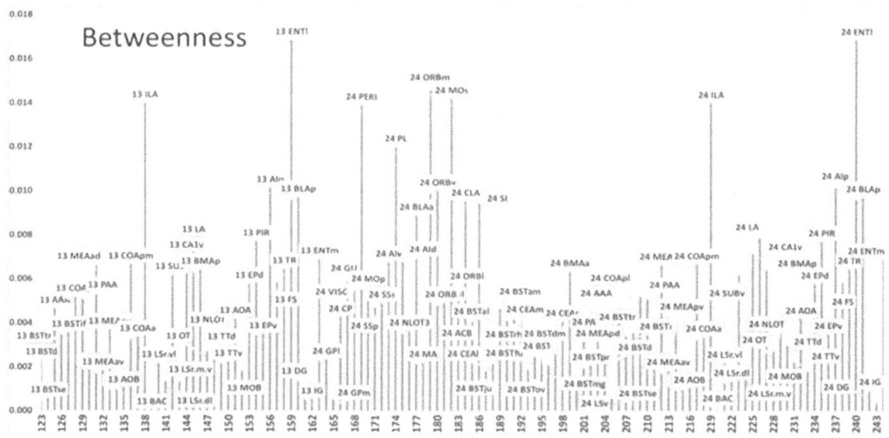

Fig. 5. Node betweenness measure as defined in [10] for nodes 123–244.

References

1. Avena-Koenigsberger, A., Misic, B., Sporns, O.: Communication dynamics in complex brain networks. Nat. Rev. Neurosci. **19**, 17–33 (2018). https://doi.org/10.1038/nrn.2017.149
2. Ball, M.O., Colbourn, C.J., Provan, J.S.: Network reliability. In: Handbooks in Operations Research and Management Science, vol. 7, pp. 673–762 (1995). Chapter 11
3. Barabási, A.-L.: Network Science. Cambridge University Press, Cambridge (2016)
4. Blondel, V.D., Guillaume, J.-L., Lambiotte, R., Lefebvre, E.: Fast unfolding of communities in large networks. J. Stat. Mech.: Theory Exp. P10008 (2008). https://sci-hub.tw/https://iopscience.iop.org/article/10.1088/1742-5468/2008/10/P10008/pdf
5. Chaudhary, A.K., Warner, L.A.: Introduction to social network research: brokerage typology, AEC535, Agricultural Education and Communication Department (2018)
6. Fortunato, S., Hric, D.: Community detection in networks: a user guide. Phys. Rep. **659**(11), 1–44 (2016)
7. Girvan, M., Newman, M.E.J.: Community structure in social and biological networks. Proc. Natl. Acad. Sci. U.S.A. **99**(12), 7821–7826 (2002)
8. van den Heuvel, M.P., Sporns, O.: Network hubs in the human brain. Trends Cogn. Sci. **17**(12), 683–696 (2013). https://doi.org/10.1016/j.tics.2013.09.012
9. Kuikka, V.: Influence spreading model used to community detection in social networks. In: Cherifi, C., Cherifi, H., Karsai, M., Musolesi, M. (eds.) Complex Networks & Their Applications VI. COMPLEX NETWORKS 2017. Studies in Computational Intelligence, vol. 689, pp. 202–215. Springer, Cham (2018)
10. Kuikka, V.: Influence spreading model used to analyse social networks and detect sub-communities. Comput. Soc. Netw. **5**, 12 (2018). https://doi.org/10.1186/s40649-018-0060-z

11. Kuikka, V.: A general method for detecting community structures in complex networks. In: Cherifi, H., Gaito, S., Mendes, J., Moro, E., Rocha, L. (eds.) Complex Networks and Their Applications VIII. COMPLEX NETWORKS 2019. Studies in Computational Intelligence, vol. 881. Springer, Cham (2019). https://doi.org/10.1007/978-3-030-36687-2_19
12. Jeub, L.G.S., Sporns, O., Fortunato, S.: Multiresolution consensus clustering in networks. Sci. Rep. **9**, 3259 (2018). https://doi.org/10.1038/s41598-018-21352-7
13. Lancichinetti, A., Fortunato, S.: Community detection algorithms: a comparative analysis. Phys. Rev. E **80**, 056117 (2009)
14. Swanson, W.S., Hahn, J.D., Jeub, L.G.S., Fortunato, S., Sporns, O.: Subsystem organization of axonal connections within and between the right and left cerebral cortex and cerebral nuclei (endbrain). PNAS **115**(29), E6910–E6919 (2018)

Network-Based Approach for Modeling and Analyzing Coronary Angiography

Babak Ravandi[1(✉)] and Arash Ravandi[2]

[1] Network Science Institute, Center for Complex Network Research, Northeastern University, Boston, MA 02115, USA
bravandi@northeastern.edu, bk.ravandi@gmail.com

[2] Division of Orthopeadic Rheumatology, Friedrich-Alexander University Erlangen-Nuremberg, Waldkrankenhaus Erlangen, 91054 Erlangen, Germany
arash.ravandi@fau.de
http://bravandi.net

Abstract. Significant intra-observer and inter-observer variability in the interpretation of coronary angiograms are reported. This variability is in part due to the common practices that rely on performing visual inspections by specialists (e.g., the thickness of coronaries). Quantitative Coronary Angiography (QCA) approaches are emerging to minimize observer's error and furthermore perform predictions and analysis on angiography images. However, QCA approaches suffer from the same problem as they mainly rely on performing visual inspections by utilizing image processing techniques. In this work, we propose an approach to model and analyze the entire cardiovascular tree as a complex network derived from coronary angiography images. This approach enables to analyze the graph structure of coronary arteries. We conduct the assessments of network integration, degree distribution, and controllability on a healthy and a diseased coronary angiogram. Through our discussion and assessments, we propose modeling the cardiovascular system as a complex network is an essential phase to fully automate the interpretation of coronary angiographic images. We show how network science can provide a new perspective to look at coronary angiograms.

Keywords: Complex networks · Coronary heart disease · Angiography · Quantitative coronary angiography · Coronary network · Complex systems

1 Introduction

Coronary Heart Disease (CHD) is a major cause of disability and death in developed countries. Although over the past four decades CHD mortality rates have declined worldwide, CHD remains responsible for one-third of all deaths in people over age of 35 [31]. Invasive coronary angiography is the current gold standard to determine the presence, location, and stage of coronary artery disease as

H. Barbosa et al. (Eds.): *Complex Networks XI*, SPCOM, pp. 170–181, 2020.
https://doi.org/10.1007/978-3-030-40943-2_15

well as to follow-up with the patients after therapeutic procedures [26]. However, potential observer error from performing visual analysis of Coronary Angiograms (CAs) has been estimated to be over 35% [6]. Quantitative Coronary Angiography (QCA) approaches are emerging to minimize the observer error and further perform predictions and analysis on angiography images [13]. Feyter et al. [8] classified the limitations of QCA approaches to three categories: patient related, technique related, and methodology related. The technique and methodology related limitations such as X-Ray contrast and calibration limitations are significantly improved due to immense technical improvements in the medical imaging techniques and the advances in machine learning [5,17,32]; notably, 3D Reconstruction of coronary angiography from 2D images [3,4]. However, the main limitation of QCA approaches remains on the patient related limitations particularity the procedural constraints and capturing physiological characteristics such as side branches and bifurcations, hemodynamic assessment, and vasomotion that are technically difficult to measure [10,12,29]. Hence, several angiography phenomena can lead QCA approaches toward over or underestimation of parameters such as extensive calcium deposits, acute or chronic thrombus, and slow flow [9]. Due to these limitations, QCA approaches lack sufficient accuracy to be employed for clinical purposes [13]. We believe the missing key is dynamization of QCA; that can be achieved by utilizing the structural characteristics of the cardiovascular coronary tree as a complex network. The advantage of utilizing networks naturally arises from the way of thinking behind it, that is focusing on the relations among the entities rather than the entities themselves.

1.1 Innovation

Heart is a complex system consisting of an interconnected network of coronary arteries as the heart's blood supplier. The innovation of the proposed approach is its ability to create a collective view of the heart's coronary circulation system by capturing the structure of coronary tree. Our approach enables analyzing network structure-functions relationships, through which, we can identify hidden patterns in coronary networks. Such patterns relate to the formation or existence of conditions such as stenosis. The following summarizes contributions of this work:

- We propose a new perspective to analyze and understand coronary angiography images based on capturing the network structure of coronary tree.
- We treat the cardiovascular system as a complex system and present a showcase of three network assessments on a healthy and a diseased coronary angiogram.
- We discuss how network science can provide insights from the graph structure of coronary arteries, and ultimately paves the way to fully automate the interpretation of coronary angiography images.

1.2 Delimitations

Due to limitations to access data and patients, this research only included two case-studies. The authors acknowledge that two cases are not sufficient to derive general patterns. The authors note that the purpose of this study is not establishing statistically significant features that can distinguish and compare healthy-case and disease-case angiograms. Instead, our focus is on presenting a new perspective and modeling approach to analyze coronary angiograms. We believe the proposed model has the potential to obtain significant results if given more subjects to study.

This article is organized as follows: Sect. 2 provides a brief overview of employing network science in modeling and understanding biological systems. In Sect. 3, we introduce our modeling approach by conducting three network-based assessments on two CA. Lastly, we discuss our vision in Sect. 4.

2 Complex Networks and Biological Systems

In recent years, there has been growing interest in complex, self-organizing networks often employed to model the dynamics and structure of complex systems [22]. These are dynamical networks of diffusely interconnected components. Their behavior is a manifestation of the behavior of the individual components and a reflection of the structural connections between these components. Examples of complex dynamic graphs abound in nature, from the microscopic cellular level where cells synchronize to perform their functions (heart beating [19] and neural graphs [25]) to large ecological graphs that respond to perturbations through very slow evolutionary behaviors [20].

Lusis and Weiss [18] provided a comprehensive review of the advances achieved by employing network science to investigate the cardiovascular system and diseases from the molecular level (genes and proteins). They showed system-based approaches are likely to play an important role in understanding the higher-order interactions that lead to formation of diseases such as heart failure, atherosclerosis, cardiac hypertrophy, and arrhythmias. Moreover, Dashtbozorg et al. [5] proposed an automated graph-based approach to classify the retinal blood vessels. Their study was able to label retinal blood vessels with up to 89% accuracy. In another study, Estrada et al. [7] proposed a graph-theoretic framework to classify the retinal blood vessels. Their approach obtained an accuracy level up to 93.5%. Furthermore, West et al. [30] introduced a general model of the circulatory systems as space-filling fractal networks. Their model derives the well known biological scaling relationship (i.e., $\textit{metabolic-rate} \propto \textit{body-mass}^{3/4}$) shedding light on the evolution of biological systems. The above studies demonstrate the practicality and advantages of analyzing the graph structure of the circulatory system by modeling the blood vessels as complex networks.

3 Proposed Approach and Case Study

In this section, we propose our model by presenting a case study for both healthy and diseased CAs. The case study concentrates only on the Left Coronary

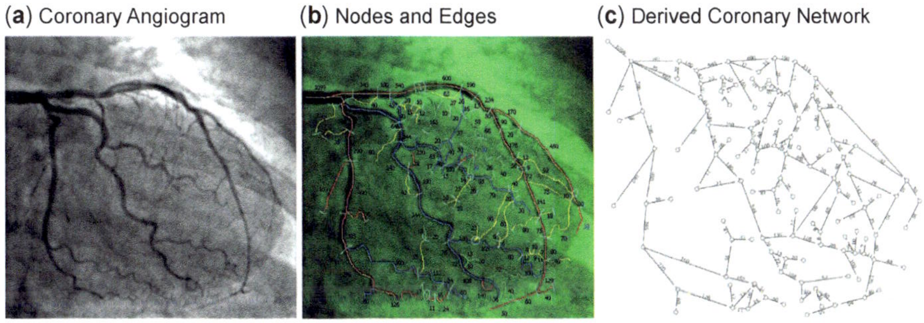

Fig. 1. An example of network creation process

Arteries (LCA). We label an angiogram as diseased if a stenosis exists in the LCA. However, without loss of generality, the proposed model is naturally extendable to integrate all cardiac vessels and provide a complete map of heart coronary arteries. Figure 1 illustrates a CA and the process to derive a network of coronary vessels. A network consists of a set of nodes (representing a system entities) and a set of edges (capturing a relationship between those entities). In the proposed model, a node represents an intersection between vessels, and a weighted edge represents a vessel. The weight of an edge is calculated by multiplying the diameter of a vessel by its length to capture the special characteristics of vessels. Two steps were taken to create the coronary networks in this work: (1) identify the intersections of vessels (i.e., nodes), and (2) measure the length and diameter of each sub-vessel between the identified nodes and calculate the weights of edges. We employed graphical filters to magnify the vessels as presented in Fig. 1(b) and manually conducted these steps. However, without loss of generality, one can fully automate the network creation process by employing the variety of tools developed for performing visual inspections on angiography images [2,3,10,28]. Figure 1(c) presents the created weighted coronary network.

3.1 Healthy and Diseased Coronary Networks

Figure 2 presents angiograms for both a healthy and a diseased heart alongside their corresponding coronary networks. The healthy angiogram is collected from [1] and the source of diseased angiogram is in [14]. In the diseased angiogram, a stenosis is marked by the green arrow. The healthy-case and disease-case angiograms in Fig. 2 are not related to each other. Our goal is to utilize the CAs in Fig. 2 to introduce our modeling approach. The global network characteristics [22] of healthy-case and disease-case networks are summarized in Table 1, in the following five columns: (1) *number of nodes* represents the number of intersections between the vessels, (2) *number of edges* represents the number of vessels, (3) *average degree* presents the average number of connections of the nodes, (4) *average clustering coefficient* captures the degree of connectedness among neighbors of nodes, and (5) *diameter length* of a network presents the length of

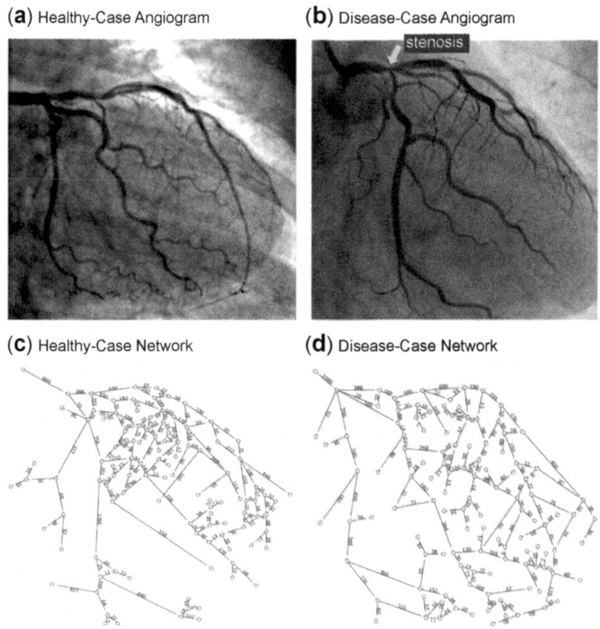

Fig. 2. Healthy and disease-case coronary angiograms and their networks.

Table 1. Coronary network characteristics

Network	Number of nodes	Number of edges	Average degree	Average clustering coefficient	Diameter length
Healthy-case	115	140	2.4348	0.099	23
Disease-case	109	138	2.5321	0.063	24

the longest shortest path between all combinations of nodes. At the first glance, the average clustering coefficient of the disease-case network is relatively smaller than the healthy-case by 36%.

3.2 Network Visualization

Visualizations of networks may provide insights on their structure and patterns of connections. Figure 3 illustrates the derived coronary networks in a circular layout and the thickness of edges represents the edge weights. Also, the green boxes in Fig. 3 mark Λ-branches as illustrated by Fig. 3(c). A Λ-branch consists of a single parent node that only has two children who are not connected to any other nodes (i.e., the coronary tree leaves). Also, the parent node must only have a single additional connection other than its children.

The disease-case network has several more Λ-branches compared to the healthy-case network. This indicates blood is not being properly supplied to

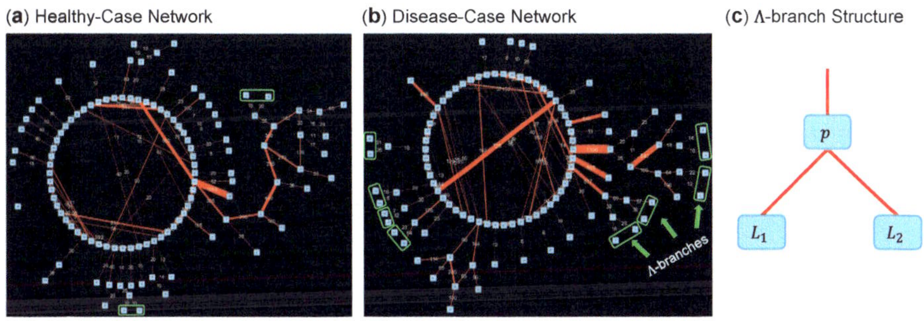

Fig. 3. Coronary networks visualizations. Thickness of edges indicate the vessel's diameter times their length and the green boxes mark Λ-branches.

the diseased heart. The abundance of Λ-branches could reflect the Neovascularization phenomenon [21], which happens when the blood is not being properly supplied and the heart starts creating new vessels. These vessels can be observed in Fig. 2(b) where many small vessels are emerged from the main arteries. In the next section, we show how to systematically capture this behavior by analyzing the degree distribution of coronary networks.

3.3 Assessment of Degree Distribution

The degree distribution of a network represents the distribution of connections among nodes. In the coronary networks, the degree distribution presents the extent, in which, vessels are connected to each other. Blood flows in a fixed direction in human's cardiovascular system. Hence, we employ directed edges to capture the direction of blood flow. Figure 4 illustrates a directed Λ-branch with the degrees of its nodes. For a given node, the *total-degree* indicates its number of connections, the *in-degree* indicates the number of connections to the node, and the *out-degree* indicates the number of connections from the node. In this paper, the weights of edges are not used in the assessment of degree distribution.

Figure 5 presents the in-degree, out-degree, and total-degree distributions of the healthy-case and disease-case networks. At the first glance, there is no significant difference between the degree distributions of the coronary networks. However, a significant difference is observed by comparing the quartile-degree distributions of the healthy-case and disease-case networks, which is presented

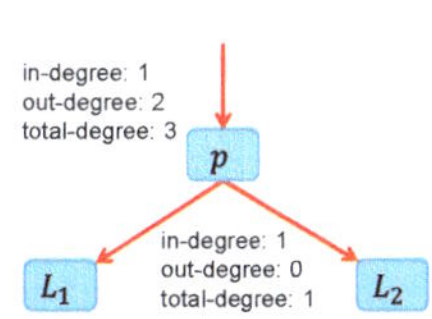

Fig. 4. Directed Λ-branch structure.

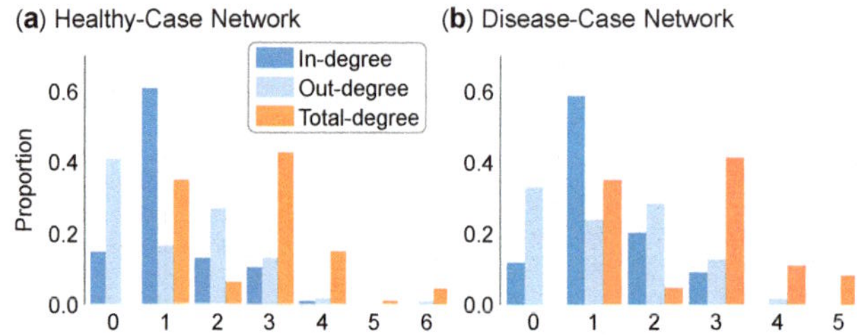

Fig. 5. Degree distributions of coronary networks.

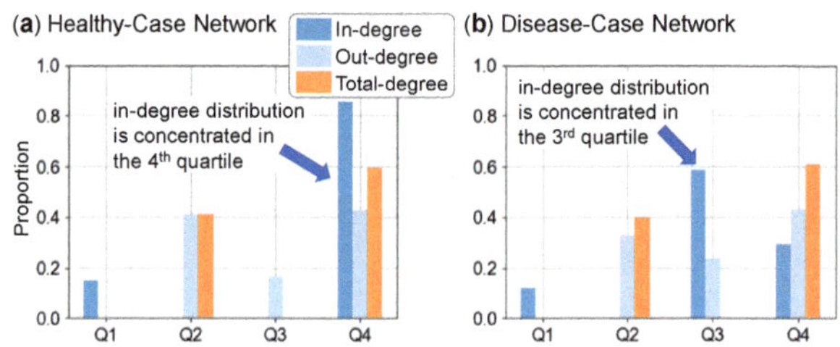

Fig. 6. Quartile degree distributions of coronary networks.

in Fig. 6. In the healthy-case network, most nodes are concentrated in the fourth quartile for all three degree distributions. However, in the disease-case network, the concentration of in-degree distribution is shifted to the third quartile. This shift is due to the abundance of directed Λ-branches in the disease-case network. To conclude, the patterns of connections in coronary networks could provide insights on the condition of the cardio vascular system. For example, the analysis of degree-distribution could be used to determine the extent, in which, a heart is trying to create new vessels to overcome inefficient blood circulation.

3.4 Assessment of Network Integration

The efficiency of a network is the measurement of how efficiently it exchanges information. In transportation networks, this measurement corresponds to the efficiency of patrons commuting in terms of time and distance. We can utilize the patterns of connections in structure of systems to infer their functional efficiency [15]. The assessment of integration in the coronary networks corresponds to quantifying the efficiency of blood circulation in the cardiovascular system.

Figure 7 provides three measures of network integration: *shortest-paths length*, *routing-efficiency*, and *search-information*. The *shortest-paths length* provides the least number of hops (i.e., edges) that needs to be taken to navigate from any source node to any destination node [22]. Figure 7(a) and (d) present the lengths of shortest-paths between all pairs of nodes.

The *routing efficiency*, also known as global efficiency enables to quantify how cost-efficient a particular network is, where the cost depends on the weight

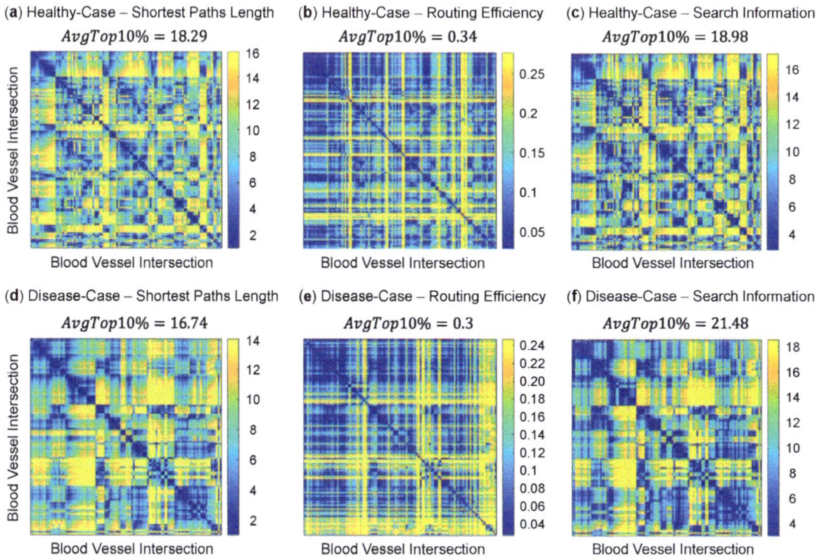

Fig. 7. Analysis of network integration between all pairs of nodes.

of edges (i.e., weighted shortest paths are used) [11]. Hence, this assessment enables to include the vessel's diameter and length (i.e., weight of edges) in quantifying the efficiency of blood circulation. For all pairs of nodes, we present this measurement in Fig. 7(b) and (e).

Lastly, the *search-information* quantifies the amount of information needed for a walker to perform an efficient routing (i.e., quantify accessibility or hiddenness). That is, how much information is needed for a walker to walk on a shortest path when the walker randomly travels between the nodes [24,27]. Figure 7(c) and (f) present this measurement between all pairs of nodes in the healthy-case and disease-case coronary networks.

Through the assessment of network integration, we observe that the healthy-case network requires less information to find efficient routes. In other words, shortest paths are less hidden to the random walker in the healthy-case network compared to the disease-case (i.e., smaller *search-information*). This observation indicates the measurement of *search-information* could be used as a feature to classify healthy and diseased coronary networks.

3.5 Assessment of Controllability

The controllability of complex networks is the study of controlling the state of networks from any initial value to a desired value in finite time via stimulating a set of key nodes called *driver nodes*. Efficient algorithms are developed to identify driver nodes in complex networks [16,23]. Most control scenarios are interested in identifying a minimum number of driver nodes needed to control a

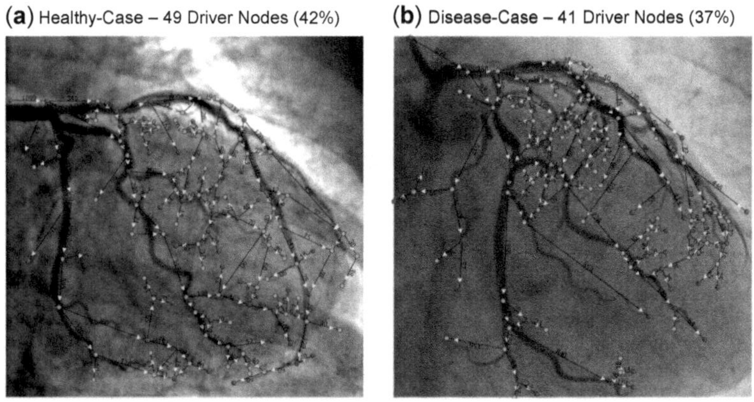

Fig. 8. The controllability assessment (driver nodes are marked with red color).

system. In coronary networks, this is analogous with controlling the flow of blood by modifying the flow that can pass through each node (arteries' intersections). Figure 8 presents the driver nodes (marked red) for both coronary networks. The weights of edges are not utilized in the assessment of controllability.

Intuitively, being easy to control (for cardiovascular systems) might be taken as a sign for healthiness. However, having a small percentage of driver nodes in a coronary network indicates a small number of malfunctions can perturb the whole system. Hence, a healthy network with a high percentage of driver nodes is more resilient to malfunctions. Figure 8 shows the disease-case network has less driver nodes (37%) compared to the healthy-case network (42%).

4 Discussion and Conclusion

The predominant methods to identify cardiovascular conditions primarily focus on analyzing the visual properties of coronary arteries (e.g., the diameter of arteries). For instance, Soroushmehr et al. [28] proposed a QCA approach to assist the diagnosis of CAs. Their approach is primarily based on the visual properties of coronary arteries (e.g., thickness of the arteries) and it can be extended by employing network science. In addition to employing the visual properties of CAs, our proposed approach enables to analyze the dynamics of cardiovascular system. Moreover, Andrikos et al. [3] introduced a novel approach for 3D reconstruction of CAs as illustrated in Fig. 9. Their approach can be naturally utilized to automate the process of network construction from CAs.

The proposed modeling approach provides the basis for development of a new systematic methodology to study the cardiovascular system and automate the diagnosis of coronary network pathology. An advantage of such a methodology is introducing new features based on network measurements such as the routing efficiency and controllability. For instance, these features could be used for the early detection of cardiovascular pathology by training machine learning

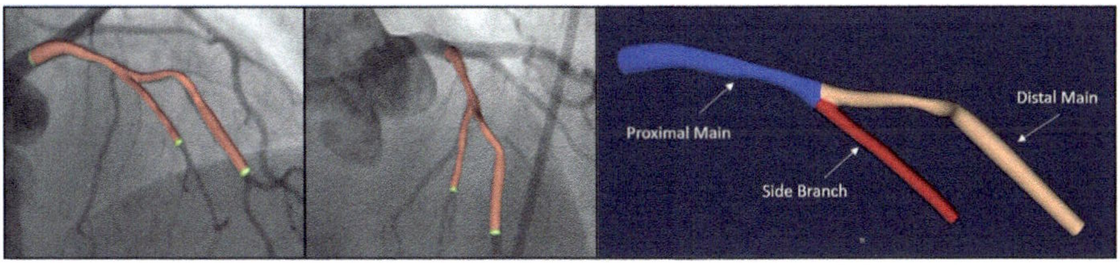

Fig. 9. 3D reconstruction of coronary angiograms (courtesy Andrikos et al. [3]).

classifiers and developing network-based diagnostic methods. Similarly, the proposed approach can improve the accuracy of procedure follow-ups such as the early detection of revascularization after stent implantation. Another important advantage of developing a systematic methodology is minimizing human error that accounts for a significant observer error [6].

Furthermore, non-invasive coronary angiography such as Computed Tomography Angiography (CTA) are already of significant value in the diagnostic procedure of patients. Our modeling approach can enhance the current literature on computer-based approaches for the interpretation of CTA images [17, 32].

The proposed network-based approach paves the way to apply the whole arsenal of network science tools on analyzing and classifying the CAs. However, the authors acknowledge that a rigorous study with more than two CAs should be done to further formalize and validate this approach.

Acknowledgements. The authors acknowledge Professor Joaquín Goñi, School of Industrial Engineering at Purdue University, West Lafayette, USA and Dr. Sophoclis Sophocleous, Pulmonology Resident in Bethanien Hospital, Solingen, Germany for their help and guidance on this paper. We like to thank Mr. Javad Darivandpour, Ph.D. candidate in the Department of Computer Science at Purdue University, West Lafayette, USA for his constructive criticism of the manuscript. We would also like to show our gratitude to Dr. Saied Ravandi for his pearls of wisdom with us during the course of this research.

References

1. What is coronary angiography. http://www.qmedicine.co.in/tophealthtopics/A/Angiography-Coronary.html
2. Andrikos, I.O., Sakellarios, A.I., Siogkas, P.K., Rigas, G., Exarchos, T.P., Athanasiou, L.S., Karanasos, A., Toutouzas, K., Tousoulis, D., Michalis, L.K., Fotiadis, D.I.: A novel hybrid approach for reconstruction of coronary bifurcations using angiography and OCT. In: 2017 39th Annual International Conference of the IEEE Engineering in Medicine and Biology Society (EMBC), pp. 588–591, July 2017
3. Andrikos, I.O., Sakellarios, A.I., Siogkas, P.K., Tsompou, P.I., Kigka, V.I., Michalis, L.K., Fotiadis, D.I.: A novel method for 3D reconstruction of coronary bifurcation using quantitative coronary angiography. In: Lhotska, L., Sukupova, L., Lacković, I., Ibbott, G.S. (eds.) World Congress on Medical Physics and Biomedical Engineering 2018, pp. 191–195. Springer, Singapore (2019)

4. Bourantas, C.V., Kourtis, I.C., Plissiti, M.E., Fotiadis, D.I., Katsouras, C.S., Papafaklis, M.I., Michalis, L.K.: A method for 3D reconstruction of coronary arteries using biplane angiography and intravascular ultrasound images. Comput. Med. Imaging Graph. **29**(8), 597–606 (2005)
5. Dashtbozorg, B., Mendonça, A.M., Campilho, A.: An automatic graph-based approach for artery/vein classification in retinal images. IEEE Trans. Image Process. **23**(3), 1073–1083 (2014)
6. DeRouen, T.A., Murray, J.A., Owen, W.: Variability in the analysis of coronary arteriograms. Circulation **55**(2), 324–328 (1977)
7. Estrada, R., Allingham, M.J., Mettu, P.S., Cousins, S.W., Tomasi, C., Farsiu, S.: Retinal artery-vein classification via topology estimation. IEEE Trans. Med. Imaging **34**(12), 2518–2534 (2015)
8. de Feyter, P.J., Serruys, P.W., Davies, M.J., Richardson, P., Lubsen, J., Oliver, M.F.: Quantitative coronary angiography to measure progression and regression of coronary atherosclerosis, value, limitations, and implications for clinical trials. Circulation **84**(1), 412–423 (1991)
9. Fleming, R.M., Kirkeeide, R.L., Smalling, R.W., Gould, K., Stuart, Y.: Patterns in visual interpretation of coronary arteriograms as detected by quantitative coronary arteriography. J. Am. Coll. Cardiol. **18**(4), 945–951 (1991)
10. Garrone, P., Biondi-Zoccai, G., Salvetti, I., Sina, N., Sheiban, I., Stella, P.R., Agostoni, P.: Quantitative coronary angiography in the current era: principles and applications. J. Intervent. Cardiol. **22**(6), 527–536 (2009)
11. Goñi, J., Avena-Koenigsberger, A., Velez de Mendizabal, N., van den Heuvel, M.P., Betzel, R.F., Sporns, O.: Exploring the morphospace of communication efficiency in complex networks. PLOS ONE **8**(3), 1–10 (2013)
12. Green, N.E., Chen, S.Y.J., Hansgen, A.R., Messenger, J.C., Groves, B.M., Carroll, J.D.: Angiographic views used for percutaneous coronary interventions: a three-dimensional analysis of physician-determined vs. computer-generated views. Cathet. Cardiovasc. Interv. **64**(4), 451–459 (2005)
13. Kern, M.J., Cutlip, D.: Quantitative coronary angiography: clinical applications, November 2017. https://www.uptodate.com/contents/quantitative-coronary-angiography-clinical-applications. Gordon M Saperia (ed.) UpToDate. Assessed Feb 2019
14. Kim, T.J., Kim, J.K., Park, B.M., Song, P.S., Kim, D.K., Kim, K.H., Seol, S.H., Kim, D.I.: Fatal subacute stent thrombosis induced by guidewire fracture with retained filaments in the coronary artery. Korean Circ. J. **43**(11), 761–765 (2013)
15. Latora, V., Marchiori, M.: Efficient behavior of small-world networks. Phys. Rev. Lett. **87**, 198701 (2001)
16. Liu, Y.Y., Slotine, J.J., Barabási, A.L.: Controllability of complex networks. Nature **473**(7346), 167 (2011)
17. Lossau, T., Nickisch, H., Wissel, T., Bippus, R., Schmitt, H., Morlock, M., Grass, M.: Motion artifact recognition and quantification in coronary CT angiography using convolutional neural networks. Med. Image Anal. **52**, 68–79 (2019)
18. Lusis, A.J., Weiss, J.N.: Cardiovascular networks. Circulation **121**(1), 157–170 (2010)
19. Makowiec, D.: The heart pacemaker by cellular automata on complex networks. In: Cellular Automata, pp. 291–298 (2008)
20. Milo, R., Shen-Orr, S., Itzkovitz, S., Kashtan, N., Chklovskii, D., Alon, U.: Network motifs: simple building blocks of complex networks. Science **298**, 824–827 (2002)
21. Moreno, P.R., Purushothaman, K.R., Sirol, M., Levy, A.P., Fuster, V.: Neovascularization in human atherosclerosis. Circulation **113**(18), 2245–2252 (2006)

22. Newman, M.: The structure and function of complex networks. SIAM Rev. **45**(2), 167–256 (2003)
23. Ravandi, B., Mili, F., Springer, J.A.: Identifying and using driver nodes in temporal networks. J. Complex Netw. **7**, 720–748 (2019)
24. Rosvall, M., Grönlund, A., Minnhagen, P., Sneppen, K.: Searchability of networks. Phys. Rev. E **72**, 046117 (2005)
25. Rubinov, M., Sporns, O.: Complex network measures of brain connectivity: uses and interpretations. Neuroimage **52**(3), 1059–1069 (2010)
26. Ryan, T.J.: The coronary angiogram and its seminal contributions to cardiovascular medicine over five decades. Circulation **106**(6), 752–756 (2002)
27. Sneppen, K., Trusina, A., Rosvall, M.: Hide-and-seek on complex networks. Europhys. Lett. (EPL) **69**(5), 853–859 (2005)
28. Soroushmehr, S., Bafna, A., Shashank, S., Brahmajee, N., Ward, K., Najarian, K.: Abstract 15456: computer-assisted diagnosis of coronary angiography. Circulation **132**(suppl_3), A15456–A15456 (2015)
29. Tomasello, S.D., Costanzo, L., Galassi, A.R.: Quantitative coronary angiography in the interventional cardiology. In: Kiraç, S.F. (ed.) Advances in the Diagnosis of Coronary Atherosclerosis, chap. 14. IntechOpen, Rijeka (2011)
30. West, G.B., Brown, J.H., Enquist, B.J.: A general model for the origin of allometric scaling laws in biology. Science **276**(5309), 122–126 (1997)
31. Wilson, P.W., Douglas, P.S.: Epidemiology of coronary heart disease, January 2018. https://www.uptodate.com/contents/epidemiology-of-coronary-heart-disease. Brian C Downey (ed.) UpToDate. Accessed Feb 2019
32. Wolterink, J.M., van Hamersvelt, R.W., Viergever, M.A., Leiner, T., Išgum, I.: Coronary artery centerline extraction in cardiac CT angiography using a cnn-based orientation classifier. Med. Image Anal. **51**, 46–60 (2019)

Connecting Neural Reconstruction Integrity (NRI) to Graph Metrics and Biological Priors

Elizabeth P. Reilly[1(✉)], Erik C. Johnson[1], Marisa J. Hughes[1], Devin Ramsden[2(✉)], Laurent Park[2], Brock Wester[1], and Will Gray-Roncal[1]

[1] Johns Hopkins University Applied Physics Laboratory, Laurel, MD 20723, USA
Elizabeth.Reilly@jhuapl.edu
[2] Johns Hopkins University, Baltimore, MD 21218, USA
dramsde1@jhu.edu

Abstract. We previously introduced the Neural Reconstruction Integrity (NRI) metric as a measure of how well the connectivity of the brain is measured in a neural circuit reconstruction, which can be represented as a graph or network. While powerful, NRI requires ground truth data for evaluation, which is conventionally obtained through time-intensive human annotation. NRI is a proxy for graph-based metrics since it focuses on the pre- and post-synaptic connections (or in and out edges) at a single neuron or vertex rather than overall graph structure in order to satisfy the format of available ground truth and provide rapid assessments. In this paper, we study the relationship between the NRI and graph theoretic metrics in order to understand the relationship of NRI to small world properties, centrality measures, and cost of information flow, as well as minimize our dependence on ground truth. The common errors under evaluation are synapse insertions and deletions and neuron splits and merges. We also elucidate the connection between graph metrics and biological priors for more meaningful interpretation of our results. We identified the most useful local metric to be local clustering coefficient, while the most useful global metrics are characteristic path length, rich-club coefficient, and density due to their strong correlations with NRI and perturbation errors.

Keywords: Neural Reconstruction Integrity · Graph metrics · Connectomics · Evaluation

This material is based upon work supported by the Office of the Director of National Intelligence (ODNI), Intelligence Advanced Research Projects Activity (IARPA), via IARPA Contract No. 2017-17032700004-005 under the MICrONS program. The views and conclusions contained herein are those of the authors and should not be interpreted as necessarily representing the official policies or endorsements, either expressed or implied, of the ODNI, IARPA, or the U.S. Government. The U.S. Government is authorized to reproduce and distribute reprints for Governmental purposes notwithstanding any copyright annotation. We also would like to acknowledge the support of the CIRCUIT 2018 initiative at JHU/APL. Distribution Statement A - Approved for public release; Distribution is unlimited.

H. Barbosa et al. (Eds.): *Complex Networks XI*, SPCOM, pp. 182–193, 2020.
https://doi.org/10.1007/978-3-030-40943-2_16

1 Introduction

Neuroscientists are working towards mapping the brain using imaging techniques such as electron microscopy, magnetic resonance imaging, and light microscopy. New large-scale results are emerging that process these imaging volumes to create a network representation of brain connectivity. High resolution imaging, such as electron microscopy, allows the reconstruction of graphs where each node is an individual neuron and each edge is a synapse, which may allow novel insights into disease and into biological foundations for information processing. These results typically contain many errors, as they are the output of imperfect processing pipelines consisting of computer vision algorithms and human decisions at scale. Additionally, ground truth data is difficult to obtain in large quantities since its creation involves a great deal of human time and effort. Thus, the community is faced with the problem of evaluating large brain graph reconstructions with small ground truth datasets. One aspect of creating high fidelity brain graph reconstructions is to ensure the connectivity of the graph is accurate, or verifying that the underlying structure for information flow is accurately captured. This includes checking if synapses are correctly associated with their pre- and post-synaptic neurons in the data to create information pathways in the reconstruction.

The Neural Reconstruction Integrity (NRI) metric [24], is a connectivity metric that operates on a ground truth brain graph and a reconstructed brain graph where the edges have been spatially aligned. It has many computational benefits and has been shown to be sufficiently sensitive to common reconstruction errors such as synapse deletions and insertions and neuron splits and merges. More specifically, the NRI is an f1 score of how many intracellular paths between two synapses (edges) are preserved during the reconstruction process. In graph theoretic language, we look at all pairs of edges incident on a single vertex. For each pair of edges, the ground truth is used to assess whether the edges are correctly associated on the same vertex in the reconstruction. The NRI can be constructed on a single neuron without requiring extensive graph structure in the ground truth data. This is desirable as ground truth data is expensive to collect, requiring a good deal of time and several human annotators, even when using more scalable approaches such as focused proofreading [22]. Because the NRI is not calculated using between-node information and rather acts as a proxy for more common graph metrics requiring extensive graph structure, it is unknown whether the NRI captures the same structural errors as more widely-used graph metrics, especially at the scale of an entire reconstruction. It is further desirable to analyze a neural circuit reconstruction when extensive ground truth is not available, necessitating the identification of other metrics that are correlated to NRI and common graph errors and that can be calculated without ground truth. In this research, we identify local and global graph metrics that are related to real biological priors. We then compare these graph metrics with NRI scores to understand their correlation on synthetic neural circuit data.

2 Methods

2.1 Background and Related Work

Generating a Brain Graph. In recent years, Electron Microscopy (EM) technology such as serial TEM (Transmission Electron Microscopy) [29] have enabled nanometer resolution imaging of neural tissue. More recent paradigms such as block-face Scanning Electron Microscopy (SEM) [7] and Focused Ion Beam SEM [19] have improved the community's ability to collect carefully aligned volumes of neural tissue. These imaging technologies have allowed the generation of synapse-level graphs, denoted by $G = (V, E)$, where individual neurons are vertices and directed edges are synapses. We use V and E to denote the set of vertices and edges respectively. If there is a synapse from v_i to v_j, then $(v_i, v_j) \in E$. The generation of graphs from EM data remains an intensive process despite recent progress in the field. After data collection, alignment, and artifact removal, terabytes or petabytes of data are stored for analysis.

Assessing Graph Fidelity. There are several existing approaches for evaluating the segmentation of the resulting graph. Some, which originated from the image segmentation community, include the Rand index [23], adjusted Rand index [15], and the Warping index [16]. More recent metrics from the Connectomics community include Variation of Information (VI) [21,22], a line graph-based f1 score [14], the Rand F-score [2], the Tolerant Edit Distance (TED) [12], and the Neural Reconstruction Integrity (NRI) [24]. Each of these is described in-depth in [24]. Expected run length (ERL) has also been recently used as a neural segmentation metric [17], which measures the average length a neuron is traced before an error is made. NRI is the metric we consider in this paper due to its in-depth evaluation in [24] and preliminary results on real data. It is worth noting that the above mentioned approaches require ground truth data for evaluation and thus do not serve as alternatives to the graph metrics presented in the following sections, which can be calculated without ground truth. Also note that NRI can be calculated over the entire network resulting in a single global NRI value, or it can be calculated on individual vertices or neurons in the network resulting in a local NRI value for each vertex.

Graph Theoretic Analysis of Brain Networks. There is also existing work in graph theoretic analysis of brain networks. For a review of several graph theoretic studies of both structural and functional brain graph networks, see [6]. For review of various brain graph representations, constructions, and analyses, see [11]. We consider both local (vertex specific) and global (volume level) graph metrics that capture information about small world properties (characteristic path length, global, and local clustering coefficient), centrality (rich-club coefficient, betweenness centrality, eigenvector centrality, PageRank), and the cost of information flow (edge density). Each of the selected graph metrics considered in this paper was chosen because of previous work indicating the metric is relevant to the study of brain networks and its potential biological interpretation in the brain.

2.2 Data

To analyze the relationship between NRI and graph metrics, we leverage synthetic neural circuit data, produced using NeuGen 2.0 [8]. While the NRI is designed to evaluate a neural circuit reconstruction in the presence of sparse ground truth, graph metrics need extensive connectivity information to truly evaluate the structure of a neural circuit. The majority of existing ground truth volumes do not have dense ground truth information, making them insufficient for this analysis. NeuGen 2.0, in particular, allows us to produce reasonably large synthetic neural circuits via biologically based neuron skeleton generation rules. By having skeletons, we are able to produce the types of errors we often see in real reconstructions. For instance, two neurons are merged probabilistically if their skeletons are within a certain distance of each other. A neuron experiences a split probabilistically when the diameter is below a threshold. Synapses are inserted probabilistically when two neurons are within a certain distance of each other. Finally, synapses are deleted uniformly at random. Imposing errors on the brain graph itself rather than the collection of neural skeletons generated by NeuGen 2.0 does not allow spatial information to inform the error generation. We analyze 4 datasets of similar size, which we label $NG1, NG2, NG3$, and $NG4$. The number of vertices (neurons) is 872 and the number of edges (synapses) is just over 1 million, as we were aiming to generate a large network with sufficient structure while considering computational limitations. Note that we are not considering dynamic graphs that evolve with time as collection of an anatomical connectome destroys the tissue, thus preventing collection for a single organism over time.

We apply simple perturbations described above to each dataset to simulate synapse (edge) insertions or deletions and neuron (vertex) splits and merges. In order to isolate changes in the metrics with respect to specific perturbations, we perturb the graph according to one error at a time. For more detail on the perturbation model, please see [24]. We measure the approximate level of perturbation using percent increase or decrease values of the number of edges (insertions and deletions) and the number of vertices (splits and merges)[1]. We will include these perturbation values as labels in figures as appropriate.

[1] Specifically, the levels of perturbation for edge deletions are a 1, 2, 5, 10, 20, 40, 60, or 80% decrease in edges across the graph. The levels of perturbation for edge insertions are 0.3, 0.7, 2, 4, 8, 15, 23, and 30% increase in edges across the graph. These are relatively small compared to the edge deletions because our perturbation method only allows erroneous edges to be inserted if two neurons are close to each other. The levels of perturbation for vertex splits are quite large, as neuron splits are a common error in neuroimaging segmentation tools [14]. The perturbation levels are 10, 30, 100, 300, 900, and 2800% increase in vertices across the graph. Finally, the levels of perturbation for vertex merges are 1, 3, 12, and 35% decrease in vertices across the graph.

2.3 Definitions and Biological Connections

The terms local and global may be used when referring to a neural volume or when referring to a graph, but the interpretation is similar for both. For a neural volume, unless otherwise indicated, the term *local* is used to refer to single-neuron focused analysis or to a small subset of neurons within a larger spatial volume. The term *global* refers to a network level or full volume analysis. The interpretation is similar in graph analysis. A local metric is calculated for a single vertex, though possibly using information about its neighbors, and a global metric is calculated on the entire graph.

We consider both local and global graph metrics since different neural connectivity properties may exist at each level. These metrics are organized to better understand certain aspects of the biology of the brain. For instance, to study the small world properties of the brain, we investigate characteristic path length [1], global clustering coefficient [27], and local clustering coefficient [28]. We also study hubs in the brain with the rich-club coefficient [26] and the centrality or importance of individual nodes in the network with betweenness centrality [27], eigenvector centrality [10], and PageRank [5]. Finally, we study the cost of information flow by observing graph density [27].

Small World Properties. The concept of a small-world network [28] was developed by noting that the Erdős-Rényi random graph model [9] does not exhibit properties that many real world networks have. In particular, small-world networks have small characteristic path length and large clustering coefficient. It has been shown that brain networks exhibit small world properties [3], which is the primary reason characteristic path length, global clustering coefficient, and local clustering coefficient are of particular interest. In [25], the authors point to many specific instances where small world properties of the brain have been examined under various conditions such as for attention deficit hyperactivity disorder (ADHD) patients or subjects at various stages of neural growth and development.

The characteristic path length is defined as the average distance between all pairs of vertices in a graph. Intuitively, this metric measures how efficiently information flows through a network.

The global clustering coefficient looks at the number of triangle relationships, or clustering, within a graph. This idea harkens from social network analysis where, if two people have a common friend, they are more likely to be friends. Thus, this coefficient provides insight about the extent to which there exist tight-knit communities within the graph. Formally, if a triangle is a set of three vertices, a, b, c with $\{a, b\}, \{b, c\}, \{a, c\} \in E$, and if a connected triplet is a set of three vertices a, b, c with at least two edges among them (say $\{a, b\}, \{b, c\} \in E$), then the global clustering coefficient is

$$\frac{3 * |T|}{|T'|} \tag{1}$$

where $|T|$ is the number of triangles and $|T'|$ is the number of connected triplets. Notice that this is defined in the context of undirected graphs. For directed graphs, we simply ignore the edge directions and use the same definition.

The local clustering coefficient for a vertex i is the proportion of the neighbors of i that are adjacent to each other. Formally, for a directed graph $G = (V, E)$, if N_i is the neighborhood of i (set of vertices that are adjacent to i) and M_i is the set of edges between i's neighbors, or

$$M_i = \{(e_j, e_k) : (e_i, e_j), (e_i, e_k), (e_j, e_k) \in E\} \tag{2}$$

then the local clustering coefficient of i, denoted C_i, is

$$C_i = \frac{|M_i|}{|N_i|(|N_i - 1|)} \tag{3}$$

Central Nodes and Groups of Nodes. The rich-club coefficient, a global metric, measures the density of highly connected nodes. A rich-club effect emerges when high degree nodes are adjacent to other high degree nodes, creating a "rich-club" or hub in the network. For a selected degree k, the induced subgraph on vertices of degree greater than k is calculated. Then, the density is computed. This value is the rich-club coefficient for parameter k.

In [26], van den Heuvel and Sporns use the rich-club coefficient to better understand the structure of highly connected central hubs in the brain. Their research demonstrates that the structural human brain exhibits rich-club properties. Thus, we include the rich-club coefficient in our analysis.

Three local metrics capture local centrality with the network - betweenness centrality, eigenvector centrality, and PageRank. Betweenness centrality captures the importance of a node along pathways within a graph. Examining betweenness centrality of nodes in a graph helps identify hubs and bridges within a network. Accurately capturing this value in a brain network means better understanding the high level structure of the brain and which specific neurons play the most important roles [13].

Betweenness centrality measures the number of shortest paths in a network that pass through that particular vertex. The intuition is that a vertex may not have high degree in a network but nonetheless may be critical because that vertex brings together communities.

Eigenvector centrality and PageRank both provide a ranking of neurons according to importance in the graph. In this paper, this is strictly with respect to structural properties of the graph. Eigenvector centrality has been examined on neural data, though usually on functional data [4,20]. In the following, we are interested in the sensitivity of the resulting rankings as errors are introduced and how those relate to NRI.

Eigenvector centrality is a way of ranking vertices of a graph based on the structural properties of the graph. Specifically, the principal eigenvector of the adjacency matrix is used to assign ranking values to vertices of a graph.

PageRank is the probability that a random walk on the graph will lead to a particular vertex. It is a variation of eigenvector centrality and we thus expect the results to be similar.

Information Flow. Graph density is a simple measure of physical costs of information flow and energy use within a brain network [6]. Brain graph density changes as an individual grows and develops. For these reasons and because of its simplicity, density is examined in this paper.

Graph density is the proportion of potential edges that actually exist in a graph. For instance, on 4 vertices, there are $\binom{4}{2} = 6$ possible edges in the graph. If there are 3 edges in the actual graph, then the density is $3/6 = 1/2$. In general, graph density of an undirected graph can be written as

$$2 \cdot \frac{|E|}{|V|(|V|-1)} \tag{4}$$

where $|E|$ is the number of edges and $|V|$ is the number of vertices. For a directed graph, we simply multiply by $1/2$.

3 Results

We computed each local and global metric on each NeuGen graph ($NG1, NG2$, $NG3$ and $NG4$) and for each level of perturbation within each perturbation type (edge (synapse) insertions and deletions and vertex (neuron) merges and splits). We also calculated both the local and global NRI values for each perturbed graph. The global metrics are visualized for all four datasets and the results for the four datasets are all similar. For local metrics, we demonstrate results on $NG1$. We discovered the following relationships between graph metrics and NRI.

3.1 NRI and Small World Properties

The graph metrics under consideration when studying small world properties are characteristic path length, global clustering coefficient, and local clustering coefficient where the first two are global and the third is a local metric.

The characteristic path length and global clustering coefficient exhibit opposite behaviors as perturbation levels increase. In particular, in the presence of increasing edge insertions or vertex merges, characteristic path length decreases while global clustering coefficient increases. For edge deletions and vertex splits, characteristic path length increases while global clustering coefficient decreases. In all cases, greater perturbation implies decreased NRI. Characteristic path length represents how efficiently information flows through a graph and hence, these results are as expected. Note that the relationship between characteristic path length and 1-Global NRI is a logarithmic relationship for splits (see Fig. 1a) while edge deletions result in a slow exponential increasing relationship. In general, the impact of merges and splits on the network and local metric values

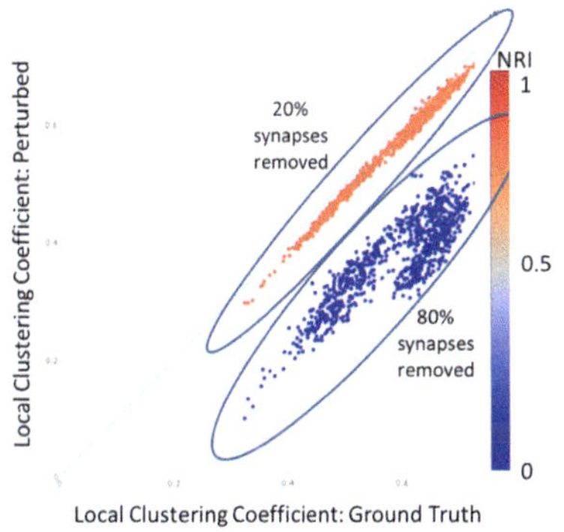

(a) NRI vs Local Clustering Coefficient when synapse deletions are made at various levels.

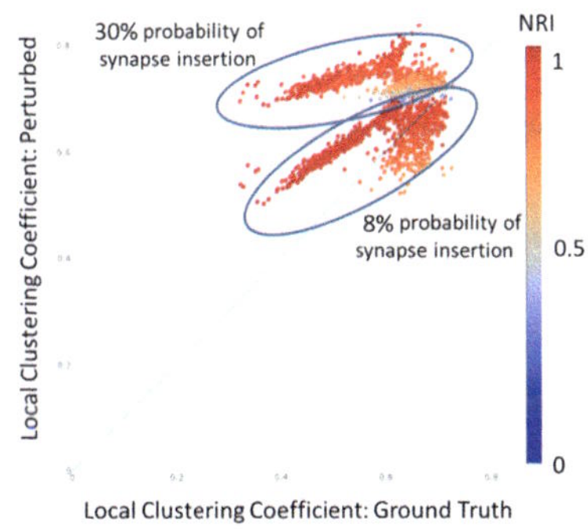

(b) NRI vs Local Clustering Coefficient when synapse insertions are made at various levels.

Fig. 1. Relationship of global NRI and two global metrics, characteristic path length and rich-club coefficient under vertex split and edge insertion perturbations respectively.

can be unpredictable (see results in Sect. 3.2). Because the characteristic path length (and other global metrics) has a nice correlative relationship with NRI and merge and split errors, this could be a powerful network level metric to use in the absence of ground truth.

The local clustering coefficient is a local metric and, when edge deletions are made, it is highly correlated with the perturbation level and NRI. This is particularly true because edge deletions are spread roughly uniformly across the volume. Figure 2a shows that the least perturbed volumes have significantly higher NRI values (red dots) and have small changes in local clustering coefficient (dots remain close to the $x = y$ line in Fig. 2a). Highly perturbed volumes have much lower NRI values (blue dots) and larger changes to local clustering coefficient (dots are further below the $x = y$ line). These relationships imply strong correlation between local clustering coefficient, NRI, and edge deletions. In the presence of edge insertions, we see that the local clustering coefficient tends to increase, though not necessarily directly according to the perturbation/NRI level. For instance, in Fig. 2b we see that the local clustering coefficient values for the highly perturbed network (30% level) are further away from the $x = y$ line. However, the most perturbed nodes have orange color, implying lower NRI, but are closest to the $x = y$ line. Both the edge deletion and insertion perturbations suggest a strong correlation between local clustering coefficient and NRI. Local clustering coefficient could be a useful local graph metric to calculate in the absence of ground truth data.

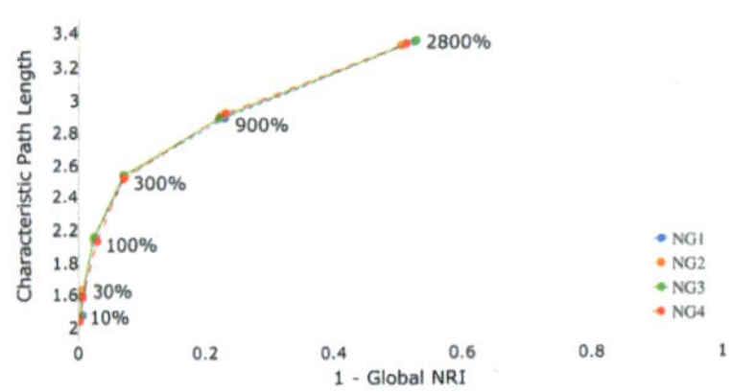

(a) 1-Global NRI vs Characteristic Path Length when vertex splits are made at various levels of perturbation.

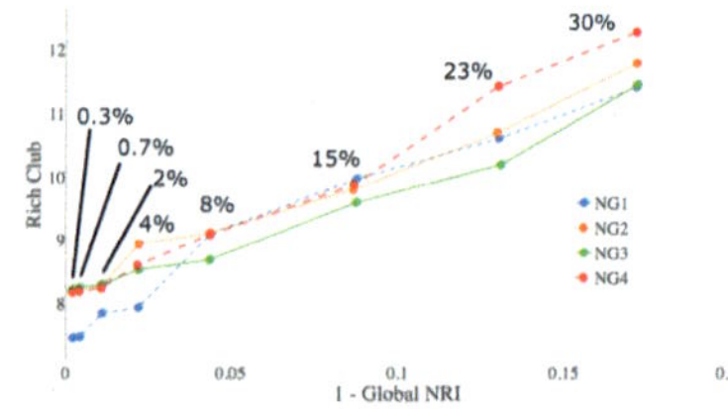

(b) 1-Global NRI vs the Rich-Club Coefficient when edge insertions are made at various levels of perturbation.

Fig. 2. Relationship of local NRI and local clustering coefficient under edge deletion and insertion perturbations.

3.2 NRI and Centrality

For measures of centrality, we computed the rich-club coefficient, which is a global metric, as well as betweenness centrality, eigenvector centrality, and PageRank. The rich-club coefficient behaves as expected. In the case of edge insertions and vertex merges, the NRI decreases as perturbations are applied, while the rich-club coefficient increases. Since edge insertions and vertex merges result in vertices with higher degree, it follows that the so-called rich-club is larger and better connected. When edges are deleted or vertices are split, the NRI still decreases, but the rich-club coefficient decreases since vertices generally become less connected as a result of these perturbations. As mentioned in Sect. 2.2, the perturbation levels for edge insertions are smaller across the board than those for edge deletions. This is reflected in the global NRI values, which, even at the highest perturbation level, is greater than 0.8. We believe this is because of the perturbation method, which requires neurons to be close in physical space in order to insert an erroneous edge. The result is that edges are inserted more frequently for certain neuron pairs. This may also explain why the rich-club coefficient increases so drastically even under small perturbation (See Fig. 1b). The rich-club coefficient exhibits strong correlations with NRI and sensitivity to edge insertion and deletion errors implying this could be a complementary global metric to characteristic path length mentioned above.

Betweenness centrality increases for most vertices in the graph as synapse removals increase. This is especially noticeable when the local NRI degrades greatly. This makes sense as betweenness centrality can represent the ability of a node to control flow of information within a network. If there are fewer pathways overall within a network, then a node that still lives on many shortest paths will increase in importance. The fact that the increase occurs across several nodes may suggest that there is a fair bit of redundancy in the pathways within the network and thinning out the network yields more powerful individual nodes.

Eigenvector centrality creates a ranking of importance among the vertices. As expected, this ranking is impacted more when the perturbation value is higher. This is true even in the case of synapse insertions where the NRI is not heavily degraded by the perturbations. The largest and least predictable changes in the ranking occur when splits and merges occur, resulting in very large local errors and significant fluctuation in eigenvector centrality values. Additionally, the highest impacted vertices don't necessarily end up with the smallest ranking. Because of the unpredictability of the eigenvector centrality for merges and splits, this metric does not promise to provide insight when the NRI is not available for comparison. PageRank has similar behavior to eigenvector centrality.

3.3 NRI and Information Flow

Density is a simple metric that speaks to the cost of information flow within a network. The impact of perturbations on edge density and NRI are as expected. Specifically, as the percentage of synapse removals or neuron splits increases, the density and NRI both decrease in a seemingly linear and exponential relationship respectively. As the percentage of synapse insertions increases, the density increases and NRI decreases. Due to its strong correlation with NRI, plus its simplicity and explainability, density would be a useful metric for analysis in the absence of ground truth.

4 Discussion

In our investigation of the relationship between NRI and local and global graph metrics, we found that, on synthetic NeuGen data, we consistently identify correlations that match our intuition given the metric formulations, which gives important insight as to how graph metrics used on brain networks relate to this key reconstruction quality metric and common errors. In particular, the local clustering coefficient, characteristic path length, rich-club coefficient, and density are metrics that are correlated with NRI. There are at least two benefits to these results. First, they indicate that, while NRI is a connectivity metric that acts as a proxy for graph metrics, it still appears to capture some of the same key structural properties. Second, from previous work, we know that the NRI is sensitive to common errors seen in connectome reconstructions. It follows that correlated graph metrics can be used in unsupervised evaluation where ground truth is not available and NRI cannot be calculated. One next step might be to use real image data to reproduce a brain graph at various parameter settings, using a tool such as those described in [14,18], to see if these graph metrics can inform the reconstruction process to determine optimal algorithm parameters and ensure reconstruction quality. It would also be beneficial to obtain a better understanding of what "normal" values of these metrics look like for real brain data, allowing anomaly detection in the presence of errors and lack of ground truth data. This would improve analysis in future studies of brain graph properties built on imperfect reconstructions.

To further expand upon the work in this paper, analysis of the computational feasibility based on network size would ensure scalability of metric calculations. Future work will also explore the analysis of neuron and synapse (or vertex and edge) attributes (e.g., functional signal, cell types, synapse strength) and other graph properties applicable for a particular research question (e.g., modularity, correlation).

Finally, note that any given perturbation model could result in different emergent graph structure and properties, such as small worldness. As our perturbation model is performed on neural skeletons in order to simulate real world reconstruction errors, an analysis of how these perturbations could influence the analyzed graph metrics in an unintended manner has not been performed.

References

1. Albert, R., Barabási, A.L.: Statistical mechanics of complex networks. Rev. Mod. Phys. **74**(1), 47 (2002)
2. Arganda-Carreras, I., Turaga, S.C., Berger, D.R., Cireşan, D., Giusti, A., Gambardella, L.M., Schmidhuber, J., Laptev, D., Dwivedi, S., Buhmann, J.M., et al.: Crowdsourcing the creation of image segmentation algorithms for connectomics. Front. Neuroanat. **9**, 142 (2015)
3. Bassett, D.S., Bullmore, E.: Small-world brain networks. Neuroscientist **12**(6), 512–523 (2006)
4. Binnewijzend, M.A., Adriaanse, S.M., Van der Flier, W.M., Teunissen, C.E., de Munck, J.C., Stam, C.J., Scheltens, P., van Berckel, B.N., Barkhof, F., Wink, A.M.: Brain network alterations in Alzheimer's disease measured by eigenvector centrality in fMRI are related to cognition and CSF biomarkers. Hum. Brain Mapp. **35**(5), 2383–2393 (2014)
5. Brin, S., Page, L.: The anatomy of a large-scale hypertextual web search engine. Comput. Netw. ISDN Syst. **30**(1–7), 107–117 (1998)
6. Bullmore, E., Sporns, O.: Complex brain networks: graph theoretical analysis of structural and functional systems. Nat. Rev. Neurosci. **10**(3), 186 (2009)
7. Denk, W., Horstmann, H.: Serial block-face scanning electron microscopy to reconstruct three-dimensional tissue nanostructure. PLoS Biol. **2**(11), e329 (2004)
8. Eberhard, J.P., Wanner, A., Wittum, G.: Neugen: a tool for the generation of realistic morphology of cortical neurons and neural networks in 3D. Neurocomputing **70**(1–3), 327–342 (2006)
9. Erdos, P., Rényi, A.: On the evolution of random graphs. Publ. Math. Inst. Hung. Acad. Sci **5**(1), 17–60 (1960)
10. Fletcher, J.M., Wennekers, T.: From structure to activity: using centrality measures to predict neuronal activity. Int. J. Neural Syst. **28**(02), 1750013 (2018)
11. Fornito, A., Zalesky, A., Breakspear, M.: Graph analysis of the human connectome: promise, progress, and pitfalls. Neuroimage **80**, 426–444 (2013)
12. Funke, J., Klein, J., Moreno-Noguer, F., Cardona, A., Cook, M.: TED: a tolerant edit distance for segmentation evaluation. Methods **115**, 119–127 (2017)
13. Gong, G., He, Y., Concha, L., Lebel, C., Gross, D.W., Evans, A.C., Beaulieu, C.: Mapping anatomical connectivity patterns of human cerebral cortex using in vivo diffusion tensor imaging tractography. Cereb. Cortex **19**(3), 524–536 (2008)
14. Gray Roncal, W.R., Kleissas, D.M., Vogelstein, J.T., Manavalan, P., Lillaney, K., Pekala, M., Burns, R., Vogelstein, R.J., Priebe, C.E., Chevillet, M.A., et al.: An automated images-to-graphs framework for high resolution connectomics. Fronti. Neuroinform. **9**, 20 (2015)

15. Hubert, L., Arabie, P.: Comparing partitions. J. Classif. **2**(1), 193–218 (1985)
16. Jain, V., Bollmann, B., Richardson, M., Berger, D.R., Helmstaedter, M.N., Briggman, K.L., Denk, W., Bowden, J.B., Mendenhall, J.M., Abraham, W.C., et al.: Boundary learning by optimization with topological constraints. In: 2010 IEEE Computer Society Conference on Computer Vision and Pattern Recognition, pp. 2488–2495. IEEE (2010)
17. Januszewski, M., Kornfeld, J., Li, P.H., Pope, A., Blakely, T., Lindsey, L., Maitin-Shepard, J., Tyka, M., Denk, W., Jain, V.: High-precision automated reconstruction of neurons with flood-filling networks. Nat. Methods **15**(8), 605 (2018)
18. Johnson, E.C., Wilt, M., Rodriguez, L.M., Norman-Tenazas, R., Rivera, C., Drenkow, N., Kleissas, D., LaGrow, T.J., Cowley, H., Downs, J., et al.: Toward a reproducible, scalable framework for processing large neuroimaging datasets. BioRxiv, p. 615161 (2019)
19. Knott, G., Marchman, H., Wall, D., Lich, B.: Serial section scanning electron microscopy of adult brain tissue using focused ion beam milling. J. Neurosci. **28**(12), 2959–2964 (2008)
20. Lohmann, G., Margulies, D.S., Horstmann, A., Pleger, B., Lepsien, J., Goldhahn, D., Schloegl, H., Stumvoll, M., Villringer, A., Turner, R.: Eigenvector centrality mapping for analyzing connectivity patterns in fMRI data of the human brain. PLoS ONE **5**(4), e10232 (2010)
21. Nunez-Iglesias, J., Kennedy, R., Parag, T., Shi, J., Chklovskii, D.B.: Machine learning of hierarchical clustering to segment 2D and 3D images. PLoS ONE **8**(8), e71715 (2013)
22. Plaza, S.M.: Focused proofreading to reconstruct neural connectomes from EM images at scale. In: Deep Learning and Data Labeling for Medical Applications, pp. 249–258. Springer (2016)
23. Rand, W.M.: Objective criteria for the evaluation of clustering methods. J. Am. Stat. Assoc. **66**(336), 846–850 (1971)
24. Reilly, E.P., Garretson, J.S., Gray Roncal, W.R., Kleissas, D.M., Wester, B.A., Chevillet, M.A., Roos, M.J.: Neural reconstruction integrity: a metric for assessing the connectivity accuracy of reconstructed neural networks. Front. Neuroinformatics **12**, 74 (2018)
25. Toga, A.W., Clark, K.A., Thompson, P.M., Shattuck, D.W., Van Horn, J.D.: Mapping the human connectome. Neurosurgery **71**(1), 1–5 (2012)
26. Van Den Heuvel, M.P., Sporns, O.: Rich-club organization of the human connectome. J. Neurosci. **31**(44), 15775–15786 (2011)
27. Wasserman, S., Faust, K.: Social Network Analysis: Methods and Applications, vol. 8. Cambridge University Press, Cambridge (1994)
28. Watts, D.J., Strogatz, S.H.: Collective dynamics of 'small-world' networks. Nature **393**(6684), 440 (1998)
29. Zheng, Z., Lauritzen, J.S., Perlman, E., Robinson, C.G., Nichols, M., Milkie, D., Torrens, O., Price, J., Fisher, C.B., Sharifi, N., et al.: A complete electron microscopy volume of the brain of adult drosophila melanogaster. Cell **174**(3), 730–743 (2018)

Social Media Analysis

Twitter Watch: Leveraging Social Media to Monitor and Predict Collective-Efficacy of Neighborhoods

Moniba Keymanesh[1(✉)], Saket Gurukar[1], Bethany Boettner[1], Christopher Browning[1], Catherine Calder[2], and Srinivasan Parthasarathy[1]

[1] The Ohio State University, Columbus, OH 43210, USA
{keymanesh.1,gurukar.1,boettner.6,browning.90,parthasarathy.2}@osu.edu
[2] University of Texas at Austin, Austin, TX 78712, USA
calder@austin.utexas.edu

Abstract. The occurrence of criminal violence is uneven across urban communities. Sociologists often associate this variation, with the concept of collective efficacy. The collective efficacy of a neighborhood is defined as social cohesion among neighbors combined with their willingness to intervene on behalf of the common good. Sociologists measure collective efficacy by conducting survey studies designed to measure individuals' perception of their community. In this work, we employ the curated data from a survey study (ground truth) and examine the effectiveness of substituting costly survey questionnaires with proxies derived from social media. We enrich a corpus of tweets mentioning a local venue with several linguistic and topological features. We then propose a pairwise learning to rank model with the goal of identifying a ranking of neighborhoods that is similar to the ranking obtained from the ground truth collective efficacy values. In our experiments, we find that our generated ranking of neighborhoods achieves 0.77 Kendall tau-x ranking agreement with the ground truth. Overall, our results are up to 37% better than the baselines.

Keywords: Collective efficacy · Social network analysis · Learning to rank

1 Introduction

Understanding occurrence of crime and disorder in cities is important for public health, policy, and governance. However, occurrence of criminal violence is uneven across the neighborhoods [1,2]. Sociologists such as Morenoff [3], Sampson [4,5], Browning [6], and Kronhauser et al. [7] associate the spatial variation of disorder to the organizational characteristics of the neighborhood. An important measure of such disorder is collective efficacy [4]. *Collective efficacy* is defined as

Electronic supplementary material The online version of this chapter (https://doi.org/10.1007/978-3-030-40943-2_17) contains supplementary material, which is available to authorized users.

H. Barbosa et al. (Eds.): *Complex Networks XI*, SPCOM, pp. 197–211, 2020.
https://doi.org/10.1007/978-3-030-40943-2_17

"social cohesion and trust among neighbors combined with the joint willingness to intervene on behalf of the common good" [4]. The computation of neighborhoods'[1] collective efficacy traditionally requires conducting expensive surveys; usually requiring funding on the order of hundreds of thousands of dollars [8]. Moreover, changes to collective efficacy over time [9], due to policy shifts (e.g. through neighborhood gentrification efforts) require additional surveys, further exacerbating this cost.

Sociologists and government agencies typically use collective efficacy to "order" neighborhoods with respect to neighborhood safety perception and social cohesion among residents [4]. Essentially one may model this as a ranking problem. Concretely the key question we seek to answer in this paper is: "Given the social media data about neighborhoods, can we rank them such that the ranked list is close to the ranked list of neighborhoods ordered by collective efficacy – thereby saving on the cost of expensive surveys?". Our approach, a first of its kind study at a city-scale, seeks to characterize neighborhood collective efficacy by levering spatially conditioned linguistic features extracted from social media. These features are related to the type of urban activity, language use, visible signs of crime and anti-social behavior reported on such media, familiarity of residents with one another, and public mood of the neighborhood. We lever additional sociological, and spatial features and develop a simple pairwise learning to rank model. We empirically show the effectiveness of our model on a real world city-scale dataset, with ground truth values of collective efficacy computed from a traditional survey-based study (details in Sect. 3). Additionally, we conduct a comprehensive analysis of the predictive power of specific features in the ranking task to better understand the relative importance of individual features. In terms of broader impacts, such ideas can be used as a cost-effective early warning mechanism to monitor the transformations of the neighborhoods and prioritize the resources.

2 Background and Related Work

The behavior of users on Twitter has been used in the past to assist prediction of criminal violence [10–15]. The link between social unrest and the user's online activity on Twitter has been studied by [16]. Moreover, Twitter has been employed to study the online behavior of gang members [17] and to measure the population at risk, considering violent crime [18]. Several studies have used Twitter to study the trust relations [19] among online users. Researchers have also leveraged Twitter data for studying social disorganization by evaluating entropy of individuals' opinion about soccer teams [20]. Although the concepts of trust, crime, and social disorganization are related to collective efficacy, to the best of our knowledge estimating individuals' perception of their neighborhood using social media data has not been addressed till now.

[1] We define neighborhoods as the census blocks groups due to the availability of geographic detail combined with decennial census data for each unit. The terms "neighborhood" and "block group" are used interchangeably in the paper.

3 Data Collection

AHDC Study: The Adolescent Health and Development in Context (AHDC) study is a longitudinal data collection effort in a representative and diverse urban setting. The study area is a contiguous space in Columbus, Ohio. In the first wave of the study, 1403 residents were asked a series of questions about their neighborhood and routine activity locations. Questions specifically focused on informal social control items measuring the participant's perception of the social climate in the area at and around each location and in the neighborhood. Participants reported agreement with the following questions: 1- whether people on the streets can be trusted? 2- whether people are watching what is happening on the street?, and 3- whether people would come to the defense of others being threatened? Responses ranged from 1 ("strongly disagree") to 5 ("strongly agree"). This step resulted in roughly 9000 location reports (4031 unique locations) nested within 567 block groups. In order to achieve the collective efficacy value of each neighborhood, we aggregated individual responses to the three social control items at the report level. Then, we aggregated report-level results for each block group. Let $y_{ijk} \in \{1, .., 5\}$ denote response to the k-th item on the j-th report for a spatial location in neighborhood i. The ground truth collective efficacy of neighborhood n_i is indicated with $C(n_i)$ and is computed using the following equation:

$$C(n_i) = \frac{\sum_{j=1}^{N_i} \sum_{k=1}^{M_j} y_{ijk}}{M_j \times N_i}$$

In the equation above, N_i indicates the number of reports for neighborhood n_i. $M_j \in \{1, .., 3\}$ indicates number of items responded in report j. We normalize the scores in range of 0 to 1 and use it as the ground truth for our study. This methodology is aligned with the traditional measurement approach employed to compute collective efficacy at neighborhood level [21–23]. Note that while we adopt a similar ground truth model [23], we lever survey reports from individuals who *both* reside within *and frequently visit* a particular neighborhood. Concomitantly, the social media postings included in our study includes postings from both individuals that reside within and frequently visit a particular neighborhood. In order to increase the reliability of the aggregation we only include the

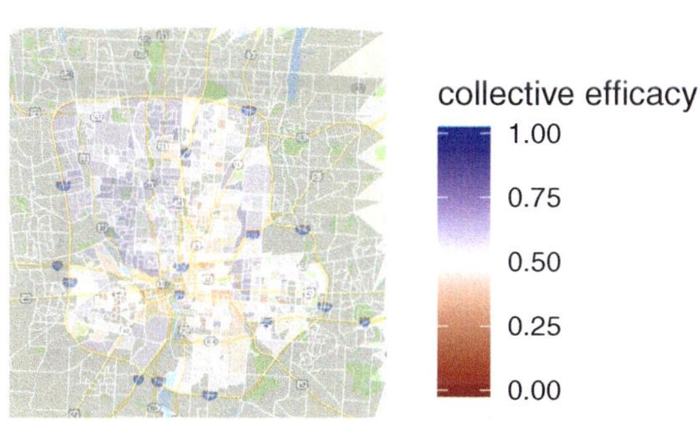

Fig. 1. Collective efficacy map of Columbus, Ohio

neighborhoods having at least 5 reports. Figure 1 shows a collective efficacy map of Columbus.

Twitter Data: To capture the informal language of local citizenry focused on local venues and localities, more than 50 million publicly available tweets were collected from the accounts of 54k Twitter users who identified their location as Columbus. These users were identified through Snowball sampling [24]. We excluded the tweets that did not contain a mention of locations within our study area. For doing so, we used a publicly available location name extractor LNEx [25]. LNEx extracts location entries from tweets, handles abbreviations, tackles appellation formation and metonomy pose disambiguation problems given a gazetteer and the region information. Open Street Map gazetteer was used and region was set to Columbus. In some cases locations extracted by LNEx are ambiguous. In our study, we exclude tweets containing ambiguous location entities. This pruning step resulted in 4846 unique locations that were spotted in 545k tweets and were mapped to 424 neighborhoods. For more details on our data collection and pruning steps see Appendix A.1 of the supplementary material.[2]

4 Methodology

As mentioned earlier, the goal of our study is to rank the neighborhoods based on features extracted from tweets such that the ranked list is similar to the list of neighborhoods ordered by collective efficacy. Hence, we formulated our problem as pairwise learning to rank task [26].

4.1 Definitions and Problem Formulation

First, we define the terms related to ordinal ranking. *Tied objects* refer to the set of two or more objects that are interchangeable in ranking with respect to the quality under consideration [27]. In our study, the neighborhoods with significantly small difference in their ground truth value of collective efficacy are considered *tied*. Ties are defined based on a threshold on the difference of collective efficacy values. Note that in this case, ties are intransitive by definition. Meaning that a tie relationship between neighborhoods n_a and n_b; and n_b and n_c does not imply that neighborhoods n_a and n_c are tied. This constraint will be reflected in the way we define the ranking matrix and will be discussed in Sect. 4.5.

Next, we formalize our ranking task and our proposed approach. Our data consists of $\{t_1, t_2, ..., t_c\}$ where t_i is the set of tweets associated with neighborhood n_i. We denote the collective efficacy of neighborhood n_i with $C(n_i)$. The goal of our framework is to automatically generate a permutation of neighborhoods $(f(n_1)f(n_2).....f(n_m))$ where f is the ordering function that maps each neighborhood to its position such that the mapped position of neighborhood is

[2] Supplementary material is provided in https://github.com/senjed/TwitterWatch.

close to its true position based on collective efficacy values. Formally the ranking task is defined as $\{f(n_i) < f(n_j) \mid \forall n_i, n_j \text{ if } C(n_i) \leq C(n_j)\}$. In order to generate a ranking of the neighborhoods, we first predict the local rank of all pairs of neighborhoods n_i and n_j. In this scenario, there can be three cases for any pair; n_i comes before n_j, n_i comes after n_j, or n_i and n_j are interchangeable in the ranking. Thus, we formulate our local ranking task as a 3-class classification task. We then use the local rankings to generate the global ranking. Details of this process are discussed in Sect. 4.4. Next, we explain the features used in this study to characterize the neighborhoods.

4.2 Features

To solve the pairwise ranking task, we represent a pair of neighborhoods with the tweets associated with them. We compute two types of features: (I) features that are computed for each neighborhood and (II) features that are computed for a pair of neighborhoods. To generate the feature vector of a pair of neighborhoods, we first concatenate feature vector of each of the neighborhoods and then add the pairwise features to the feature vector. These features are explained in detail in the following subsections.

TF-IDF of Crime Related Words: "Broken Windows" [28] is a well-known theory in criminology. The basic formulation of this theory is that visible signs of crime creates an urban environment that encourages further crime and disorder [29,30]. Under the broken windows theory, a disordered environment, with signs of broken windows, graffiti, prostitutes, and excessive litter sends the signal that the area is not monitored and that criminal behavior has little risk of detection. Such a signal can potentially draw offenders from outside of the neighborhood. On the basis of this theory, we used a lexicon of crime[3] as a proxy for visible signs of crime and disorder in neighborhoods. This lexicon contains words that people often use while talking about crime and disorder. TF-IDF captures the importance of a term in a document. With this in mind, we employed TF-IDF to capture the content surrounding the location entity in a tweet. For more details of preprocessing see the Appendix A.2 of the supplementary material.

Distribution of Spatio-Temporal Urban Activities Using Topic Modeling: Casual, superficial interaction and the resulting public familiarity engender place-based trust among residents and ultimately the expectation of response to deviant behaviour [31]. Identifying the activities that individuals conduct in a city is a non-trivial step to understanding the ecological dynamics of a neighborhood such as the potential for street activity and public contact. Following the same methodology as in [32] we applied Latent Dirichlet Allocation (LDA) [33] to tweets associated with a given neighborhood to identify the main activity types at each area. The number of topics in a set of tweets is an important prior parameter in LDA model. We used Rate of Perplexity Change (RPC) [34] to

[3] https://github.com/sefabey/fear_of_crime_paper.

evaluate the topic model and determine the optimal number of topics. We varied the number of topics from 10 to 150 and observed that RPC is maximized at 70 topics. Thus, we trained the LDA model with 70 topics on a subset of 5M tweets collected from user profiles. For more details see Appendix A.3 of the supplementary material.

Document Embeddings: In order to represent the semantics of tweets associated to a neighborhood with a fixed-length feature vector, we used Doc2vec [35], an unsupervised framework that learns continuous distributed vector representations for pieces of texts. Details on training the doc2vec model can be found in Appendix A.4.

Sentiment Distribution: Sentiment analysis, has been used by researchers for quantifying public moods in the context of unstructured short messages in online social networks [36]. We also characterize the neighborhoods in our study using the mood of the tweets mentioning a venue located inside the neighborhood. As reported in [37] the existing methods for sentiment analysis vary widely regarding their agreement; meaning that depending on the choice of sentiment analysis tool, same content could be interpreted very differently. Thus, we use a combination of several methods to make our framework more robust to the limitations of each method. We used five of the best methods for sentiment analysis [37] including Vader [38], Umigon [39], SentiStrength [40], Opinion Lexicon [41], and Sentiment140 [42]. We applied the methods on each tweet and normalized the values. Next, we categorized the observed sentiment values in 4 bins and reported the distribution of tweets sentiment for each neighborhood. For more details see Appendix A.5 of the supplementary material.

Spatial Distance: In order to represent the spatial relationship of the neighborhoods, we computed the geodesic distance between the center points of each pair of neighborhoods. We then normalize the distance values using min-max normalization.

Common Users: Frequent interaction and the resulting public familiarity engender place-based trust among residents [31]. For a pair of neighborhoods we assume that the greater the number of users that tweeted about both neighborhoods, the higher is the level of the public familiarity of the residents and the more similar are the neighborhoods in terms of collective efficacy. Thus, for each pair we computed the number of users that tweeted about both of the neighborhoods. Then we divided this value by the total number of users that tweeted about at least one of the neighborhoods in the pair.

4.3 Model

In this section we discuss our ranking task and model architecture. We use a pairwise approach to generate a ranking of neighborhoods with respect to their collective efficacy. The goal here is to use the extracted features for generating a permutation which is close to the ranking of neighborhoods if sorted by collective efficacy values. In the pairwise approach, the ranking task is transformed into a pairwise classification problem. In our case, given representations of a pair of neighborhoods $<n_a, n_b>$ the goals is to predict if n_a should be ranked higher than n_b or n_a should come later in the ranking. In the first case, a value $+1$ is the label to be predicted and in the latter case the value -1 is assigned as the true label. We consider a label value of 0 for a pair of tied neighborhoods since we do not want to move one of them higher or lower in the list with respect to the other one. We then use this local ordering to generate a global ordering of the neighborhoods. We employed different classifiers for the local ordering task including a neural ranker which is a feed-forward neural network. Extensive experiments were conducted to evaluate the effect of the model architecture as well as the predictive power of the features. Our experimental setup and the results are provided in Sects. 5 and 6.

4.4 Ordering

We train the local ranker model for each pair of the neighborhoods $<n_i, n_j>$ and their corresponding local rank label $r_{ij} \in \{-1, +1, 0\}$. For each pair, we also include another training instance $<n_j, n_i>$ as the input and $-r_{ij}$ as the ground truth value. Given the set of tweets associated with neighborhoods in our study, we rank the neighborhoods as follows: for every pair $<n_i, n_j>$ we first extract features from tweets of neighborhood n_i and neighborhood n_j then we compute the pairwise features including the spatial distance, and normalized common users count for each pair. We concatenate all the features for every pair mentioned in Sect. 4.2. The model then predicts the local ranking for each pair of neighborhoods using the feature representation of each pair. Let $R(n_i, n_j)$ be the predicted local rank value of neighborhoods n_i and n_j. In order to get the global rank of the neighborhoods, we compute the final score $S(n_i)$ for all neighborhoods by computing:

$$S(n_i) = \sum_{n_i \neq n_j \in N} R(n_i, n_j)$$

Then we rank the neighborhoods in decreasing order of these scores. The lower the score the lower the degree of collective efficacy of the neighborhood. Similar ranking setup has been used in [43–45] for substitution ranking.

4.5 Evaluation

We evaluated the accuracy of our model by measuring the agreement between the generated ranking and the ground truth ranking. As mentioned in Sect. 4.1, the ranked list of neighborhoods has non-transitive ties. The quality of predicted ranking in this setting can be computed using τ_x rank correlation coefficient [46].

τ_b is another metric that is used for measuring ranking consensus, however [46] uncovers fundamental issues with the usage of τ_b metric in the presence of ties.

The τ_x Rank Correlation Coefficient. Let A be a ranking of n objects. Then [46] defines a *weak ordering* A of n objects using the $n \times n$ score matrix. Element a_{ij} of this matrix is defined as follows:

$$a_{ij} = \begin{cases} 1 & \text{if object } i \text{ is ranked ahead of or tied with object } j \\ -1 & \text{if object } i \text{ is ranked behind object } j \\ 0 & \text{if } i = j \end{cases}$$

The τ_x rank correlation coefficient between two weak orderings A and B is computed by the dot product of their score matrices.

$$\tau_x(A, B) = \frac{\sum_{i=1}^{n} \sum_{j=1}^{n} a_{ij} b_{ij}}{n(n-1)}$$

We further evaluate our proposed framework by computing the τ_x ranking correlation between the generated ranking and the ground truth ranking. It is important to note that the cumulated gain-based metrics [47] such as *Discounted Cumulated Gain (DCG)* and the normalized version of it (NDCG) widely used in information retrieval literature for examining the retrieval results are not appropriate to evaluate our framework. The main reason being these measures penalize the ranking mistakes more on the higher end of the ranking while devaluing late retrieved items. However, such an objective does not work for our context - mistakes in ranking on the higher end should be penalized the same as the mistakes in the middle or end of the list. Thus, employing a measure of ranking agreement is a more appropriate way to evaluate our model.

5 Experiments

In this section, we empirically evaluate our hypothesis that "one can leverage social media data to quantify collective efficacy of neighborhoods".

5.1 Dataset and Empirical Setup

We lever the survey data of the AHDC study and tweets collected from residents' Twitter accounts (for details see Sect. 3). We sorted the neighborhoods based on the number of tweets collected for each of them and used the top 40% of neighborhoods in this list in our experiments in Sects. 6.1 and 6.2. This list contains 157 neighborhoods that were mentioned in 3,047 tweets on average. Our results considering all the neighborhoods as well as experiments on effect of data availability is shared in Sect. 6.3. The information on the count of block groups in each set as well as statistics on number of collected tweets is reported in Table 1. For more information on the distribution of collective efficacy values

Table 1. Tweet count statistics at each set of top k% neighborhoods; sorted based on their tweet count. The maximum number of tweets in each set of top k% neighborhoods is 98,951.

% of top tweeted neighborhoods	Neighborhood count	Min.	Median	Mean	Standard deviation of collective efficacy
20	78	490	1394.5	5903.5	0.1826
40	157	110	473	3047.7	0.1821
60	235	39	197	2058.8	0.1962
80	314	14	110	1546.9	0.2038
100	393	1	63	1237.2	0.2063

in each group see the Appendix A.6 of the supplementary material. We learn our proposed learning to rank model based on 90/10 train/test split of collected tweets. The train/test split also maintains the temporal order where train split is treated as current tweets while test split is treated as future tweets.

5.2 Baselines for the Ranking Task

Following are the baselines for the ranking task:

- **Venue count**: Neighborhoods were sorted by the number of venues located in them that were mentioned in the tweets.
- **Population**: Neighborhoods were sorted by total population of them. The values are extracted from the 2013 report of the United States census bureau[4].
- **Tweet count**: We sort the neighborhoods by number of tweets that mentioned a venue located in them.
- **User count**: We sort the neighborhoods by number of users that tweeted about a venue located in them.
- **Random**: We generate 100 random permutations of the neighborhoods and report the average τ_x.

5.3 The Classifier for the Local Ranking Task

Since we rely on pairwise learning to rank, we experiment with below classifiers for our local ranking task. The parameters are tuned using grid search with cross-validation parameter set to 5 and scoring function set to ‘f1’.

- **Logistic Regression (LR)**: The estimator penalty is set to ‘L1’ and the inverse of regularization strength is set to 0.1.
- **Support Vector Machine (SVM)**: The kernel is set to ‘rbf’, the penalty parameter C is set to 1, and the gamma kernel coefficient for rbf is set to 0.1.
- **Random Forest (RF)**: The number of estimators is set to 200. The minimum number of samples required to be at a leaf node is set to 5, and the function to measure the quality of a split is set to ‘gini’.

[4] https://www.census.gov.

- **Multi-layer Perceptron (MLP)**: We use a feed-forward neural network with 3 hidden layers and 100 units at each hidden layer, and a task-specific output layer. We use cross entropy loss and Adam algorithm [48] for optimization.

As discussed in Sect. 4.1 we define the tied neighborhoods as the ones having a significantly small difference in collective efficacy value. Tied neighborhoods are considered interchangeable in the ranking. We define the ties based on a threshold on collective efficacy difference. We compute the standard deviation of the collective efficacy value of the neighborhoods in our study and define our threshold based on different coefficients of this value. We vary the coefficient from 0 to 1 with 0.2 increments and evaluate the ranking consensus using a ranking correlation metric discussed in Sect. 4.5. For more detail on the number of tied neighborhoods in each set see Appendix A.7. The results are discussed in Sect. 6.

6 Results

6.1 Ranking Performance

In this section, we present the results of ranking agreement of the permutation generated by our framework using 4 different classifiers when the most informative combination of features discussed in Sect. 4.2 were used. More specifically, we used doc2vec, distribution of topics, distribution of sentiment, normalized

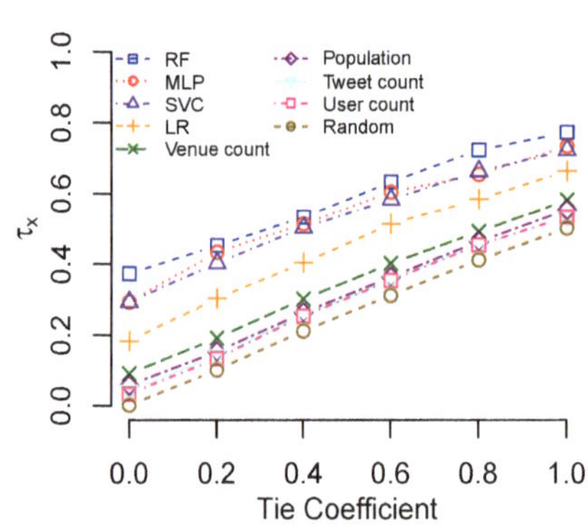

Fig. 2. Ranking performance of our model and the baselines. We used 4 classifiers for local ranking module. The x axis indicates the coefficient that is multiplied by standard deviation to make the tie threshold. The standard deviation of the ground truth collective efficacy for the 157 block groups included in this experiment is 0.18.

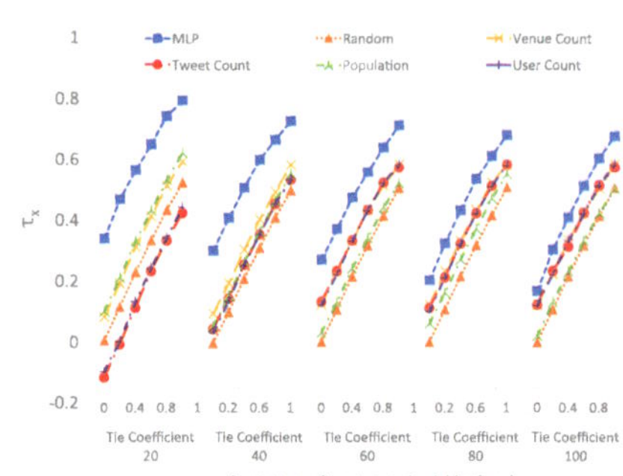

Fig. 3. Ranking performance of our framework and the baselines on different sets of neighborhoods. The sets are defined based on the number of collected tweets. Tie threshold is computed by multiplying the standard deviation of collective efficacy by the tie coefficient. The standard deviation of each set is reported in Table 1.

common user count, and spatial distance to characterize each pair of neighborhoods in our study. As shown in the Fig. 2, our framework even when used with a linear classifier such as logistic regression outperforms the baselines by at least 20%. Also, it can be seen that random forest closely followed by multi-layer perceptron is consistently giving better ranking correlation results in comparison to other classifiers.

6.2 Model Drill Down

In this section we discuss our experiments related to the effect of each feature discussed in Sect. 4.2 on ranking. To determine the best content feature, we experimented with features in this group namely TF-IDF of crime lexicon, topic distribution of urban activities, and doc2vec on the top 40% highly tweeted neighborhoods. We performed experiments with all different combinations of our 3 content factors. Each content feature is enabled in 3 combinations and disabled in 3 corresponding paired combinations. Each factorial experiment was conducted using 3 classifiers for the local ranking module including random forest, multi layer perceptron, and logistic regression. We repeated this process for 6 tie coefficients. Tie coefficients varied from 0 to 1 with an interval of 0.2. This resulted in 54 ($3 \times 3 \times 6$) experiments in which a content feature is enabled and 54 experiments in which a content feature is disabled. We observed that adding doc2vec increases the ranking performance and this boost is statistically significant (Wilcoxon signed-rank test with p-value < 0.001) [49]. However, this was not the case for two other content features. The box plot of these experiments is shared in Appendix A.8 of the supplementary material. Next, we examine the impact of additional features along with doc2vec on the ranking performance. In each experiment we computed the ranking correlation of the generated ranking with the ground truth ranking with tie coefficient values ranging from 0 to 1 with interval of 0.2. The results are presented in Table 2. As it can be seen from the Table, regardless of choice of feature combination or tie coefficient our model consistently outperforms the baselines. For summarizing the ranking correlation results, we rely on **AUC-ERC**. AUC-ERC is area under the curve of graph created by plotting tie coefficients against τ_x. From Table 2, we see that models 7 to 11, show better AUC-ERC score than the baselines. Model number 10 achieves the highest AUC-ERC score. We see that model 11 which includes all the features is not the best performing model. We conjecture that this behaviour is due to the over-fitting of the model on the training set. The TF-IDF feature has a dimensionality of 100 and the classifier might learn a function to predict the local rank between pair of neighborhoods based on few crime lexicon words (e.g. gun, shooting) in the training set. However, the test set might not contain those words on which classifier learned the function thereby resulting in wrong prediction. To summarize, using Model 10 as our proposed model, our generated ranking of neighborhoods achieves 0.77 Kendall tau-x ranking agreement with the ground truth ranking. Our results are between 20% to 37% better than the baselines depending on choice of the tie threshold.

Table 2. The comparison of Kendall τ_x and AUC-ERC of the baselines and our proposed framework. Top 40% of highly tweeted block groups were included in this experiment. Random forest was used for local ordering task. The best performance for each tie coefficient is boldfaced.

ID	Models	0	0.2	0.4	0.6	0.8	1	AUC-ERC
1	Random	0	0.09	0.20	0.30	0.40	0.49	0.25
2	Coordinates	−0.04	0.05	0.17	0.27	0.37	0.47	0.22
3	User count	0.03	0.13	0.25	0.35	0.45	0.53	0.292
4	Tweet count	0.04	0.14	0.25	0.35	0.45	0.53	0.295
5	Population	0.05	0.15	0.26	0.36	0.46	0.55	0.306
6	Venue count	0.09	0.19	0.3	0.4	0.49	0.58	0.342
7	Doc2vec + Sentiment	0.3539	0.4504	0.5256	0.6347	0.713	0.7635	0.5764
8	Doc2vec + Sentiment + Common Users	0.3698	**0.461**	**0.5497**	0.6043	0.7033	0.7642	0.5770
9	Doc2vec + Sentiment + Common Users + Topics	0.368	0.4367	0.5425	**0.6412**	0.6988	0.7647	0.5771
10	Doc2vec + Sentiment + Common Users + Topics + Distance	**0.3748**	0.4565	0.5388	0.6322	**0.7207**	**0.7735**	**0.5844**
11	Doc2vec + Sentiment + Common Users + Topics + Distance + Tfidf	0.3686	0.4597	0.5207	0.6294	0.6957	0.7666	0.5746

6.3 Effect of Data Availability

In this section, we explored to what extent the result of ranking consensus is related to the amount of data we have for each neighborhood. With this in mind, we solved the ranking tasks for different set of block groups. These sets are introduced in Sect. 5.1. We used our ranking framework with MLP as the classifier to rank each set. As indicated in Fig. 3 the more the amount of data we have for the neighborhoods in our study, the higher is the ranking consensus of the generated ranking and the ground truth ranking.

7 Conclusion

In this paper, we focused on the problem of costly computation of collective efficacy values for the neighborhoods. With the help of extensive experiments, we showed that this problem can be addressed by leveraging the social media data. Our proposed framework allows frequent and less costly access to collective efficacy values of the neighborhoods. In the future, we plan to leverage data from other sources (e.g., additional social forums and census) to improve our model. Additionally, we plan to explore the ego-net of users on social media and weigh high importance to tweets of users who are more familiar with a particular neighbourhood. Our proposed framework can act as an early warning system to capture the transformations in the neighborhoods' composition. This potentially can assist regulators and policymakers to prioritize resources, monitor neighborhood safety, and upkeep.

Acknowledgements. This work was supported by NIH-1R01 HD088545-01A1. Any opinions, findings, and conclusions in this material are those of the author(s) and may not reflect the views of the NIH.

References

1. Chainey, S., Ratcliffe, J.: GIS and crime mapping (2013)
2. Weisburd, D., Bruinsma, G.J., Bernasco, W.: Units of analysis in geographic criminology: historical development, critical issues, and open questions. In: Putting Crime in Its Place (2009)
3. Morenoff, J.D., Sampson, R.J., Raudenbush, S.W.: Neighborhood inequality, collective efficacy, and the spatial dynamics of urban violence. Criminology **39**, 517–560 (2001)
4. Sampson, R.J., Raudenbush, S.W., Earls, F.: Neighborhoods and violent crime: a multilevel study of collective efficacy. Science **277**, 918–924 (1997)
5. Sampson, R.J., Groves, W.B.: Community structure and crime: testing social-disorganization theory. AJS **94**, 774–802 (1989)
6. Browning, C.R., Cagney, K.A., Boettner, B.: Neighborhood, place, and the life course. In: Handbook of the Life Course (2016)
7. Kornhauser, R.R.: Social sources of delinquency: an appraisal of analytic models (1978)
8. Couper, M.P.: New developments in survey data collection. Ann. Rev. Sociol. **43**, 121–145 (2017)
9. Hipp, J.R.: Collective efficacy: how is it conceptualized, how is it measured, and does it really matter for understanding perceived neighborhood crime and disorder? JCJ **46**, 32–44 (2016)
10. Wang, X., Brown, D.E., Gerber, M.S.: Spatio-temporal modeling of criminal incidents using geographic, demographic, and Twitter-derived information. In: ISI (2012)
11. Wang, X., Gerber, M.S., Brown, D.E.: Automatic crime prediction using events extracted from twitter posts. In: SBP-BRiMS (2012)
12. Gerber, M.S.: Predicting crime using Twitter and kernel density estimation. DSS **61**, 115–125 (2014)
13. Aghababaei, S., Makrehchi, M.: Mining social media content for crime prediction. In: WI (2016)
14. Williams, M.L., Burnap, P., Sloan, L.: Crime sensing with big data: the affordances and limitations of using open-source communications to estimate crime patterns. BJC **52**, 320–340 (2017)
15. Bendler, J., Brandt, T., Wagner, S., Neumann, D.: Investigating crime-to-Twitter relationships in urban environments-facilitating a virtual neighborhood watch (2014)
16. Compton, R., Lee, C., Lu, T.C., De Silva, L., Macy, M.: Detecting future social unrest in unprocessed twitter data: "emerging phenomena and big data". In: ISI (2013)
17. Patton, D.: Gang violence, crime, and substance use on Twitter: a snapshot of gang communications in detroit. In: SSWR (2015)
18. Malleson, N., Andresen, M.A.: The impact of using social media data in crime rate calculations: shifting hot spots and changing spatial patterns. CaGIS **42**, 112–121 (2015)

19. Vedula, N., Parthasarathy, S., Shalin, V.L.: Predicting trust relations within a social network: a case study on emergency response. In: WebSci (2017)
20. Pacheco, D.F., Oliveira, M., Menezes, R.: Using social media to assess neighborhood social disorganization (2017)
21. Bandura, A.: Self-efficacy. The Corsini encyclopedia of psychology (2010)
22. Paskevich, D.M., Brawley, L.R., Dorsch, K.D., Widmeyer, W.N.: Relationship between collective efficacy and team cohesion: conceptual and measurement issues. Group Dyn.: Theory Res. Pract. **3**, 210 (1999)
23. Sampson, R.J., Morenoff, J.D., Earls, F.: Beyond social capital: spatial dynamics of collective efficacy for children. Am. Sociol. Rev. **64**, 633–660 (1999)
24. Goodman, L.A.: Snowball sampling. Ann. Math. Stat. **32**, 148–170 (1961)
25. Al-Olimat, H.S., Thirunarayan, K., Shalin, V., Sheth, A.: Location name extraction from targeted text streams using gazetteer-based statistical language models. In: COLING (2018)
26. Liu, T.Y., et al.: Learning to rank for information retrieval. Found. Trends Inf. Retrieval **3**, 225–331 (2009)
27. Kendall, M.G.: The treatment of ties in ranking problems. Biometrika **33**, 239–251 (1945)
28. Wilson, J.Q., Kelling, G.L.: Broken windows. Atlantic Monthly (1982)
29. Skogan, W.: Disorder and decline: the state of research. JRCD **52**, 464–485 (2015)
30. Welsh, B.C., Braga, A.A., Bruinsma, G.J.: Reimagining broken windows: from theory to policy. J. Res. Crime Delinquency **52**, 447–463 (2015)
31. Jacobs, J.: The Death and Life of Great American Cities. Vintage, New-York (1961)
32. Fu, C., McKenzie, G., Frias-Martinez, V., Stewart, K.: Identifying spatiotemporal urban activities through linguistic signatures. Comput. Environ. Urban Syst. **72**, 25–37 (2018)
33. Blei, D.M., Ng, A.Y., Jordan, M.I.: Latent Dirichlet allocation. JMLR **3**, 993–1022 (2003)
34. Zhao, W., Chen, J.J., Perkins, R., Liu, Z., Ge, W., Ding, Y., Zou, W.: A heuristic approach to determine an appropriate number of topics in topic modeling. In: BMC (2015)
35. Le, Q., Mikolov, T.: Distributed representations of sentences and documents. In: ICML (2014)
36. Bertrand, K.Z., Bialik, M., Virdee, K., Gros, A., Bar-Yam, Y.: Sentiment in New York City: a high resolution spatial and temporal view. arXiv preprint arXiv:1308.5010 (2013)
37. Ribeiro, F.N., Araújo, M., Gonçalves, P., Gonçalves, M.A., Benevenuto, F.: SentiBench-a benchmark comparison of sentiment analysis methods. EPJ Data Sci. **5**, 23 (2016)
38. Gilbert, C.H.E.: Vader: a parsimonious rule-based model for sentiment analysis of social media text. In: ICWSM (2014)
39. Levallois, C.: Umigon: sentiment analysis based on terms lists and heuristics. In: SemEval (2013)
40. Thelwall, M., Buckley, K., Paltoglou, G., Cai, D., Kappas, A.: Sentiment strength detection in short informal text. JASIST **61**, 2544–2558 (2010)
41. Hu, M., Liu, B.: Mining and summarizing customer reviews. In: SIGKDD (2004)
42. Go, A., Bhayani, R., Huang, L.: Twitter sentiment classification using distant supervision. CS224N Project Report, Stanford (2009)
43. Glavaš, G., Štajner, S.: Simplifying lexical simplification: do we need simplified corpora? In: ACL (2015)

44. Paetzold, G., Specia, L.: Lexical simplification with neural ranking. In: EACL (2017)
45. Maddela, M., Xu, W.: A word-complexity lexicon and a neural readability ranking model for lexical simplification. In: ACL (2018)
46. Emond, E.J., Mason, D.W.: A new rank correlation coefficient with application to the consensus ranking problem. J. Multi-Criteria Decis. Anal. **11**, 17–28 (2002)
47. Järvelin, K., Kekäläinen, J.: Cumulated gain-based evaluation of IR techniques. TOIS **20**, 422–446 (2002)
48. Kingma, D.P., Ba, J.: Adam: a method for stochastic optimization. ICLR (2015)
49. Wilcoxon, F., Katti, S., Wilcox, R.A.: Critical values and probability levels for the Wilcoxon rank sum test and the Wilcoxon signed rank test. In: Selected Tables in Mathematical Statistics (1970)

A Longitudinal Analysis of Vocabulary Changes in Social Media

Harith Hamoodat[1(✉)], Firas Aswad[1(✉)], Eraldo Ribeiro[1(✉)], and Ronaldo Menezes[2(✉)]

[1] Department Computer Engineering and Sciences, Florida Tech, Melbourne, USA
{hhamdon2013,faswad2013}@my.fit.edu, eribeiro@fit.edu

[2] Department of Computer Science, University of Exeter, Exeter, UK
r.menezes@exeter.ac.uk

Abstract. The vocabulary size of a language indicates the evolution of the language. The way people use their vocabulary in social media has changed, especially with the appearance of pictorial representations of ideas (e.g., emojis, memes). The adoption of emojis in the last few years motivated us to look into possible effects on vocabulary sizes in social media and maybe understand a little more about language evolution. In this paper, we do a longitudinal analysis of the vocabulary size used in social media for 14 different cities in the USA for a period of 2010–2015. We are especially interested in the relationship between vocabulary and education attainment. We computed the size of the vocabulary for each of the cities over time and compared that to the emoji usage for the same period. We found that emoji usage increases with time. Interestingly, the average size of the vocabulary behaves erratically with increases in the first two years, then reductions from 2012–2014, and then increases again in 2015. We investigated two factors that could be related to such pattern in vocabulary usage: (i) increase of reliance on emojis instead of words, which is negatively correlated with the growth of the vocabulary; (ii) increase the educational attainment, which shows a positive correlation with the increase of vocabulary for a specific time and place.

Keywords: Language usage · Vocabulary of social media · Emoji · Educational attainment

1 Introduction

Languages evolve according to the needs of its speakers; such evolution may be driven by factors such as new technologies, new products, and the incorporation of words from other languages. The development of language may happen to any part, including sounds, grammar, and vocabulary [10,21]. The English language vocabulary has changed over the last millennium [8,16] but this change is quite slow and hard to notice from year to year. Compared to how things happened in the past, current change happens faster than we expect because

H. Barbosa et al. (Eds.): *Complex Networks XI*, SPCOM, pp. 212–221, 2020.
https://doi.org/10.1007/978-3-030-40943-2_18

of new technologies such as television, radio, or the Internet, and overall world globalization. Recently, the Internet has played an essential role in accelerating this process because of the use of social media and texting.

The increased use of social media affects the way we use English (and other languages) on a daily basis [14]. Social-media messages are shorter due to length restrictions, which led to the introduction of several new acronyms. The acronyms are now part of our colloquial language (e.g., LOL, LMAO, OMG). Furthermore, users on social media adopted emojis to communicate thoughts and feelings in a visual and condensed way [4].

Researchers argue that the emojis are becoming a new global language [6,7, 30], while others claim they are just pictures that are naturally combined with plain text creating a new form of language [2,3]. The reality is that the wide adoption of emojis is reflected in the length of texts being written playing an important role in language development [1,4].

The development of the languages may occur either by adding or removal words and phrasal constructions. In this paper, we focus on how vocabulary changed in social media. We attempt to investigate why the development of vocabulary size happens and what we should expect next in this development, our efforts can help companies be prepared for the future of texting. Several factors might influence the development of vocabulary; we focused on two of them and how they correlated with the people's vocabulary. First, is the frequency of emojis in tweets, while the second is the educational attainment for different regions in the United States. While many studies on vocabularies development focused on lexical, phonetic, spelling, semantic, and syntactic perspectives [18, 19,22,28], our study examines the size of language vocabulary over time.

Many researchers have investigated the evolution of languages from a different point of view. For example, Michel et al. [23] used a corpus of around 5 million books. They argue that the number of words in the English lexicon increased from 544,000 to 1,022,000 in the last century; the ratio increased 70% during the past 50 years. They exploit the changes on word frequency as a measure for the quantitative investigation of cultural and linguistic phenomena.

The evolution of vocabulary has been studied using different types of data sources—A popular data source Google Books N-gram is used in many works. Gulordava and Baroni [11] used two periods of 4 years from the 1960s and 1990s to examine the semantic changes using the co-occurrence matrix of words and their context. They suggested that the method of measuring the semantic is suitable for detecting the semantic evolution of words during the time. Kim et al. [17] presented a model to identify the changes in specific words at a certain time. They used a corpus from 1900–2009 to train the model to obtain word vector for a particular year. Moreover, Popescu and Strapparava [27] presented a methodology that can predict future changes in the distribution of words in the Google N-gram corpus and their relationships with emotion words.

Another source of data is social media. Moise et al. [24] studied language development using data extracted from social media. They processed 10TB of the Twitter dataset using Spark DataFrame to examine the mobility of languages derived from Geo-located tweets. Next, they investigated a temporal and spatial evaluation of language using techniques such as density-based clustering and Self-Organizing Maps. As a result of using the Twitter dataset, they were able to detect real-world events and tourism trends.

This paper organized as follows. In Sect. 2, we discuss the methods used to collect, pre-process, normalize, and standardize the data. In Sect. 3, we present the results and demonstrate that our strategy seems to be a good indicator of evolution vocabulary in social media for a short period. We conclude our work in Sect. 4 with some final thoughts about this work and possible future avenues.

2 Methodology

2.1 Data Curation

Our study was carried out using two datasets, one related to socio-economics and another collected from a social micro-blog site. The first dataset contains the educational attainment for regions in the United States of America as recorded by the American Community Survey (ACS) for the period of 2010–2015. The educational attainment for a city is a ratio of the people who graduated from high school or higher to the total population. The dataset is the fraction of the population aged 25 years and older. This population age is matched with the three largest fractions of Twitter users aged (25–34, 35–44, and 55–64) [15], which already included in the dataset of the educational attainment.

The second dataset contains 569 million geo-located messages (i.e., tweets) from the Twitter microblogging platform. Around 17 million users sent the tweets during the five years from May 2010 to July 2015. Tweets were geographically spread among 58 cities in 32 countries (Fig. 1). Here, each tweet contains the following six attributes: user id, latitude, longitude, text, date-time, and language tag. We selected only tweets written in English.

To obtain a useful set of tweets, we performed several data-cleaning tasks on the dataset. First, we removed any tweets missing attributes as well as duplicated tweets. We also removed tweets that appeared to be generated by automated senders (i.e., bots); we assume bots were in place if we saw high-frequency tweets sent in less than two-second intervals. Moreover, we had to remove any tweets sent by users who are traveling faster an acceptable speed (i.e., users moving faster than a typical airplane speed of 750 km/h); the user speed was estimated by dividing the distance (latitude and longitude) between two tweets over their time difference between the same tweets. Last, we removed any numbers, special characters, links, function words/stop words (e.g., a, is, was), and punctuation from the text of the tweets.

2.2 Data Sampling

We tested 14 different regions in the United States to cover a variety of demographic areas. The number of tweets in our dataset varies in terms of years and

Fig. 1. The Twitter dataset used was collected for 58 world cities from around the world (shown as red circles) [12].

Table 1. Number of users and tweets for each year in the dataset before and after removing spurious tweets.

	2010	2011	2012	2013	2014	2015
Before data cleaning						
Users	404,139	1,591,575	4,157,311	6,970,707	8,343,348	6,357,308
Tweets	4,160,014	20,703,647	50,609,869	107,347,885	192,852,312	193,259,271
After data cleaning						
Users	395,733	1,563,298	4,152,449	6,961,825	8,334,160	6,346,505
Tweets	3,640,600	18,111,598	46,844,592	102,898,323	179,256,310	178,663,575

cities (Tables 1 and 2). To minimize bias when calculating vocabulary size in regions with a high number of tweets, we took a sample of tweets for each year in a city. The methodology of choosing the sample is implemented by determining the size of the sample, and the number of iterations runs. This determination was performed based on the value of the beta parameter extracted from Heaps' law [13] in Eq. 1.

$$V_R(n) = Kn^{\beta}, \tag{1}$$

where V_R is the number of distinct words (vocabulary) in a text of size n, and K and β are parameters determined experimentally.

To specify the smallest size of the sample that represents the data, we calculate the beta value for different sample sizes starting from 1,000 to 10,000 steps by 1,000. We repeat the computation 100 times and use Knee detection

Table 2. The 14 regions used in this study and the range of the number of tweets for each region between the years of 2010 to 2015. The range represents the minimum and maximum number of tweets in the years of study.

Region	Range	Region	Range	Region	Range
Atlanta	51K–2.04M	Miami	27K–1.42M	San-Diego	25K–1.11M
San-Francisco	77K–2.33M	Dallas	35K–2.95M	Phoenix	23K–1.16M
Chicago	55K–2.39M	Boston	43K–2.08M	Houston	27K–2.20M
Philadelphia	90K–4.69M	Los-Angeles	121K–5.85M	Detroit	34K–2.23M
New-York	188K–6.77M	Washington D.C	88K–3.03M		

[29] to find a stable point for the size of the sample. Next, we use beta value once more to find the minimum number of samples that may represent the data. We tested a different number of samples ranging from 1 to 10. For each number, we repeated the computations 100 times and used the variance value to find the number of samples. This normalization retains enough information regarding the growth of the vocabulary in a region. Based on the calculations above, we found that the size of the sample appears to be between 3,000 and 4,000 tweets. While for the number of samples, it must be larger than four. Consequently, we used five samples of 4,000 tweets[1] [12].

We calculated the vocabulary index of a region for a particular sample size $V_r(s)$ using Eq. 2 to compare the samples of cities. In essence, $V_r(s)$ is the average value of the proportion of distinct words in a Twitter sample to the total number of words in the same sample for a given region, and for a one-year interval is given by:

$$V_r(s) = \frac{1}{s} \sum_{k=1}^{s} (U_k/N_s), \tag{2}$$

where U_k is the number of distinct words in a random sample of 4,000 tweets, N_s is the total number of words in the same sample for region r, while s is the number of samples chosen in each region which is five samples.

3 Results and Discussion

A language's vocabulary size increases with time. Socio-economic and education backgrounds also are significant factors affecting the size of the vocabulary. According to the American Community Survey (ACS), the educational attainment increased by a small number each year for 14 cities (Fig. 2). The level of educational attainment positively correlated with the size of vocabulary for the same region. It is a phenomenon described in our previous work [12] and also, supported by other works [5, 9, 26].

We expect to see the value increase on the ratio of vocabulary each year (due to the positive relationship between the level of vocabulary with the educational

[1] For more information regarding the sampling method, please read [12].

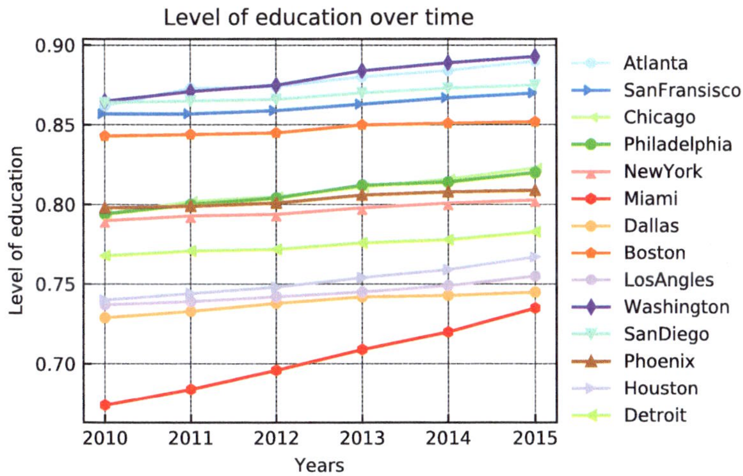

Fig. 2. Temporal effect on the level of educational attainment for the 14 cities in the United States.

attainment), especially after observing the increase in the educational attainment as described in Fig. 2. On the one hand, this expectation was correct for the years 2010, 2011, and 2012. We also assume the year 2015 is valid in terms of vocabulary ratio because it is higher than the year of 2014. Regardless, the value is less compared with 2012. On the other hand, the result for the years of 2013 and 2014 was not as expected. Figure 3 shows that in 2010–2012, the vocabulary ratio value increased in all the 14 cities. Later of 2012, it starts decreasing until 2014, and steady with tend to be increasing a bit in 2015.

In 2012, the use of emojis in social media, especially on Twitter, started been noticed compared to the previous years. We began to observe 1 or 2 emojis per 10 tweets in 2012, and less than one emoji per 20 tweets in the preceding years. Later, using emojis significantly increased from 0.3 to 0.7 emoji per tweet in years (2013 and 2014), and increased continuously in 2015 (Fig. 4). Accordingly, the increase of using emojis may affect the tweet text in terms of the number of words and the size of the vocabulary. Recently, we have seen users start to use emojis along with words in their tweets or messages.

Consequently, we presume there is a relation between the ratio of vocabulary and the using of emoji. We calculated the Pearson correlation between the ratio of emoji usage and vocabulary size. Due to the availability of emoji in the dataset, we performed the calculation to the years of 2012–2015. It is worth mentioning that we dropped the years of 2010–2011 due to the lack of emoji in the dataset. Our finding shows a negative correlation between the ratio of vocabulary and emoji usage for all regions under the test, as shown in Fig. 5. The correlation varied from −0.863 for the Dallas region to −0.972 for San-Diego. Our results show a significant *p-value* for most regions such as Houston 0.037, and San-Francisco 0.046. Also, we noticed that the *p-value* is not sufficient for some

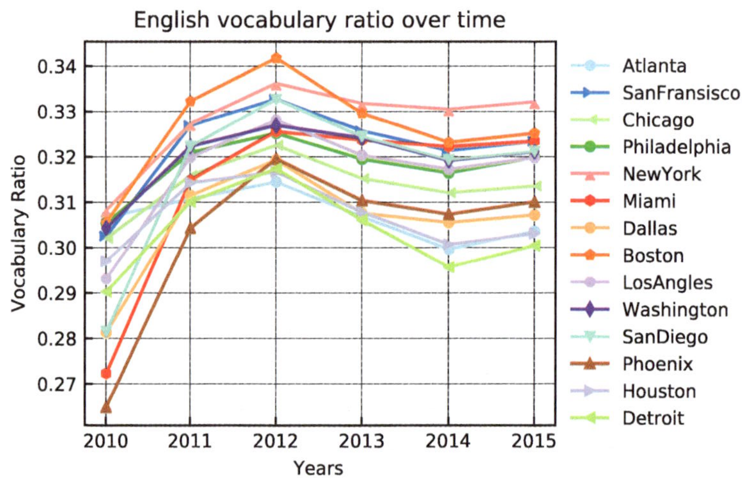

Fig. 3. Temporal effect on the level of English vocabulary ratio for the 14 regions in the United States in the time period from 2010 to 2015. The x-axis represents the years and the y-axis represents the ratio of the unique words to the total number of words in the same sample.

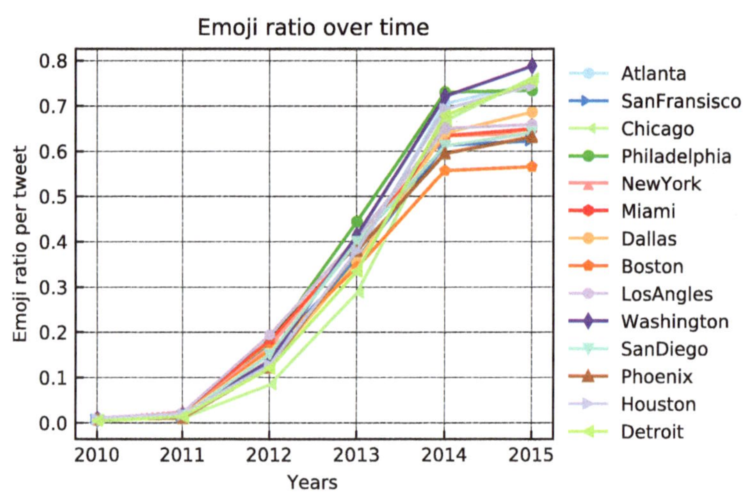

Fig. 4. The ratio of emoji per tweet from 2010 to 2015 for 14 regions in the United States. The behavior looks indistinguishable for all the regions under the test which reflect the reality of using emojis in balance, from the perspective of the number of emoji used.

cities, such as Dallas 0.136 and Miami 0.102 because of the smallness of the sample (4 points only). The sample size strongly influences the *p-value* of a test; the value fails to be significant in a small sample, which can be significant in a larger sample [25].

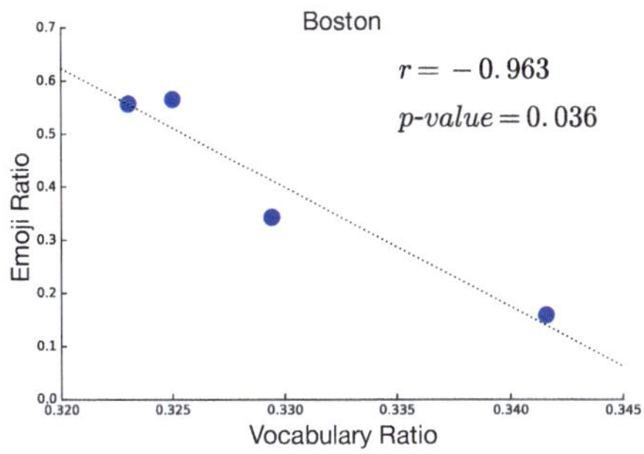

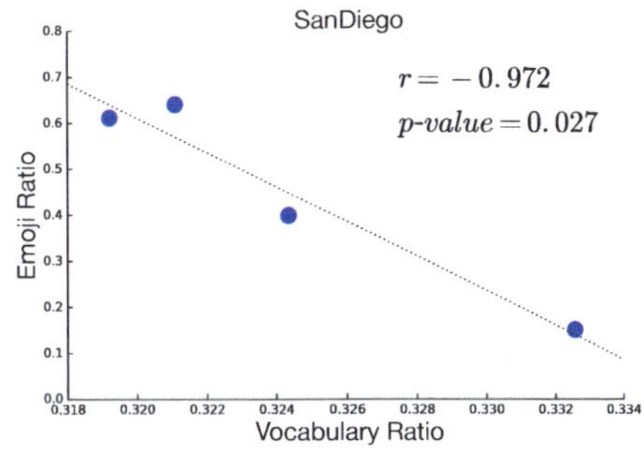

Fig. 5. Sample of two regions shows the correlation between the emoji ratio and the vocabulary ratio for the years 2012 to 2015 with a significant *p-values*. The region of Boston with $r = 0.963$ on the left side. While the right side shows the correlation for the region of San-Diego with $r = 0.972$.

Interestingly, we noticed that the size of the English tweets decreased by 24% for some regions such as Miami and Dallas and 12% in the Atlanta region. Moreover, cities with low-level educational attainment seem to have a high ratio of shrinkage on the size of the tweet. Interestingly for Miami, even though it has the highest increasing rate of educational attainment among the 14 cities, as shown in Fig. 2, yet it has the lowest value of education.

4 Conclusion

We showed that the text on social media after 2012 uses fewer words and a smaller vocabulary size year after year. We found that the reason behind that is the increases usage of emoji instead of regular words in users' tweets. The correlation is negative between emoji usage and size of vocabulary for the 14 regions in our dataset. This fact contradicts the expectation of the natural behavior of language vocabulary, which is increased over time and positively correlated with the educational attainment of a region.

Our method may predict the future of the text in social media, which could have more emoji images than words. This result opens several research avenues, such as to classify the languages according to the temporal effect of using emoji on different languages.

Our results suggest that social media texts nowadays can be treated as plain text for two reasons: (i) the diversity of subjects, which then leads to having a richer vocabulary; (ii) the time associated with tweets that assist the text to be more orderly and very useful to study several behaviors of temporal studies. This feature of social media will not be applicable in the future, for our best knowledge, because of the changes in text structure.

In the future, we plan to apply our method to include more regions and several languages for different periods to study the effect of time on the level of vocabulary. Additionally, we will consider the impact of tourism on vocabulary growth. Finally, we will examine how vocabulary size changes over certain seasons, given that temporal aspects may influence how people communicate.

Acknowledgement. The authors would like to thank Bruno Gonçalves for providing the Twitter dataset from his work [20]. The data was invaluable to us. Also, Harith Hamoodat and Firas Aswad would like to thank the Ministry of Higher Education and Scientific Research (MoHESR, Iraq), the Northern Technical University, and the University of Mosul for financial support.

References

1. Alshenqeeti, H.: Are emojis creating a new or old visual language for new generations? A socio-semiotic study. Adv. Lang. Lit. Stud. **7**(6), 56–69 (2016)
2. Barbieri, F., Ballesteros, M., Saggion, H.: Are emojis predictable? arXiv preprint arXiv:1702.07285 (2017)
3. Barbieri, F., Kruszewski, G., Ronzano, F., Saggion, H.: How cosmopolitan are emojis? Exploring emojis usage and meaning over different languages with distributional semantics. In: Proceedings of the 24th ACM International Conference on Multimedia, pp. 531–535. ACM (2016)
4. Barbieri, F., Ronzano, F., Saggion, H.: What does this emoji mean? A vector space skip-gram model for Twitter emojis. In: LREC (2016)
5. Becker, W.: Teaching reading and language to the disadvantaged-what we have learned from field research. Harv. Educ. Rev. **47**(4), 518–543 (1977)
6. Cheung, R.: How emojis became the modern world's status symbols-and how they've crossed from messaging apps to real life. South. China Morning Post **2** (2017). https://www.scmp.com/lifestyle/article/2083504/how-emojis-became-modern-worlds-status-symbols-and-how-theyve-crossed
7. Cohn, N.: Will emoji become a new language. BBC Future (2015)
8. Crystal, D.: The Cambridge Encyclopedia of the English Language. Ernst Klett Sprachen, Stuttgart (2004)
9. Graves, M.F.: Chapter 2: Vocabulary learning and instruction. Rev. Res. Educ. **13**(1), 49–89 (1986)
10. Gray, R.D., Atkinson, Q.D., Greenhill, S.J.: Language evolution and human history: what a difference a date makes. Philos. Trans. Roy. Soc. B Biol. Sci. **366**(1567), 1090–1100 (2011)
11. Gulordava, K., Baroni, M.: A distributional similarity approach to the detection of semantic change in the Google Books Ngram corpus. In: Proceedings of the GEMS 2011 Workshop on Geometrical Models of Natural Language Semantics, pp. 67–71 (2011)
12. Hamoodat, H., Ribeiro, E., Menezes, R.: Social media vocabulary reveals education attainment of populations. In: International Workshop on Complex Networks, pp. 157–168. Springer (2019)
13. Herdan, G.: Type-Token Mathematics, vol. 4. Mouton (1960)
14. Indrajith, I., Varghese, T.: Language into "lang": Whatsapp imprints on teenagers. ACME Int. J. Multidiscip. Res. **6**(1), 71–82 (2018)
15. Clement, J.: Distribution of Twitter users in the united states as of September 2018, by age group, August 2019. https://www.statista.com/statistics/192703/age-distribution-of-users-on-twitter-in-the-united-states/. Accessed 9 Aug 2019

16. Keller, R.: On Language Change: The Invisible Hand in Language. Routledge, London (2005)
17. Kim, Y., Chiu, Y.I., Hanaki, K., Hegde, D., Petrov, S.: Temporal analysis of language through neural language models. arXiv preprint arXiv:1405.3515 (2014)
18. Kroch, A.S.: Syntactic Change. na (2001)
19. Labov, W.: The social motivation of a sound change. Word **19**(3), 273–309 (1963)
20. Lamanna, F., Lenormand, M., Salas-Olmedo, M.H., Romanillos, G., Gonçalves, B., Ramasco, J.J.: Immigrant community integration in world cities. PLoS ONE **13**(3), e0191612 (2018)
21. Lieberman, P.: On the Origins of Language: An Introduction to the Evolution of Human Speech. Macmillan, New York (1975)
22. Masterson, J.J., Apel, K.: The spelling sensitivity score: noting developmental changes in spelling knowledge. Assess. Eff. Interv. **36**(1), 35–45 (2010)
23. Michel, J.B., Shen, Y.K., Aiden, A.P., Veres, A., Gray, M.K., Pickett, J.P., Hoiberg, D., Clancy, D., Norvig, P., Orwant, J., et al.: Quantitative analysis of culture using millions of digitized books. Science **331**(6014), 176–182 (2011)
24. Moise, I., Gaere, E., Merz, R., Koch, S., Pournaras, E.: Tracking language mobility in the Twitter landscape. In: 2016 IEEE 16th International Conference on Data Mining Workshops (ICDMW), pp. 663–670. IEEE (2016)
25. Moore, D.S., McCabe, G.P., Craig, B.A.: Introduction to the practice of statistics, New York (2012)
26. Nagy, W.E., Anderson, R.C.: How many words are there in printed school English? Read. Res. Q. **19**(3), 304–330 (1984)
27. Popescu, O., Strapparava, C.: Behind the times: detecting epoch changes using large corpora. In: Proceedings of the Sixth International Joint Conference on Natural Language Processing, pp. 347–355 (2013)
28. Radinsky, K., Agichtein, E., Gabrilovich, E., Markovitch, S.: A word at a time: computing word relatedness using temporal semantic analysis. In: Proceedings of the 20th International Conference on World Wide Web, pp. 337–346. ACM (2011)
29. Satopaa, V., Albrecht, J., Irwin, D., Raghavan, B.: Finding a "kneedle" in a haystack: detecting knee points in system behavior. In: 2011 31st International Conference on Distributed Computing Systems Workshops, pp. 166–171. IEEE (2011)
30. Thompson, C.: The emoji is the birth of a new type of language (no joke). Wired **24**, 16 (2016)

Communities of Human Migration in Social Media: An Experiment in Social Sensing

Firas Aswad[1(✉)], Harith Hamoodat[1(✉)], Eraldo Ribeiro[1(✉)], and Ronaldo Menezes[2(✉)]

[1] Department Computer Engineering and Sciences, Florida Tech, Melbourne, USA
{faswad2013,hhamdon2013}@my.fit.edu, eribeiro@fit.edu

[2] Department of Computer Science, University of Exeter, Exeter, UK
r.menezes@exeter.ac.uk

Abstract. Migration has been key to the success of humans as it can improve both the lives of migrants and the economy of destination regions. However, it can also overstretch the resources of hosting regions. Therefore, an accurate assessment of immigration data is important to both immigrants and governments. In this work, we detect and relate communities of migration by analyzing discussions from social media and compare them with official immigration records. Our goal is two-fold. First, it measures the agreement between the official migration numbers and popular perception. Second, the measure of individuals' feelings from social-media data may be used by governments to guide immigration policy-making. This paper contributes to the growing body of knowledge in social sensing; it shows that people's perspectives on migration, as reflected in social media, is mismatched with reality.

Keywords: Human migration · Community structure · Social sensing

1 Introduction

Migration has always impacted human development [7,10,11]. Migration can benefit both migrants and the places to which they migrate to [6]. Yet, it can also affect the economy, demographics, and political stability of host countries [12]. Thus, governments must plan the allocation of resources and public services while taking into account the impact of immigration.

The accurate assessment of immigration flows is an essential part of government planning and policymaking. Governments often rely on immigration data from agencies and organizations such as the International Passenger Survey[1], the Organization for Economic Co-Operation and Development (OECD)[2], and

[1] https://www.ons.gov.uk.

[2] http://www.oecd.org.

H. Barbosa et al. (Eds.): *Complex Networks XI*, SPCOM, pp. 222–232, 2020.
https://doi.org/10.1007/978-3-030-40943-2_19

the United Nations' International Organization for Migration[3]. However, these sources may not contain enough detail (e.g., migrants vs. international students) or may not be updated frequently enough to ensure correct decisions. Also, traditional measurements and collection of immigration flows can be time-consuming and expensive [8].

Social sensing, or the inference of the world state from data from social media information, may provide an effective alternative to traditional methods of immigration-data assessment. Recent developments on social sensing have led to the modeling of many human behaviors, including epidemics, health, city science, environment, and education [1,2,9,15,16]. Similar sensing approaches may be useful to the modeling of human migration. Moreover, the understanding of immigration from social sensing can be critical to policymakers as a decision-support tool based on real-time information.

Data from social media has been used to understand immigrant integration in urban environments [14] and also flows of migrants between countries [3]. Some studies attempt to track, analyze, and predict human displacement and migration. Zagheni et al. [23] used geolocated data from Twitter users in countries belonging to the OECD to estimate the flow of migrants. Because the OECD countries do not reflect the Twitter users' distribution, the study has limitations of representing the status world-wide. Thus, they proposed a difference-in-differences approach to reduce selection bias when it comes to checking trends in out-migration rates for individual countries.

To study human migration, Kikas et al. [13] used data from Skype to explain human migration. They used socio-economic indicators such as GDP and correlated them with a network built from Skype. On the one hand, they found a low correlation between World Bank migration data and Skype migration data. On the other hand, the correlation in Europe was robust when compared to the migration information obtained from EUstat (Euskal Estatistika Erakundea).

Mendieta et al. [17] collected geolocated tweets from a sample of users (travelers) for 21 months. They focused on the flow of migrants to Ecuador in the time of the Pope's visit in 2015. Their findings show that the destination of local and foreign travelers from social-network data highly correlated with migration data from official records.

At a global level, Messias et al. [18] investigated clustering in human migration by introducing a new model consisting of a triad of countries. Here, they argued that relying on the flow of pairs of countries is insufficient to determine clusters of countries. Messias et al. collected data from Google+ users, and extracted the geolocations for about 14% of the users, classifying them based on the number of countries they lived.

Another recent work by Lamanna et al. [14] has demonstrated the notion that immigrants are deeply connected to their original home countries. Here, they used five years of social-media data collected from Twitter to quantify the spatial integration of immigrant communities in 53 major cities. They built a

[3] https://www.iom.int.

bipartite spatial-integration network to quantify the spatial segregation of each immigrant community in each city.

Considering the importance of communities in the analysis of immigration, we propose a method to identify *immigration communities* and determine whether the communities from social media data reflect the data from official sources. We compare migration communities for two datasets, first obtained from official sites and second from social media. Here, we used the Jaccard similarity coefficient to show how well one dataset reflects the other. Our finding shows that people's views on migration, as reflected in social media, is mismatched with reality.

This paper organized as follows. Section 2 describes the methods used to collect and pre-process the data; it also describes the community detection algorithm used in this paper. Section 3 shows the Jaccard score among the communities in the two datasets and presents the main findings. Finally, we conclude our work in Sect. 4.

2 Datasets and Methods

Our study was carried out using two datasets. The first dataset contains a set of keywords related to migration extracted from Twitter messages sent between November 2017 and October 2018. The messages were written in 8 different languages. The second dataset was obtained from the United Nations Department of Economic and Social Affairs [21]. It contains migrants' source and destination countries, and metadata associated with the immigrants (e.g., gender).

2.1 Twitter Dataset

This dataset was collected using the Twitter Streaming API. We gathered tweets containing keywords related to migration (e.g., immigrant, emigrant, refugee, displacement) and for the eight most-common languages on Twitter (i.e., English, Japanese, Spanish, Malay, Portuguese, Arabic, French, and Turkish) [19]. In total, the dataset has some 81 million tweets for the period between November 2017 and October 2018. Most tweets in our dataset are in English (about 57 million), and come from locations worldwide. A small subset of approximately 64,000 tweets are geo-located, which enabled us to know their source countries. The distribution of geo-located tweets is shown in Fig. 1. It shows areas of wide Twitter adoption. Although the figure does not show all the tweets in our dataset, we expect the distribution of all the tweets to spread across many countries, except for countries restricting Twitter access (e.g. China).

Next, we built the network, and we used off-the-shelf Network-Science tools to analyze it. The network nodes represent the world countries, and the links are the flow estimate for the number of immigrants from the origin to the host (destination) country. We extracted country names from the tweets by searching for 249 countries listed in the *pycountry* Python package [20] and their alternative translations in the eight languages aforementioned.

The network links are implemented from co-occurrences of countries' names in a single tweet, i.e., if a tweet mentions at least two names of countries, we link

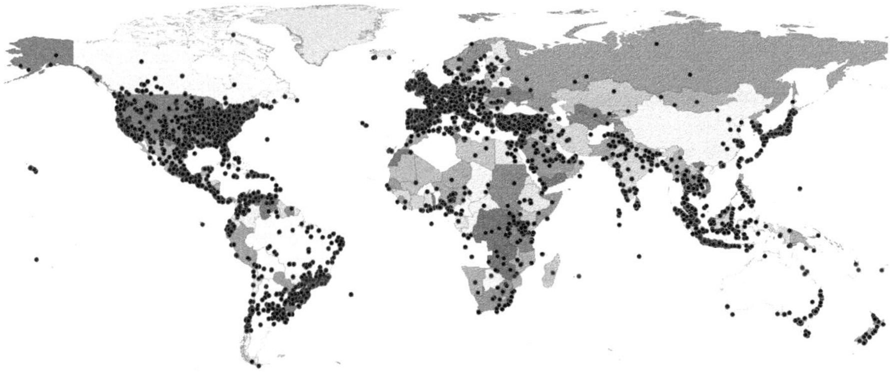

Fig. 1. Geo-located tweets spatial distribution around the world. There are approximately 64 thousand tweets depicted here and this sample shows us that our tweets are well distributed except for places that restrict Twitter access (e.g. China).

them in the network. For instance, given a tweet *"At least 380 Latin migrants have died this year, many of them drowning while trying to cross the U.S.-Mexico border"*, the link *[United States–Mexico]* is added to the network. If a single tweet mentions three or more countries, we linked them to one another as a clique. For instance, for a tweet *"Syria plans for Turkey and Germany includes fighting smugglers to curb illegal migration"*, the links *[Syria–Turkey]*, *[Syria–Germany]*, *[Turkey–Germany]* are added to the network. Inconsistencies regarding country names were dealt with by using alternative spellings, e.g., we used *UK*, *U.K*, *U.K.* when detecting mentions of the *United Kingdom.*

We assume that the order countries' names appear in a tweet is irrelevant in this study. While the direction is important in a migration network, the social-media data does not carry enough information to allow us to infer the direction of migration reliably. Thus, network links are presented as undirected edges. Furthermore, when adding a new link between already connected countries, the link weight increases by 1; as a result, we end up with a weighted network.

2.2 The UN Dataset on Migration

International migration data was obtained from the UN Department of Economic and Social Affairs, which provides accumulated data about international migration every five years and sometimes every two years. In this work, we worked with the data for 2015 and 2017. Next, we subtracted between the two years and took the absolute value of the results. This technique allows us to identify the flow of migrants.

The UN network was created using a similar approach we used for the Twitter network; countries are the nodes, and the links are weighted by the flow of immigrants between the two countries. The nature of the UN data is directed (i.e., source and destination are known). Thus, we converted the network from directed to undirected to compare it with the Twitter network by adding the flows in both directions to get the undirected weight between the two countries. We tested other approaches for defining the weights (i.e., subtraction of flows, maximum of both directions) and, by using correlation analysis, we found that the methods are equivalent [3].

2.3 Community Detection

In network science, the ability to explore and analyze the structure of a set of nodes in a graph by looking at the pieces of subgraphs that may have similar properties is called community analysis. In this work, we used community detection to provide insights into overall immigrants' behavior; that is, how they move between subsets of countries. Community detection enables us to reveal these structures which have corresponding meaning in the real-world phenomena the network represents. There are several approaches to implement community detection [22], but in general, a community represents a set of nodes whose density of links within the set is larger than the density of links from members of the set to the nodes outside the set.

A fast and popular community detection method is the Louvain algorithm by Blondel et al. [5]. It works by optimizing a heuristic objective function known as *modularity*, which captures the density of in-community links compared to the links of nodes in the community to nodes outside the community. The algorithm maximizes the modularity by applying the vertex mover. Each vertex examines all the possible moves to a neighborhood community with increased modularity.

3 Results

This paper analyzes communities of migration; we calculated the node degree and weighted degree distributions for both the Twitter and UN networks. This is a similar structural analysis to that described in our previous work [3], but this time for a larger dataset. In the migration networks, the node degree can be seen as a metric that gives us information about the diversity of immigrants in countries. The weighted degree represents the number of migrants moving between countries.

We used the log-likelihood ratio to fit the node degree and weighted degree distributions. The stretched exponential distribution achieved the best fit for the node degree distributions (Fig. 2) while the truncated power-law distribution was the best fit for the weighted-degree distributions (Fig. 3).

Next, we used a community detection algorithm to uncover the network structure exposed by the density of links in the migration network. Recall that a community is defined as a group with a high density of edges within each group but having sparser connections with other groups in a network. Thus, we want to examine whether the community structure for the data extracted from Twitter has similarities at the community level with the UN dataset.

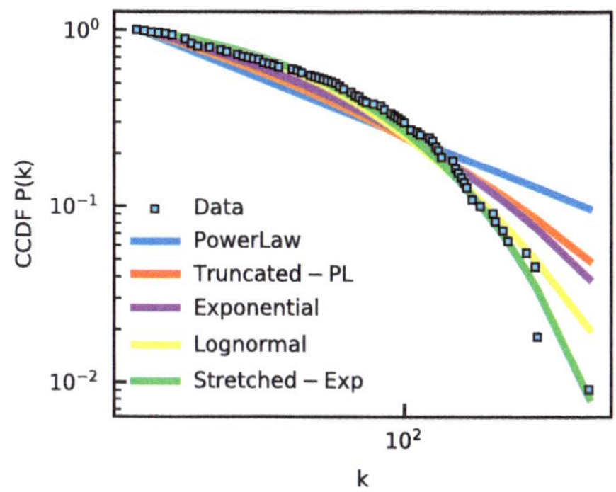

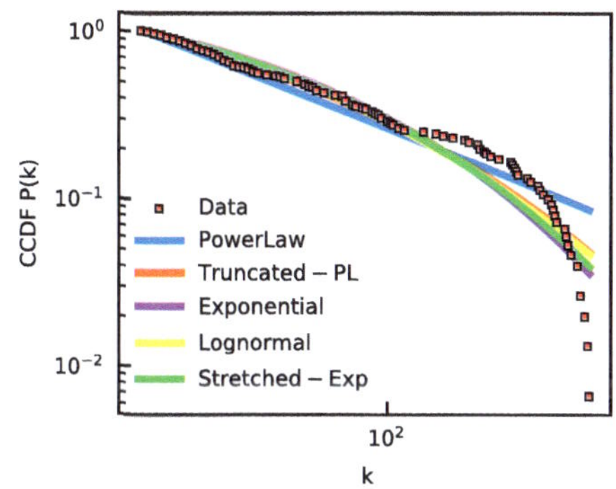

Fig. 2. Node degree distributions. Twitter network (left) and UN network (right). We used the log-likelihood ratio to fit the distribution. The stretched exponential distribution achieved the best fit for both datasets.

We applied the Louvain method, and the results of the execution showed nine communities for each dataset. To measure the similarity between these communities and comprehend if they reflect each other, we treated each community as a set of countries. Then, we used *Jaccard Similarity* to compute the similarity between the sets. Figure 4 shows the world countries' communities map, the color represent the nine communities for the datasets.

The Jaccard similarity coefficient J uses the concept of the cardinality of the intersection over the cardinality of the union. The coefficient values range from 0 to 1, where 1 indicates identical communities, and 0 indicates completely different communities. $J(c_t, c_u)$ is defined as:

$$J(c_t, c_u) = \frac{|c_t \bigcap c_u|}{|c_t \bigcup c_u|}, \tag{1}$$

where c represents the set of nodes in the community, while t and u represent whether c is referring to the Twitter and the UN sets respectively.

Because of the number of communities in each network, we have to compare all the sets (communities) in the Twitter (t) and UN (u) networks; all pairs are compared to find the best match, shown as a matrix in Fig. 5 (top). The highest value of the comparison is adopted as the "matched" pairs of communities. Thus, the community pair (c_t, c_u) in the network t and u is chosen based on the maximum Jaccard score $J_{\max}(c_t, c_u)$ for all possible pairs of communities, defined as:

$$J_{\max}(c_t, c_u) = \max J(c_t, c_u), \quad \forall t \in T, u \in U, \tag{2}$$

where T and U represent all sets of communities for the Twitter and UN networks respectively. Equation 2 ensures that all pairs are compared. The results of this calculation can be shown in Fig. 5 (bottom).

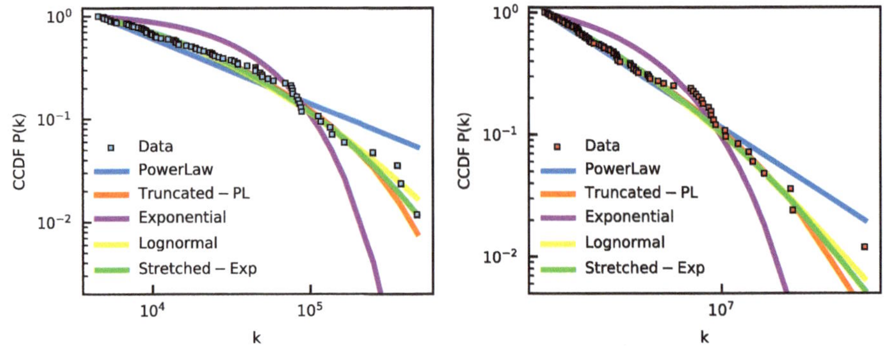

Fig. 3. Weighted degree distributions. Twitter network (left) and UN network (right). We used the log-likelihood ratio. The truncated power-law distribution achieved the best fit for both datasets.

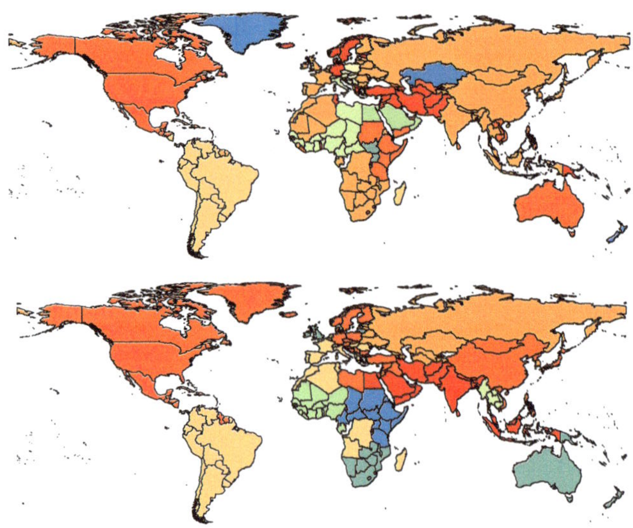

Fig. 4. Communities map for Twitter (top) and the UN (bottom). The countries' colors represent communities.

Finally, we calculate the average Jaccard similarity J_{avg} for the entire paring considering all the matched pairs found according to Eq. 2 to obtain the final value of Jaccard of the entire community matching, which is given by:

$$J_{\text{avg}} = \frac{1}{m} \sum_{t,u=1}^{m} J_{\max}(c_t, c_u), \tag{3}$$

where m represents the total number of matched pairs of communities, and $J_{\max}(c_t, c_u)$ represents the matched pair from Twitter and UN with the maximum Jaccard as defined in Eq. 2.

The results of the Jaccard score in some communities is considered high, such as 0.44 between community 6 in Twitter and community 7 in the UN, particularly when compared to other sets of communities, as shown in Fig. 5. The average value of the Jaccard similarity J_{avg} is 0.232.

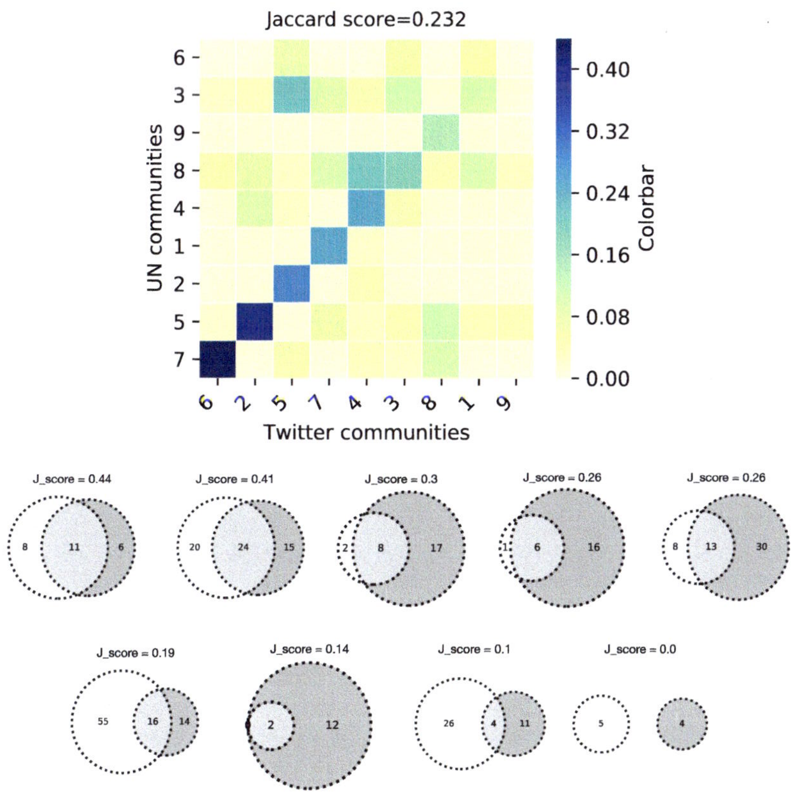

Fig. 5. The heatmap on the top shows the nine communities for Twitter and the UN associated with the values of the Jaccard similarities for Twitter × UN. The Venn diagrams on the bottom represent the nine maximum values of the Jaccard score for the comparison among the communities.

This analysis shows that people's perception about immigration on social media *does not* reflect the reality reported by the UN statistics given that J_{avg} is relatively low. Yet, we saw that some pairings are close to reality when it comes to countries that share a border or that are geographically close to each other. Some of these countries experience significant immigration. For instance, in the same communities on both datasets include the United States of America, Canada, Mexico, Anguilla, Belize, Bermuda, Cuba, Cayman Island, Dominica, Dominican Republic, Guatemala, Guadeloupe, Honduras, Haiti, Saint Kitts, St Lucia, Puerto Rico, El Salvador, and St Vincent. Also, we saw that most countries in Africa formed communities that agree with the ground truth (official records), such as Niger, Nigeria, Ghana, Benin, Burkina Faso, Cote d'Ivoire, Guinea, Gambia, Guinea-Bissau, Equatorial Guinea, and Senegal.

4 Conclusion

In this paper, we built two networks about migration; one from Twitter and another from official UN data. Then, we extracted the degree and weighted degree distributions in both networks. The degree gives the diversity of immigration, whereas the weighted degree indicates the number of people who immigrated.

The main contribution of this paper is to compare the community structure for both networks to understand if social sensing leads to information that reflects the reality of the migration. We used the Louvain method to find the communities for both networks; the results gave us nine communities for each dataset. Later, we examined the similarity among the sets of the communities in the datasets using the Jaccard coefficient similarity. The results of the comparison among the communities of the two datasets yielded a similarity of 0.232. Regardless of such low value, we can see a high Jaccard score for some of the communities in particular for countries with a high level of migration such as the USA and Mexico but also for local aspects of migration such as in some of the countries in Africa.

In sum, the difference of the Jaccard score among the communities in the results between the two datasets showed us that the public has a mismatched view of reality but slightly correct when it comes to some of the countries.

Our comparison in this paper is quite simple because by using the Jaccard similarity, we only compare the communities from the point of view of nodes but not edges. This means that the internal structure of the communities is not considered. As future work, we intend to look at methods for network comparison such as the one introduced by Bagrow and Bollt [4]. Their approach could be used to compare the structure of the communities (by looking at communities as separate networks) but indeed even compare the entire Twitter and UN networks.

Acknowledgments. Firas Aswad and Harith Hamoodat would like to thank the Ministry of Higher Education and Scientific Research (MoHESR, Iraq), the University of Mosul, and the Northern Technical University for financial support.

References

1. Anastasi, G., Antonelli, M., Bechini, A., Brienza, S., D'Andrea, E., De Guglielmo, D., Ducange, P., Lazzerini, B., Marcelloni, F., Segatori, A.: Urban and social sensing for sustainable mobility in smart cities. In: 2013 Sustainable Internet and ICT for Sustainability (SustainIT), pp. 1–4. IEEE (2013)
2. Arthur, R., Boulton, C.A., Shotton, H., Williams, H.T.: Social sensing of floods in the UK. PLoS ONE **13**(1), e0189327 (2018)

3. Aswad, F.M.S., Menezes, R.: Refugee and immigration: Twitter as a proxy for reality. In: The Thirty-First International Flairs Conference (2018)
4. Bagrow, J.P., Bollt, E.M.: An information-theoretic, all-scales approach to comparing networks. Appl. Netw. Sci. **4**(1), 45 (2019)
5. Blondel, V.D., Guillaume, J.L., Lambiotte, R., Lefebvre, E.: Fast unfolding of communities in large networks. J. Stat. Mech Theory Exp. **2008**(10) (2008). https://doi.org/10.1088/1742-5468/2008/10/P10008
6. Borjas, G.J.: The economic benefits from immigration. J. Econ. Perspect. **9**(2), 3–22 (1995)
7. Coppel, J., Dumont, J.C., Visco, I.: Trends in immigration and economic consequences. Technical report (2001)
8. Findlay, A., Gould, W.T.: Skilled international migration: a research agenda. Area **21**(1), 3–11 (1989)
9. Hamoodat, H., Ribeiro, E., Menezes, R.: Social media vocabulary reveals education attainment of populations. In: International Workshop on Complex Networks, pp. 157–168. Springer (2019)
10. Hanson, G.H.: Immigration and economic growth. Cato J. **32**, 25 (2012)
11. Hanson, G.H.: The Economic Logic of Illegal Immigration. Council on Foreign Relations, New York (2007)
12. Kapur, D.: Political effects of international migration. Annu. Rev. Polit. Sci. **17**, 479–502 (2014)
13. Kikas, R., Dumas, M., Saabas, A.: Explaining international migration in the skype network: the role of social network features. In: Proceedings of the 1st ACM Workshop on Social Media World Sensors, pp. 17–22. ACM (2015)
14. Lamanna, F., Lenormand, M., Salas-Olmedo, M.H., Romanillos, G., Gonçalves, B., Ramasco, J.J.: Immigrant community integration in world cities. PLoS ONE **13**(3), e0191612 (2018)
15. Madan, A., Cebrian, M., Lazer, D., Pentland, A.: Social sensing for epidemiological behavior change. In: Proceedings of the 12th ACM International Conference on Ubiquitous Computing, pp. 291–300. ACM (2010)
16. Madan, A., Moturu, S.T., Lazer, D., Pentland, A.S.: Social sensing: obesity, unhealthy eating and exercise in face-to-face networks. In: Wireless Health 2010, pp. 104–110. ACM (2010)
17. Mendieta, J., Suárez, S., Vaca, C., Ochoa, D., Vergara, C.: Geo-localized social media data to improve characterization of international travelers. In: 2016 Third International Conference on eDemocracy & eGovernment (ICEDEG), pp. 126–132. IEEE (2016)
18. Messias, J., Benevenuto, F., Weber, I., Zagheni, E.: From migration corridors to clusters: the value of Google+ data for migration studies. In: Proceedings of the 2016 IEEE/ACM International Conference on Advances in Social Networks Analysis and Mining, pp. 421–428. IEEE Press (2016)
19. MIT Technology Review: Most-used languages on Twitter, December 2013. https://www.statista.com/statistics/267129/most-used-languages-on-twitter. Accessed 20 Nov 2017
20. Theune, C.: Pycountry python package, August 2019. https://pypi.org/project/pycountry/. Accessed 18 Aug 2019
21. United Nations Department of Economic and Social Affairs: Population division, international migration. https://www.un.org/en/development/desa/population/migration/data/estimates2/estimates17.asp. Accessed 2017

22. Yang, Z., Algesheimer, R., Tessone, C.J.: A comparative analysis of community detection algorithms on artificial networks. Sci. Rep. **6** (2016). Article number: 30750. https://www.nature.com/articles/srep30750
23. Zagheni, E., Garimella, V.R.K., Weber, I., et al.: Inferring international and internal migration patterns from Twitter data. In: Proceedings of the 23rd International Conference on World Wide Web, pp. 439–444. ACM (2014)

Demographic Analysis of Music Preferences in Streaming Service Networks

Lidija Jovanovska[1,2], Bojan Evkoski[1(✉)], Miroslav Mirchev[1], and Igor Mishkovski[1]

[1] Faculty of Computer Science and Engineering, Ss. Cyril and Methodius University, 1000 Skopje, North Macedonia
lidija.jovanovska@outlook.com, bojanevkoskl@outlook.com

[2] Jožef Stefan International Postgraduate School, 1000 Ljubljana, Slovenia

Abstract. As Daniel J. Levitin noted, music is a cross-cultural phenomenon, a ubiquitous activity found in every known human culture. It is indeed, a living matter that flows through cultures, which makes it a complex system potentially holding valuable information. Therefore, we model country-to-country interactions to reveal macro-level music trends. The purpose of this paper is twofold. Firstly, we explore the way specific demographic characteristics, such as language and geographic location affect the global community structure in streaming service networks. Secondly, we examine whether a clear flow of musical trends exists in the world by identifying countries who are prominent leaders on the music streaming charts. The community analysis shows that there is strong support for the first claim. Next, we find that the flow of musical trends is not strongly directional globally, although we were still able to identify prominent leaders and followers within the communities. The obtained results can further lead to the development of more sophisticated music recommendation systems, kindle new cultural studies and bring discoveries in the field of musicology.

Keywords: Music · Community detection · Leader-follower relationship

1 Introduction

Throughout the previous decade, the world has experienced a drastic change in the way music is listened to. While physical and digital copy sales continue to fall globally, streaming services usage is growing massively. Namely, streaming income increased by 230% between 2013 and 2017 and it continues to do so year after year by almost 25%, marking it as a period of steady development.

In regards to the fluidity of the CD music market, Ferreira and Waldfogel stated that in at least one year of continuity, 31 artists appeared concurrently on

H. Barbosa et al. (Eds.): *Complex Networks XI*, SPCOM, pp. 233–242, 2020.
https://doi.org/10.1007/978-3-030-40943-2_20

the charts in more than 18 countries on the worldwide music trade between 2001 and 2007 [4]. This suggests a high similarity of the regional musical preferences on a global level. However, digital songs are more likely to fall off the chart in the first week than CD songs, indicating a highly volatile market, resulting regularly in heterogeneous charts [5]. These findings prompted us to examine whether this volatile market possesses clear patterns that could be used for further research and applications such as developing music recommendation systems, running cultural studies and discovering more about the nature of today's music.

Our study is organized as follows: Firstly, in Sect. 2, we provide an overview of some of the most relevant studies on networks and music. Then, in Sect. 3, we describe the data taken from the global log of streaming habits recorded by Spotify. In the following Sect. 4, we describe how the music networks were constructed and investigate whether and to what extent communities are influenced by language and geographical distance when looking at the whole dataset, but also at specific genres. In Sect. 5, we present an approach for detecting leadership in the music network, by measuring the correlation between top charts of each pair of countries in the dataset. Finally, in Sect. 6 we conclude the paper with a summary of contributions and avenues for future work.

2 Related Work

The field of Network Science provides a framework for modeling interactions between entities to reveal macro-level properties that may not be noticeable at the individual level. The potential of these methods stems from the fact that the creation and transmission of cultural products are essentially network phenomena. Therefore, networks can lead to a new fundamental understanding of the complex nature of culture in music cognition. In the past decade, there has been plentiful research on music preference due to the demand for better recommendation systems in the online world. This has motivated many researchers to try to better understand the connections in music networks across the globe.

Gunaratna and Menezes analyzed a social network of Brazilian musicians to identify the artists responsible for the flow of information in the network. They also demonstrated that the network follows a power-law distribution, with prominent hierarchical characteristics only during the latter decades (1990–2010). This tendency towards a hierarchical structure can be explained as a result of the increasing availability of collaboration tools, i.e. social media [7]. Salganik et al. studied the effect of social influence on a micro-level (success of a song) and the macro-level (market inequality). They did so by creating an artificial "music market" in which a group of participants rated previously unknown songs either with or without knowledge of previous participants' choices. They discovered that social influence increased both inequality and unpredictability of success [10]. Nagy et al. provided a powerful methodology for detecting leader-follower pairs, which was previously applied in the search for the leadership hierarchy present in pigeon flocks [8]. This was further adapted by Lee and Cunningham in their research about the global flow of music on Last.fm [6], which served as an inspiration for this research.

3 Data

Streaming services provide easy access to rich and various content which can be used for thorough analysis of many properties of music, artists and its consumers. In recent times, Spotify has arguably been the most popular streaming platform with a user base of 248 million active users worldwide. Since January 2017, the Spotify Top 200 Charts are open to the public via their API. It allows easy access to song and artist data, but it also provides daily and weekly worldwide streaming top charts. They are available for each country separately, which makes them suitably formatted for our analysis.

Spotify currently provides streaming services to 79 countries, of which 65 have regular weekly Top 200 Charts. Since we were interested in finding the flow of the musical trends in the network, it was necessary to provide data for the longest time span possible (January 2017–June 2019). Considering that not every country started providing top charts from day one, the filtered dataset ultimately contained 55 countries.

The working dataset contained over 270 billion streams with a median of 1.6 billion and an average of 5 billion streams per country. It included 6810 different artists with over 31000 unique songs, which managed to climb into the Top 200 in at least one region. Each sample contained the song name, artist, Spotify URL of the song (unique), date, country, the streaming count of the song for the specific week and its position (Table 1).

Table 1. Basic dataset statistics

# Countries	# Songs	# Artists	# Weeks	Total streams	Min streams	Max streams
55	31093	6810	124	273400 mil.	Lux: 93 mil.	US: 69073 mil.

Finally, to be able to undergo genre-specific analysis, we used the genre hierarchy defined in the Free Music Archive (FMA) [3] to group all 1500 different Spotify genres into their corresponding 12 parent genres defined in FMA: blues, country, electronic, folk, hip-hop, classical, jazz, pop, reggae, rock, soul-rnb and indie/experimental (contains every genre that did not fit the rest).

4 Identifying Global Communities

To gain knowledge about the connections and the structure of the music streaming world in terms of countries, we attempted to find clusters based on language and geographic distance by analyzing the similarity of their top charts.

4.1 Streaming Matrices

For the purpose of creating a suitable mathematical representation of a country's streaming history, the data was aggregated into matrices, which we refer to as streaming matrices. These streaming matrices contain every unique song ever to appear on the Spotify top charts as a column, while each row represents one of the 55 countries. Since most of the countries do not have the same set of songs appearing on their charts, the matrices are sparse. The cells showed the total streaming count of a song for a particular country. In mathematical terms, a non-zero entry in the matrix at position i, j is a positive integer, indicating the total number of times the users from country i have streamed the song j. With this representation, each country was a vector of 31000 values (unique songs). Since there was a big difference in the number of streams between countries (the USA with over 7 billion while Peru with only 300 million) min-max normalization on each row separately was necessary (Table 2).

Table 2. Streaming matrix example

	Africa	Christmas lights	...	Mr. Brightside	Shape of you
Ireland	4112k	810k	...	6032k	15821k
...	...	...	...	...	...
USA	41639k	1823k	...	9336k	351342k

4.2 Agglomerative Clustering

The normalized streaming matrices were used to compute the agglomerative clusters and to create dendrograms by using Ward's distance method. By examining Fig. 1, it became clear that there are two major clusters: a Spanish speaking and non-Spanish speaking, which asserts the influence of language on the development of the major clusters. It is also noticeable that the Spanish speaking countries form their cluster much faster than the rest of the world. The language bond is visible in the other cluster too, where there is a strong connection between the UK and Ireland, Germany and Austria, etc. Strong geographical influences are also present. For example, even though Greek and Bulgarian come from different language families, there is a strong musical similarity between Greece and Bulgaria, due to geographical closeness. The same pattern appears with Hungary and the Czech Republic. This repeats to an even larger extent if we analyze country pairs that use similar languages. Finally, it should be noted that the cluster containing France, Italy, Turkey, and Brazil, which shows up rather late in the groupings is not a cluster at all. As we would discover later, these four countries mostly listen to local music and they have largely independent top charts from the rest of the world.

4.3 Community Networks

To be able to visualize the clustering, but also apply some other clustering techniques specific to graphs, undirected community networks were generated for

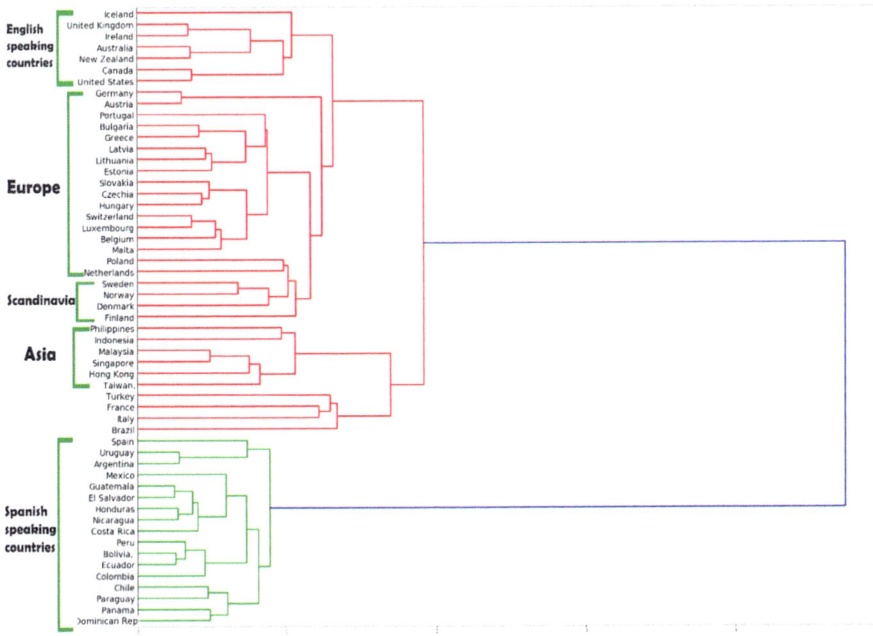

Fig. 1. Agglomerative clustering on top charts similarity

each genre separately. To form edges, the cosine similarity between each vector was measured. Its most frequently used in text analysis, to measure document similarity. However, it can be used in other scenarios which can be represented with vector notation. Afterwards, using Gephi [1], the modularity of the network was calculated [2]. It is a widely used asset in community structure analysis which measures the strength of division of the network into communities.

Figure 2 shows the community network for the pop music genre where the size of the nodes represents the streaming counts. As expected, the Spanish cluster was present and strongly interconnected, but the other smaller communities revealed much more interesting results. The next smaller clusters were eminent: the English-speaking countries, the Balkan community, Central Europe, Eastern Europe and finally, the Asian countries. Italy and Brazil remained strongly disconnected from the rest of the world, while Paraguay and Finland join them as cultures mostly devoted to domestic music. Furthermore, even though the Spanish language dominates politics and diplomacy in Paraguay, their native Guarani language is used by over 90% of the population. On the other hand, Finland's disconnectedness is also due to the uniqueness of the Finnish language as part of the Uralic language family, unlike the rest of the Scandinavian countries, who are part of the Germanic language family.

The main support of the hypothesis that language has the biggest influence on the development of musical communities is the fact that when looking at the genre which has almost no words at all - the classical genre, the communities

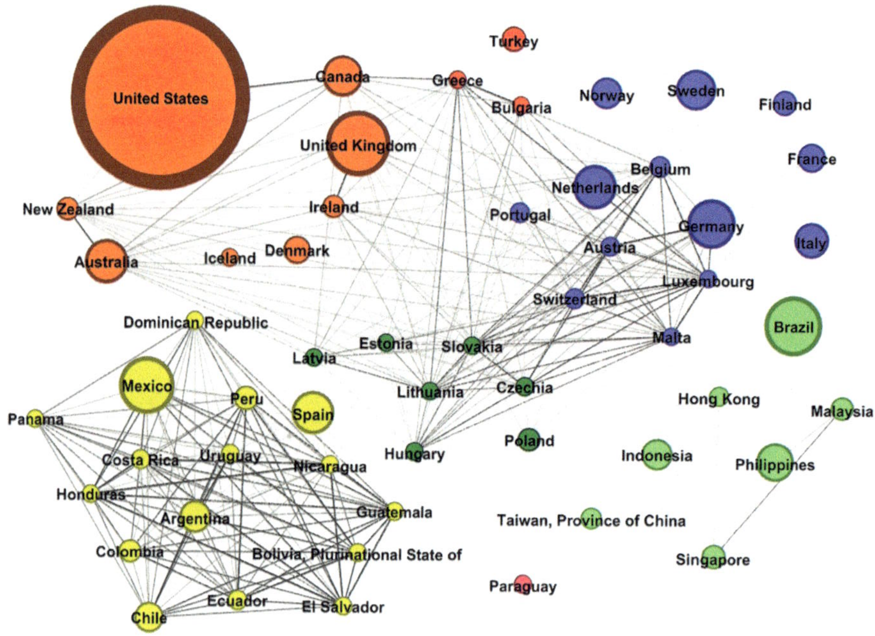

Fig. 2. Pop music top charts community network using modularity score (node size - total stream count, edge weight - cosine similarity)

that emerge are significantly more versatile in terms of language and geographical distance (Fig. 3).

5 Detecting Global Music Leaders

The next step was to delve deeper into the characteristics and the dynamics of the obtained network. In the previous section we identified the biggest communities in the network, which can be very helpful for reasons stated above, but detecting the leaders in those communities is another important aspect of acquiring knowledge about the structure of the network. Generally, leaders are considered as countries that dictate the top charts and initiate the trends, while followers are countries that adopt certain trends within a certain time lag. This led to the design of a novel approach for finding leadership in the community network.

5.1 Methodology

When looking at a pair of countries we want to find out whether country i follows country j, j follows i, or no connection exists. Since we have weekly top charts for every country, we determine the relationship by comparing charts of the pair of countries, but in different time steps. For example, if we want to determine the

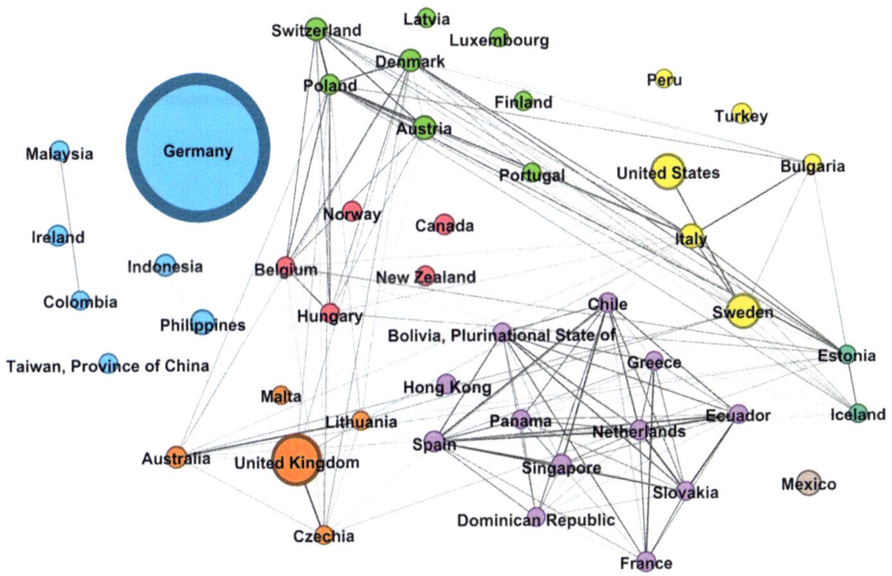

Fig. 3. Classical music top charts community network using modularity score

relationship between Sweden and Denmark, we measure the Jaccard similarity of Sweden's top charts in week t, with Denmark's top charts in week $t-1$, and vice versa for every t in the dataset. If the similarity is strong enough, we choose the higher similarity and add a directed link from the leader to the follower, e.g. from Denmark to Sweden. Hence, a simple follower score can be defined as:

$$\text{Follower\_Score}(C_i, C_j) = \sum_t \text{Jaccard}(C_i, C_j) = \sum_t \frac{C_i(t) \cap C_j(t-1)}{C_i(t) \cup C_j(t-1)}.$$

However, to make this approach more reliable, these computations are calculated for different time lags: from one week (as in the example above), to 12 weeks for each pair of countries. We then compute the global averages for every time lag separately, which we take as thresholds that help us determine the significance of the similarities. Afterward, we sum the 12 similarities for each pair and measure how many of those similarities are above their respective threshold, labeled as k in our formula. We use these counts as parameters to the similarity sums. Larger k corresponds to a stronger followership, while a smaller value means a less consistent one. These scaled similarity sums are the weight of the network edges. Note that if k is zero, the edge is removed. The follower score which we propose takes the following form:

$$\text{Follower\_Score}(C_i, C_j) = (\sum_{d-1}^{12} \sum_t \frac{(C_i(t) \cap C_j(t-d))}{(C_i(t) \cup C_j(t-d))}) \cdot k.$$

Finally, to identify if, for example, "Sweden follows Denmark" significantly more than "Denmark follows Sweden", we calculate the average difference between A and B and B and A scores for all pairs of countries. Only if this threshold is crossed, we create an edge between Sweden and Denmark, otherwise we conclude that the following is not significant enough, thus there is no leader-follower relationship.

5.2 Results

After the directed network was generated with the method described above, the well-known PageRank algorithm was used to compute the importance of the nodes [9]. The algorithm utilizes the weights of the edges and helps by enabling a better visualization of the network. In that sense, the size of the nodes in the network represents their PageRank score. As in the previous section, the modularity was measured to see if the same communities that emerged in Sect. 4 would emerge again.

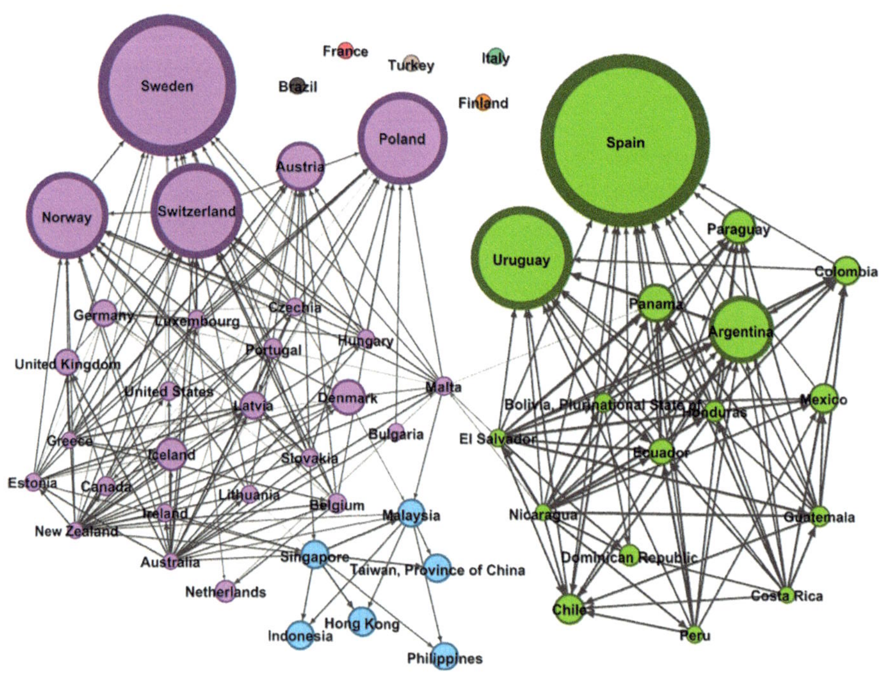

Fig. 4. Music leaders network (node size - PageRank, edge weight - lagged jaccard similarity)

As we can see in Fig. 4, the Spanish speaking countries form a strong cluster once again, with Spain as the obvious leader in this group. Interestingly enough, the other large cluster formed by the non-Spanish Western world is not what we expected to see. Even though the USA and UK are regularly two of the largest recorded music markets in the world[1], Sweden, Switzerland, and Germany are the first that recognize popular patterns and lead the trends in the weekly top charts in the streaming world.

From this analysis, we can infer that if a song becomes popular in Spain, it will most likely succeed in the other Spanish speaking countries too. On the other hand, if it succeeds in Sweden, Switzerland or Norway, it might dominate the whole Western music scene. These findings can help greatly in predicting top charts of the follower countries, but also give better recommendations to users from those regions based on the most popular music in the leader countries.

6 Conclusion

The examination of the music streaming networks revealed clear evidence that communities are formed under the influence of language and geographic location. In addition, we presented an approach for detecting countries that lead the music trends on Spotify. The results showed that global leaders in the music industry are not necessarily trendsetters in the streaming world. Furthermore, it was revealed that both similarity and leadership between countries in music streaming differ across genres.

The results from this research could be leveraged from music streaming platforms, such as Spotify, to build recommendation engines based not only on content and/or collaborative-based filtering, but will also take into account specific demographic characteristics and music genre leaders and followers. In the future, we intend to perform this analysis using song-specific features. This approach will uncover new genres which are not merely labels, but a combination of certain feature values. Music has never been more available, thus the future awaits with many new challenges in the fields of music analysis and music recommendation.

References

1. Bastian, M., Heymann, S., Jacomy, M.: Gephi: an open source software for exploring and manipulating networks. In: Third International AAAI Conference on Weblogs and Social Media (2009)
2. Blondel, V.D., Guillaume, J.L., Lambiotte, R., Lefebvre, E.: Fast unfolding of communities in large networks. J. Stat. Mech Theory Exp. **2008**(10) (2008). https://doi.org/10.1088/1742-5468/2008/10/P10008
3. Defferrard, M., Benzi, K., Vandergheynst, P., Bresson, X.: FMA: a dataset for music analysis. ArXiv abs/1612.01840 (2016)
4. Ferreira, F., Waldfogel, J.: Pop internationalism: has half a century of world music trade displaced local culture? Econ. J. **123**(569), 634–664 (2013)
5. Lao, J., Nguyen, K.H.: One-hit wonder or superstardom? The role of technology format on billboard's hot 100 performance (2016)

[1] https://en.wikipedia.org/wiki/List_of_largest_recorded_music_markets.

6. Lee, C., Cunningham, P.: The geographic flow of music. In: 2012 IEEE/ACM International Conference on Advances in Social Networks Analysis and Mining, pp. 691–695. IEEE (2012)
7. Menezes, R., Gunaratna, C., Patel, M.: Network sciences in music recommendation a case study with Brazilian music (2012)
8. Nagy, M., Akos, Z., Biro, D., Vicsek, T.: Hierarchical group dynamics in pigeon flocks. Nature **464**(7290), 890 (2010)
9. Page, L., Brin, S., Motwani, R., Winograd, T.: The PageRank citation ranking: bringing order to the web. Technical report, Stanford InfoLab (1999)
10. Salganik, M.J., Dodds, P.S., Watts, D.J.: Experimental study of inequality and unpredictability in an artificial cultural market. Science **311**(5762), 854–856 (2006)

Mobility Networks

Comparative Analysis of Store Opening Strategy Based on Movement Behavior Model over Urban Street Networks

Takayasu Fushimi(✉) and Masaya Yazaki

School of Computer Science, Tokyo University of Technology, Hachioji 192-0982, Japan
takayasu.fushimi@gmail.com, c0116270ac@edu.teu.ac.jp

Abstract. In this study, we address the problem of identifying trade-areas of facilities, such as convenience stores, gas stations, and supermarkets, based on closeness centrality and betweenness centrality while considering the target city a spatial network. When placing a new facility in a local area, locating it with high accessibility for neighboring residents can attract customers from existing facilities and expand its own trade-area. Therefore, it is important to properly grasp the trade-area of existing facilities. To this end, we consider two movement behavior models of people. In the first model, which is an existing model, the trade-area of each facility is extracted by Voronoi tessellation against its installation site, which assumes that each resident goes to the nearest facility. In the second model, which is our proposed model, it is extracted by the proportion including the facility on the shortest paths from the resident's departure point to various destinations in the network. Based on these models, we propose a selection probability that a resident selects a facility, and attempt to extract trade-area of each facility. From experimental evaluations using actual road networks and location information of the convenience stores, we confirmed that the existing model extracts the trade-areas of each store with good balance; in the proposed model, the trade-area of stores located along the main road becomes wider. By analyzing the entropy of selection probabilities, it was confirmed that the competitiveness of existing stores can be understood and can be used as an evaluation measure when determining candidate locations for new store openings.

Keywords: Centrality · Spatial network · Facility location · Trade-area

1 Introduction

In recent years, convenience stores, gas stations, supermarkets, and so forth have been located throughout the given region, and more than 50,000 convenience stores have opened in Japan. Convenience stores need to be located in a place

H. Barbosa et al. (Eds.): *Complex Networks XI*, SPCOM, pp. 245–256, 2020.
https://doi.org/10.1007/978-3-030-40943-2_21

that is easily accessible from local residents, that is, close to many residents, in order to satisfy the needs of local residents to purchase foods that are necessary for their daily lives. In addition, it must be a place that is accessible for people not only in the local resident but also on the move. To satisfy the demands of many people and maximize the profits of stores, each chain store has a strategy for opening stores. When establishing a new store, it is important to understand the trade-area of the existing stores. The trade-area of a store is equivalent to basins of attraction when the store is as an attractor.

In the field of network science, closeness centrality is a typical measure of accessibility. Closeness centrality is based on the distance to other nodes in the network, and the concept that nodes with a small average distance are important. When the road structure of an area is considered as a network, the facilities in the upper nodes in the closeness centrality ranking are said to have high accessibility from many residents (other nodes). In this study, we consider the urban street structure of a region as a network, and formalize two behavior models of people based on the concept of network centrality. The first one is a model in which each resident goes to the nearest facility as the destination. This is based on the concept of closeness centrality, and we refer to this model as the Nearest Neighbor Movement model. The second one is a model in which each resident drops by at a facility on the shortest route from the starting point to the destination. This is based on the concept of betweenness centrality, and we refer to this model as the Shortest Path Movement model. Based on these models, we propose a selection probability that a resident selects a facility, and attempt to extract trade-area of each facility. Extracting trade-areas based on selection probability has a similar flavor to overlapping community detection from a spatial network, i.e., the Nearest Neighbor Movement model is equivalent to Voronoi tessellation based on graph distance. On the other hand, trade-area based on the Shortest Path Movement model is a novel concept.

Various community extraction methods have been proposed to divide node set on a network into several subsets. Typical methods include extracting the part of the network with a relatively high link density [10,11] and cutting the link into several subgraphs [5,12]. The two trade-area-extraction methods that we focus on in this study are different from the community extraction method described above, because all nodes are divided into K groups according to the positional relationship with the K pivot nodes (facilities).

In our experiment using the actual road structure and the actual location information of the convenience store, we compare the store opening strategy from the viewpoint of movement behavaior of the target customers. Furthermore, the degree of competition among stores is quantified from the overlap of store trade-areas, and differences of the models are considered from the viewpoint of the candidate locations for new stores.

The paper is organized as follows. Section 2 describes related work. Section 3 explains an existing movement model and our proposed model. In Sects. 4 and 5, we report and discuss experimental results using real world data. Finally, Sect. 6 concludes this paper and address the future work.

2 Related Work

This section summarizes existing research on the analysis of the centrality measure for spatial networks such as road networks and the facility location problem on a network.

2.1 Centrality Analysis of Spatial Networks

There are many studies that analyzed road networks by network analysis approach [2,7,8]. Crucitti et al. analyzed the distribution of four centrality indices in a road network considering distance weights between junctions [2]. The area with similar road structure is classified by the fitting parameter and Gini coefficient of the centrality distribution.

Montis et al. analyzed multiple undirected networks with municipalities as nodes and commuter traffic between municipalities as weighted links [7]. The relationship between the degree and the clustering coefficient indicates that there is a hierarchy in the municipality and that there is a positive correlation between the centrality index and population or wealth of residents.

Park et al. evaluated the difference in the topological structure of residential areas and downtown areas by applying the centrality index to the road network and calculating its entropy [8]. Thus, centrality plays an important role in research on road network analysis. The centrality index used in existing studies is a quantification of the independent nature of individual nodes for the entire network. Since overestimation due to overlapping influences of neighboring nodes is not considered, the centrality scores of neighboring nodes tend to be similar. Therefore, simply focusing on nodes with high existing centrality scores, only nodes that are biased to a part of the network are extracted, and are not suitable as facility location candidate sites for this study. In this study, we focus on group centrality [3] in which the centrality score of the node set is quantified in consideration of the dependency between nodes.

2.2 Facility Location Problem on Networks

In a branch of operations research and computational geometry, facility location problems have been studied for many years [1,4,6,9,13,14]. Basically, facility location problems are formalized as p-center or p-median problem, the latter can be regarded as k-medoids clustering based on graph distance. These studies try to solve in polynomial or pseudo-polynomial time, for small-size graphs with a simple topological structure like tree or line. On the other hand, in this study, we attempt to grasp the trade-areas of existing facilities based on network centrality notion, for large-scale road networks extracted from actual urban streets.

In Tabata et al.'s method [13], the first node in the ranking based on proximity centrality is calculated very quickly, and that node is used as a facility location candidate site. That is, the first representative node in the greedy method of k-medoids clustering is obtained at high speed. In general, installing multiple facilities can meet the demand of more nodes (residents), and extracting the areas that each facility affects will lead to a strategy for the contents of the installation. This study is different in that the location of a new store is discussed after considering the location of the store that has already been installed (Fig. 1).

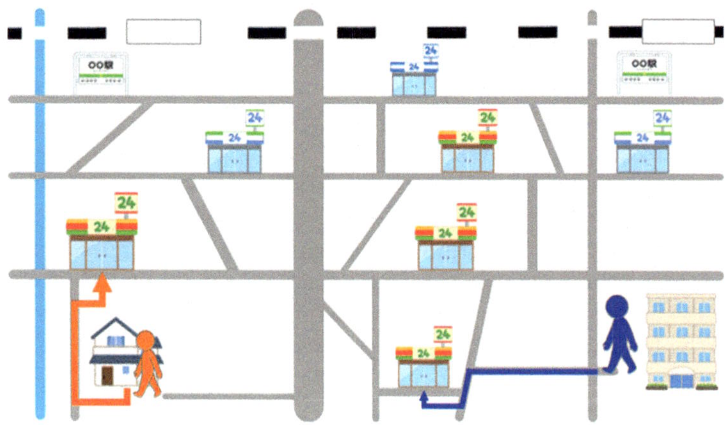

Fig. 1. NNM model. Residents go to the nearest convenience store in order to buy what they need.

3 Movement Behavior Models

In this study, we consider two types of movement behaviors: the Nearest Neighbor Movement model, which is an existing model based on closeness centrality and the Shortest Path Movement model, which is a novel model based on betweenness centrality. Now we consider the road network $G = (\mathcal{V}, \mathcal{E})$ in the target area, where $\mathcal{V}$ is a set of junction nodes, and $\mathcal{E}$ is a set of road links between junctions. For convenience, the junction node represents an inhabitant, and the set of junction nodes where facilities are installed is $\mathcal{R} \subset \mathcal{V}$, $K = |\mathcal{R}|$.

3.1 Nearest Neighbor Movement Model

The Nearest Neighbor Movement (NNM) model is a behavior model that assumes people move from the residence to the nearest facility (store). This can be said to be a behavior model whose purpose is to go to the store. In this model, residents select stores with a probability inversely proportional to distance. That is, the probability that the resident u selects the store r is defined as

$$P_{nnm}(r|u) = \frac{d(u,r)^{-\theta}}{\sum_{r' \in \mathcal{R}} d(u,r')^{-\theta}}, \tag{1}$$

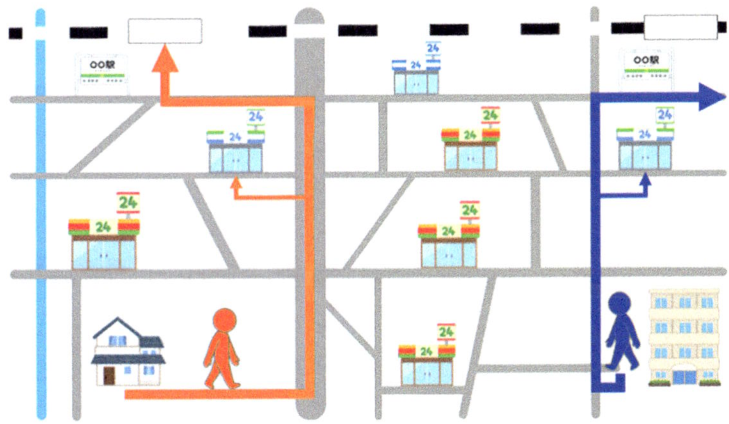

Fig. 2. SPM model. Residents drop by convenience stores on the way to the station for commuting and attending school.

where $d(u,r)$ represents the distance between the resident node u and the store node r, and θ is a parameter. As shown in the Eq. (1), since this probability is calculated based on closeness centrality, nodes with high closeness located on urban area are expected to indicate a high selection probability (Fig. 2).

The following trade areas (communities) can be extracted by finding the store with the highest selection probability for all nodes, and assuming the group of nodes that select each store as the trade area of each store.

$$\mathcal{C}(r) = \left\{ u \in \mathcal{V}; r = \underset{r' \in \mathcal{R}}{\arg\max}\, P_{nnm}(r'|u) \right\}. \tag{2}$$

Since all nodes belong to the trade area of any store, this is equivalent to Voronoi tessellation of the road network based on the graph distance.

3.2 Shortest Path Movement Model

The Shortest Path Movement (SPM) model is a behavior model that assumes people drop in at facilities (stores) on the route when traveling to destinations such as schools, workplaces, sightseeing spots, and hospitals. This model is different from the NNM model in which the store is assumed to be the destination in that it is assumed to be a stopover on the way from the starting point to the destination. In this model, a store is selected with a probability proportional to the number of times it appears on the shortest route toward various destinations. The ratio of passing through store r over the shortest path between starting node u and the various destinations t can be defined as

$$\delta_u(r) = \sum_{t \in \mathcal{V} \setminus \{u\}} \frac{\sigma_{u,t}(r)}{\sigma_{u,t}},$$

where $\sigma_{u,t}$ is the number of shortest paths from the starting node u to the destination node t and $\sigma_{u,t}(r)$ is the number of those paths that pass through the store node r. Therefore, the probability that the resident u selects the store r is defined as

$$P_{spm}(r|u) = \frac{\delta_u(r)^\theta}{\sum_{r' \in \mathcal{R}} \delta_u(r')^\theta}. \tag{3}$$

As shown in the Eq. (3), since this probability is calculated based on betweenness centrality, nodes with high betweenness located on arterial roads are expected to indicate a high selection probability.

The following communities can be extracted by finding the store with the highest selection probability for all nodes, and assuming the group of nodes that select each store as the trade area of each store.

$$\mathcal{B}(r) = \left\{ u \in \mathcal{V}; r = \arg\max_{r' \in \mathcal{R}} P_{spm}(r'|u) \right\}. \tag{4}$$

This is a novel community extraction method for spatial networks which is different from Voronoi tessellation based on graph distance.

4 Experiments

In our experimental evaluation, two cities, Hachioji and Shizuoka, are the target areas. From Open Street Map (OSM)[1], we collected the road structure in the target area and extracted all junctions and roads of each city. We then constructed a spatial network with the junctions as the nodes and the roads between the junctions as the links by following a standard formulation of road networks, such as those presented by SNAP (Stanford Large Network Dataset Collection)[2]. We also collected the location information of actual stores of the three major convenience store chains in Japan (7-Eleven, FamilyMart, Lawson) from NAVITIME[3]. Table 1 shows the number of junction nodes where convenience stores of the three companies are established and the number of normal junction nodes (Non) where no store is established. Approximately, all residents are assigned to junction nodes.

Table 1. #nodes (#stores or #junctions) in each city.

City	7-Eleven	FamilyMart	Lawson	Non	Total
Hachioji	95	76	36	12,117	12,324
Shizuoka	118	106	65	30,752	31,041

[1] https://www.openstreetmap.org/.
[2] http://snap.stanford.edu/data/index.html.
[3] https://www.navitime.co.jp/category/.

In our experiments, residents are assumed to live equally at each junction node regardless of residential area, urban area, or mountainous area. However, a more realistic analysis is possible by assigning the number of inhabitants at each junction node from the population density data. Furthermore, the parameter θ is set to 1, which is equivalent to setting the slope of the softmax function to 1.

5 Results

5.1 Qualitative Comparison of Movement Behavior Models

Figure 3 shows the results of extracted trade-areas based on two models for 8 convenience stores (Ministop) in Hachioji. The star mark in the figure represents the location of 8 convenience stores, and the color of the junction node represents the trade-area of each convenience store. Also, the black roads in the figure represent the top 3% links in the edge-betweenness centrality ranking, and are the main roads in the target area. In the results by the NNM model in Fig. 3(a), the trade areas of 8 convenience stores are extracted with good balance. Figure 3(b) shows that the cyan convenience store located along the main road has a very strong trade-area. Since the cyan convenience store faces the road with the highest edge-betweenness centrality ranking, it can be the shortest route between many nodes, so it is considered that a large trade-area has been acquired. On the other hand, although the green convenience store is located near the station, it is along a complicated road. Therefore, it is rarely on the shortest route for car users, and it is thought that only a small trade-area was obtained. The brown convenience store is located in the area where the residential complex exists, so it is considered that the residential complex has become a trade-area because residents often go along the way to the destination.

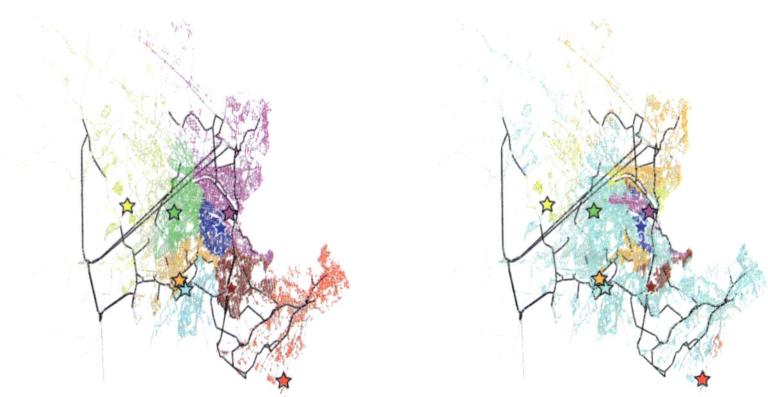

(a) NNM model (b) SPM model

Fig. 3. Extracted trade-areas of 8 stores in Hachioji

5.2 Quantitative Comparison of Extracted Trade-Areas

For each convenience store belonging to a chain of three major companies, the trade-area was extracted using Eqs. (2) and (4). Figures 4 and 5 show the visualization results of the trade-area in Hachioji and Shizuoka. The red circles, green triangles, and blue squares in the figure represent the locations of 7-Eleven, FamilyMart, and Lawson stores, respectively. The points in the figure are the junction nodes representing the residents, and the color represents the chain of the trade-area to which they belong.

From Fig. 4(a), it can be seen that there are many red nodes as a whole, that is the trade-area of 7-Eleven that has the most stores in Hachioji is wide. On the other hand, Fig. 4(b) shows that the trade-area of FamilyMart is wider than that

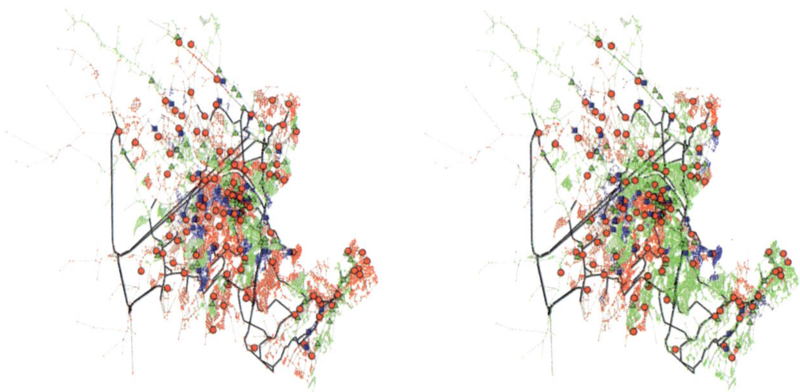

(a) NNM model (b) SPM model

Fig. 4. Extracted trade-areas of 3 chains in Hachioji

(a) NNM model (b) SPM model

Fig. 5. Extracted trade-areas of 3 chains in Shizuoka

of 7-Eleven. This is thought to be due to the fact that FamilyMart has many stores along higher-betweenness roads (main roads) and has an efficient location to acquire a larger trade-area with fewer stores. Conversely to the results for Hachioji, in Fig. 5(a) and (b), it can be seen that somewhat lots of green nodes by the NNM model and many red nodes by the SPM model are extracted.

Table 2 shows the number of nodes belonging to the trade-area of each chain. The size of trade-area of each chain was measured by summing those of stores that belong to the chain, i.e., let $\mathcal{R}_h$ be the set of stores belonging to the chain h, the size of trade-area of the chain h is calculated as

$$N_{nnm}(\mathcal{R}_h) = \sum_{r \in \mathcal{R}_h} |\mathcal{C}(r)| = \sum_{u \in \mathcal{V}} \sum_{r \in \mathcal{R}_h} \delta\left(r, \arg\max_{r' \in \mathcal{R}} P_{nnm}(r'|u)\right), \quad (5)$$

$$N_{spm}(\mathcal{R}_h) = \sum_{r \in \mathcal{R}_h} |\mathcal{B}(r)| = \sum_{u \in \mathcal{V}} \sum_{r \in \mathcal{R}_h} \delta\left(r, \arg\max_{r' \in \mathcal{R}} P_{spm}(r'|u)\right), \quad (6)$$

where $\delta(A, B)$ returns 1 if $A = B$ otherwise 0.

Similarly, we computed the size of trade-area by summing the selection probabilities of stores that belong to the chain, i.e., the size of trade-area of the chain h is calculated as

$$\tilde{N}_X(\mathcal{R}_h) = \sum_{u \in \mathcal{V}} P_X(\mathcal{R}_h|u) = \sum_{u \in \mathcal{V}} \sum_{r \in \mathcal{R}_h} P_X(r|u) \quad \text{with } X \in \{\text{nnm}, \text{spm}\}, \quad (7)$$

where $P_X(\mathcal{R}_h|u)$ stands for the probability that the stores belonging to the chain h are selected by resident u. From Table 2, it can be confirmed that compared with the results by Eqs. (5) and (6) that select a store with the highest selection probability with probability 1, the values of the trade-area sizes by Eq. (7) are somewhat smoothed. In results of either calculation method, it can be said that (1) in Hachioji, FamilyMart, which has a smaller number of stores than 7-Eleven,

Table 2. Size of trade-areas

Model	Eq.	7-Eleven	FamilyMart	Lawson	Total
(a) Hachioji					
NNM model	$N_{nnm}(\mathcal{R}_h)$	7,292.00	3,601.00	1,431.00	12,324
	$\tilde{N}_{nnm}(\mathcal{R}_h)$	5,837.71	4,416.26	2,070.03	12,324
SPM model	$N_{spm}(\mathcal{R}_h)$	3,870.00	7,238.00	1,216.00	12,324
	$\tilde{N}_{spm}(\mathcal{R}_h)$	5,197.63	5,156.51	1,969.86	12,324
(b) Shizuoka					
NNM model	$N_{nnm}(\mathcal{R}_h)$	14,678.00	10,375.00	5,988.00	31,041
	$\tilde{N}_{nnm}(\mathcal{R}_h)$	12,683.93	11,331.56	7,025.51	31,041
SPM model	$N_{spm}(\mathcal{R}_h)$	20,706.00	5,620.00	4,715.00	31,041
	$\tilde{N}_{spm}(\mathcal{R}_h)$	16,189.95	7,112.97	7,738.09	31,041

is an effective location for residents who follow the SPM model; (2) in Shizuoka, 7-Eleven, which has almost the same number of stores like FamilyMart, is an appropriate location for residents who move according to the SPM model.

From these results, FamilyMart may be executing a store opening strategy intended for residents who stop by the way of moving by car; whereas 7-Eleven may be executing a store opening strategy intended for residents who walk to nearby stores on foot. In this way, it was also suggested that it is necessary to switch the trade-area extraction method depending on what kind of behavior model the customer is supposed to be.

5.3 Entropy of Selection Probabilities

Next, we show the visualization results colored by the entropy of selection probability based on the NNM and the SPM models. For a node u, the entropy of selection probability is calculated as $E_{nnm}(u) = -\sum_{h=1}^{H} P_{nnm}(\mathcal{R}_h|u) \log P_{nnm}(\mathcal{R}_h|u)$ or $E_{spm}(u) = -\sum_{h=1}^{H} P_{spm}(\mathcal{R}_h|u) \log P_{spm}(\mathcal{R}_h|u)$, where H is the number of chains, $H = 3$ in our data, and $\sum_{h=1}^{H} P_{nnm}(\mathcal{R}_h|u)$ and $\sum_{h=1}^{H} P_{spm}(\mathcal{R}_h|u)$ hold 1. In Figs. 6 and 7, reddish nodes have higher values and bluish nodes lower values. From these figures, we can confirm that in the NNM model, the entropy is high in an area where the stores of the three chains are located in a well-balanced distance and number; in the SPM model, the entropy is high along the trunk road highlighted in solid black line and in the area between several trunk roads because the main road used for the shortest route movement depends on the destination location and depends on the probability of visiting the store chain on the main road.

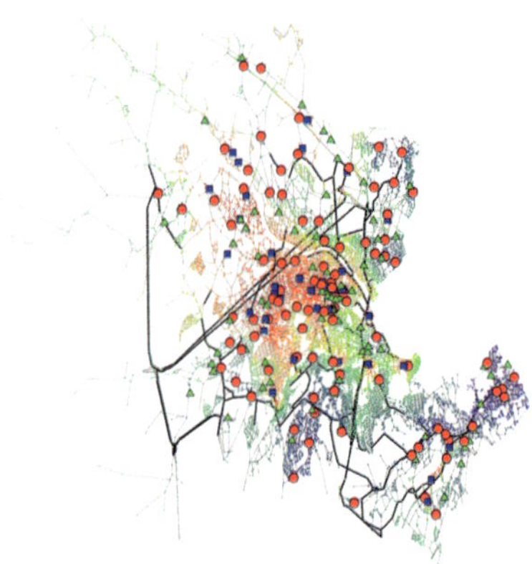

(a) NNM model

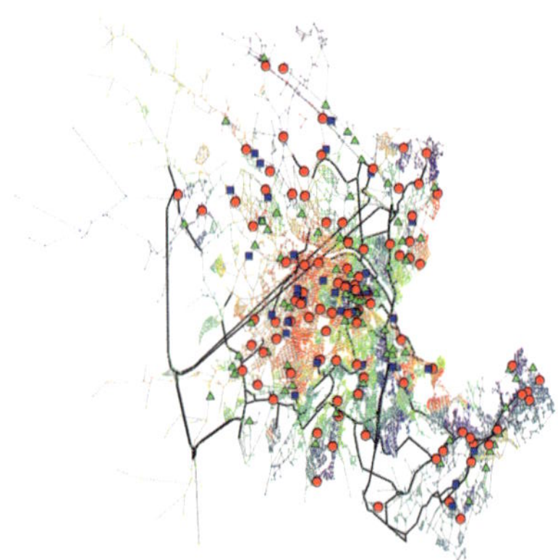

(b) SPM model

Fig. 6. Entropy of selection probability (Hachioji)

(a) NNM model (b) SPM model

Fig. 7. Entropy of selection probability (Shizuoka)

Thus, by analyzing the entropy of the selection probability, it is possible to grasp the degree of overlap of trade-areas or the degree of competition among chains. When opening a new store in an area with low entropy by many stores of the same chain, it would be undesirable to take away customers from the same chain. Conversely, by opening stores in areas with low entropy by many stores of different chains, customers from other chains can be taken away. In areas with high entropy, stores with multiple chains are open and competition is intense, so it is desirable to open stores to compensate for the power of the same chain.

6 Conclusion

Since the location of a store greatly affects the ability to attract customers and profits, it is important to properly grasp the trade-area of the store that has been installed when a new store is installed. In this study, we considered two models, a behavior model when a store is a destination, which we call the Nearest Neighbor Movement model, and a behavior model when a store is a transit point, which we call the Shortest Path Movement model. The former is based on the idea of closeness centrality, and corresponds to the traditional territory model or k-medoids clustering. The latter is a new model proposed in this paper and is based on the idea of betweenness centrality. From the experiments using the actual road network and the actual location information of convenience stores of Hachioji and Shizuoka, we confirmed that the existing model extracts the trade-areas of each store with good balance; in the proposed model, the trade-area of stores located along the main road becomes wider. By analyzing the entropy of selection probabilities, it was confirmed that the competitiveness of existing stores can be understood and can be used as an evaluation measure when determining candidate locations for new store openings. As future work, it is considered to introduce the population density of each junction node as a weight, and a high probability to the node that is likely to be the destination. In addition, by considering demographic attributes of residents, it leads to area marketing.

Acknowledgments. This material is based upon work supported by JSPS Early-Career Scientists (No.19K20417).

References

1. Agra, A., Cerdeira, J.O., Requejo, C.: A decomposition approach for the p-median problem on disconnected graphs. Comput. Oper. Res. **86**, 79–85 (2017)
2. Crucitti, P., Latora, V., Porta, S.: Centrality measures in spatial networks of urban streets. Phys. Rev. E **73**(3), 036125+ (2006)
3. Everett, M.G., Borgatti, S.P.: The centrality of groups and classes. J. Math. Sociol. **23**(3), 181–201 (1999)
4. Gimadi, E.K.: On exact solvability of the restricted capacitated facility location problem. In: Proceedings of the OPTIMA-2017 Conference, pp. 209–216 (2017)
5. Hagen, L., Kahng, A.B.: New spectral methods for ratio cut partitioning and clustering. IEEE Trans. Comput. Aided Des. Integr. Circuits Syst. **11**(9), 1074–1085 (1992). https://doi.org/10.1109/43.159993
6. Jinmei, W., Kejia, Z.: Study of facility location and allocation problem based on fuzzy graph theory. In: 2010 International Conference on Management and Service Science, pp. 1–5, August 2010
7. Montis, D.A., Barthelemy, M., Chessa, A., Vespignani, A.: The structure of interurban traffic: a weighted network analysis. Environ. Plan. **34**(5), 905–924 (2007)
8. Park, K., Yilmaz, A.: A social network analysis approach to analyze road networks. In: Proceedings of the ASPRS Annual Conference 2010 (2010)
9. Puerto, J., Ricca, F., Scozzari, A.: Extensive facility location problems on networks: an updated review. TOP **26**(2), 187–226 (2018). https://doi.org/10.1007/s11750-018-0476-5
10. Saito, K., Yamada, T., Kazama, K.: The k-Dense method to extract communities from complex networks. In: Zighed, D., Tsumoto, S., Ras, Z., Hacid, H. (eds.) Mining Complex Data, Studies in Computational Intelligence, vol. 165, pp. 243–257. Springer, Heidelberg (2009)
11. Seidman, S.B.: Network structure and minimum degree. Soc. Netw. **5**(3), 269–287 (1983)
12. Shi, J., Malik, J.: Normalized cuts and image segmentation. IEEE Trans. Pattern Anal. Mach. Intell. **22**(8), 888–905 (2000)
13. Tabata, K., Nakamura, A., Kudo, M.: An efficient approximate algorithm for the 1-median problem on a graph. IEICE Trans. Inf. Syst. **E100.D**(5), 994–1002 (2017). https://doi.org/10.1587/transinf.2016EDP7398
14. Tamir, A.: The k-centrum multi-facility location problem. Discret. Appl. Math. **109**, 293–307 (2001)

Optimisation of Signal Timings in a Road Network

Samadhi Nallaperuma[1(✉)], Shahin Jalili[1], Edward Keedwell[1], Alex Dawn[2], and Laurence Oakes-Ash[2]

[1] University of Exeter, Exeter, UK
s.n.nallaperuma@exeter.ac.uk
[2] City Science, Exeter, UK

Abstract. Road network simulation models and tools are increasingly being used for strategic and operational traffic management with the use of widely available online traffic data. The widespread use of such models raises the prospect of transport system optimisation, improving energy consumption, delays and carbon emissions. Although strategic interventions such as the building of new roads or infrastructure is costly and time-consuming, significant savings can be made through the modelling and optimisation of the operation of the network through signal timings.

Keywords: Infrastructure networks · Algorithms · Road networks

1 Introduction

With rapid economic development, fast growing populations, and an increasing number of vehicles in the urban areas, transportation networks are becoming increasingly congested. This results in wasted energy, increased delays, poor urban air quality, and increased carbon emissions. Due to the high costs of extending current transportation infrastructure [21] and space limitations in cities, it is often not a feasible choice to increase the capacity of the network by constructing new roads and traffic equipment. Hence, researchers have been focused on improving the efficiency of the current transportation networks through intelligent traffic control strategies.

It is well known that the performance of a transportation network is highly sensitive to the traffic lights' settings in various ways including mobility [25], the safety of pedestrians and vehicles [20] and environmental pollution [15]. One practical way to make a given transportation network more efficient is to apply a suitable signal control strategy. As an efficient approach for enhancing the capacity of current transportation networks, optimisation of signal timing parameters of traffic lights, such as cycle time [8,9,17], green time [3,15,29], offset time [2,15,17], and phase sequence [14,16,22], has received a great deal of attention in studies of transportation networks. From an optimisation perspective, the problem of optimum signal timings of a transportation network

H. Barbosa et al. (Eds.): *Complex Networks XI*, SPCOM, pp. 257–268, 2020.
https://doi.org/10.1007/978-3-030-40943-2_22

is highly nonlinear and non-convex, in which the modification of a timing plan of a given traffic light can significantly affect the traffic flow in the other areas of the network. The Webster method is one of the most efficient classic signal optimisation method, in which the delay time of a single junction is optimised to find a optimum cycle time formula [27]. However, this method only optimises the delay of a single junction and it does not consider the interactions of the traffic lights of multiple junctions. The precise layout and configuration of the network will determine the extent to which a globally optimised solution can outperform a locally optimised one. The number of junctions, their geographic proximity, traffic demand and the number of signal cycles will all determine the potential for interaction between signals.

To consider the interaction phenomenon of the traffic lights within the whole network, the traffic signal control problem of a transportation network can be formulated as a global optimisation problem. Classical optimisation algorithms, in which the calculation of the gradients of the objective function and constraints is inevitable, are not suitable approaches for the signal optimisation of network with numerous signalized intersections. As alternative approach for classical optimisation methods, meta-heuristic optimisation algorithms, such Genetic Algorithm (GA) [10], Particle Swarm Optimisation (PSO) [5], Ant Colony Optimization (ACO) [4], Differential Evolution (DE) [24] algorithm, and Harmony Search (HS) [7] algorithm, have been successfully applied to solve a wide range of complex highly nonlinear engineering optimisation problems. They can be easily implemented to solve an optimisation problem without gradient information about the objective function and constraints. In this paper, a meta-heuristic approach is proposed to solve the signal optimisation problem of a real-world transportation network. To measure the performance of the network, a set of objective functions are considered, including waiting time, fuel consumption, and vehicular emissions. To demonstrate the efficiency of the proposed approach, the signal timing plans of the traffic lights of the transportation network of a UK city is optimised and the obtained results are compared to those obtained from the classical Webster method. The numerical results verify the efficiency of the proposed approach to solve the signal optimisation of a real-world transportation network.

The organization of the rest of the paper is as follows. In Sect. 2, the signal optimisation of a transportation network is formulated by considering various objective functions. The modelling of a real-world transportation network of a city in United Kingdom (UK) is investigated in Sect. 3. The proposed meta-heuristic optimisation approach is discussed in Sect. 4. Experimental results of the optimisation is presented in Sect. 5. Finally, Sect. 6 concludes the paper.

2 Signal Optimisation Problem

The signal optimisation problem is an optimisation problem looking for the best signal timing plans for the traffic lights within the network in order to improve the performance of the network in terms of the waiting time, fuel consumption, number of stops, vehicular emissions, etc. In the literature there are various objectives considered for signal optimisation such as minimisation of waiting times [12,13,18,25,28], fuel consumption [2,16,19,23] and carbon emissions [2, 15,23,29]. These studies provide insights into how to improve the signal lights in a network based on the considered objective function. However, fewer studies have been carried out to understand the relative importance of different nodes of the network, different objectives and the effects from interactions between them in achieving overall network performance.

In the current study, a set of four objective functions are considered for the signal optimisation problem, which are the minimisation of carbon emissions, minimisation of total waiting times, minimisation of fuel consumption and the minimisation of the aggregated network performance. The signal optimisation problem for each of the mentioned objective functions can be formulated as follows:

- Waiting time minimisation:

$$Find : \mathbf{X} = [x_1, x_2, ..., x_n] \tag{1a}$$
$$Minimise : C_w(\mathbf{X}) = \Sigma_{j=0}^{m} W_{v_j} \tag{1b}$$

- Fuel consumption minimisation:

$$Find : \mathbf{X} = [x_1, x_2, ..., x_n] \tag{2a}$$
$$Minimise : C_f(\mathbf{X}) = \Sigma_{j=0}^{m} F_{v_j} \tag{2b}$$

- Emissions minimisation:

$$Find : \mathbf{X} = [x_1, x_2, ..., x_n] \tag{3a}$$
$$Minimise : C_e(\mathbf{X}) = \Sigma_{j=0}^{m} E_{v_j} \tag{3b}$$

- Performance optimisation:

$$Find : \mathbf{X} = [x_1, x_2, ..., x_n] \tag{4a}$$
$$Minimise : C_p(\mathbf{X}) = p\Sigma_{j=0}^{m} W_{v_j} + q\Sigma_{j=0}^{m} F_{v_j} + r\Sigma_{j=0}^{m} E_{v_j} \tag{4b}$$

In the above equations, $\mathbf{X}$ is a vector of decision variables containing the signal timing plans for the whole network, n is the number of signal parameters, $C_w(.)$, $C_f(.)$, and $C_e(.)$ represent the waiting time, fuel consumption, and emissions within the network, respectively, $C_p(.)$ indicates overall performance of the network, p, q and r represent the weighting coefficients for each of the objective functions, $\{v_1, v_2, ..., v_m\}$ represents the set of vehicles within the network and W_{v_j}, F_{v_j} and E_{v_j} represent waiting time, fuel consumption and emissions for vehicle v_j respectively. The weights for each objective can be adjusted based on user preference. The signal timing variables are the phase duration of traffic light states for each junction.

To consider the practical operation of traffic lights and safety regulations [1], the upper and lower bounds for the signal timing parameters are assumed as follows:

$$7\,\mathrm{s} \leq x_i \leq 60\,\mathrm{s}, i = 1, 2, ..., n \tag{5}$$

Other constraints based on the real-world conditions of the network could also be assumed, such as speed limits for vehicles, etc. However, in this study, we only assume the bounds of signal timing variables as the optimisation constraints.

3 Modelling a City Network with SUMO

In this study, the signal optimisation of a real-world transportation network of a city in the UK is investigated. (We call it "the benchmark city" and due to privacy regulations the actual name of the city is not mentioned in this study). The network is modeled by Simulation of Urban MObility (SUMO) software, which is a microscopic traffic simulation tool. Figure 1 shows the network configuration of the benchmark city. The open street map of the city is imported to the NETEDIT software, which is a graphical network editor for SUMO. The network has 9494 edges and 2207 nodes. In NETEDIT, the location of the traffic lights and their initial timings are defined.

For a weekday period 8:00AM–18:00PM, the realistic movements for cars, vans, lorries, and buses are used in the model, using the validated trip matrices. The Traffic Analysis Zones (TAZ) edges are used to describe sources (origins) and sinks (destinations) of trips. Pedestrian while not explicitly modelled are also considered as an all red traffic phase.

3.1 Inputs and Outputs of the Model

As it displayed in Fig. 1, the model includes 76 traffic lights with differing numbers of phases. The duration of each phase of traffic lights is treated as the inputs to the model. Hence, the number of decision variables of the optimisation model is 350. As it expressed by Eq. (5), the lower and upper bounds for these variables are set to 7 s and 60 s, respectively.

The outputs of the SUMO model are the waiting times, fuel consumption and carbon emissions for all the vehicles in the network as defined in Eqs. 1, 2, 3 respectively.

4 Evolutionary Optimisation Framework for a Road Network

The optimiser framework is built using the genetic algorithm library ParadisEO [11] which is available under an open source licence. For the Genetic Algorithm (GA) the population $P = \{X_1, X_2, ..., X_{j-1}, X_j, X_{j+1}, ..., X_\mu\}$ consists of

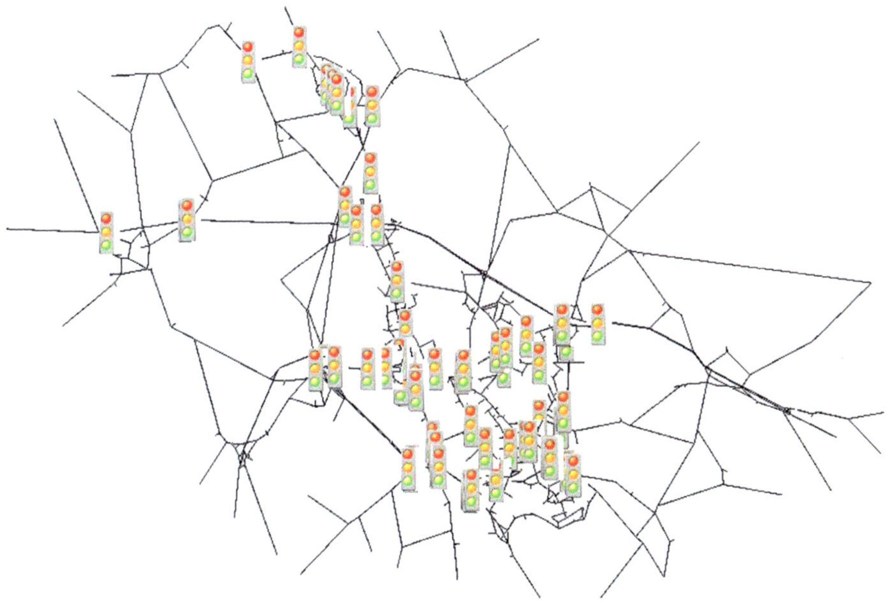

Fig. 1. Network of a benchmark city

individuals each representing a candidate solution. Such an individual is represented by a real valued vector $X_j = [x_1, x_2, ..., x_{i-1}, x_i, x_{i+1}, ..., x_n]$, where each GA gene [6] is represented by a real valued signal phase duration variable x_i. "Fitness" or the quality is evaluated through a fitness function which captures the objectives and the constraints discussed in Sect. 2. Algorithm 1 outlines the evolutionary optimisation process in general.

Algorithm 1. $(\mu + \lambda)$-EA: Evolutionary Algorithm

1) Initialise the population $P = \{X_1, X_2, ..., X_{j-1}, X_j, X_{j+1}, ..., X_\mu\}$ with μ traffic light individuals $X_j = [x_1, x_2, ..., x_{i-1}, x_i, x_{i+1}, ..., x_n]$, i.e. a vector of potential traffic light phase durations x_i.
2) Select $C \subseteq P$ where $|C| = \lambda$.
3) For each $I_1, I_2 \in C$, produce offspring I_1' I_2' by crossover and mutation. Add offspring's to P.
4) Fitness evaluation of all $I \in P$
5) Select $D \subseteq P$ where $|D| = \mu$.
6) $P := D$
7) Repeat step 2 to 6 until termination criterion is reached.

In step 1, we generate a population P with μ traffic light individuals X_js within the feasibility region defined by the bound 5. In step 2, we employ random selection to select parents to apply genetic operators. We apply genetic operators, uniform crossover and uniform mutation in step 3. In step 4, the fitness function invokes the SUMO simulation with the traffic light assignment represented by the GA individual X_j as the input and, retrieves waiting time, fuel and emissions data as the outputs at the end of the SUMO simulation as described in Sect. 3.1. In the aggregated performance case, these outputs are aggregated to one formulae for fitness evaluation as described in Eq. 4 in Sect. 2. In step 5, we use the fittest μ individuals as survivors.

5 Experiments

5.1 Impact from Signal Timings on Network Performance

For this initial experiment we consider a rather simpler set up. The goal of this experiment is to understand the effect of traffic light timings on network performance in several different aspects.

We consider simple (1+1) EA with $\mu = 1$ and $\lambda = 1$. The initial population is generated from a uniform distribution within the feasibility region as defined in the constraint 5. The algorithm is run for 500 generations, separately for each objective, namely waiting time (Eq. 1), fuel consumption (Eq. 2), carbon emissions (Eq. 3) and for the aggregated objectives with weighting of 1 (Eq. 4).

As shown in Fig. 2 during only 500 generations with the simple EA, the waiting time, fuel consumption, carbon emissions and overall network performance has been improved by 12%, 1%, 2% and 2% respectively. It is evident from these results that the performance of the road network can be improved in several aspects (waiting times, fuel consumption, carbon emissions) by changing the signal timings.

5.2 Relative Importance of Nodes

In order to identify the impact of each node (representing a junction) in the road network we conduct a set of experiments. In these experiments, we run the optimisation process changing the phase duration of each traffic light located at each junction separately and record the fitness achieved at the end of the process. Accordingly, in each algorithm run, the decision variables represent the phase durations of the traffic light for the specific junction defined by the id. The rest of the experimental set up is similar to Sect. 5.1. Figure 3 depicts these fitness values achieved by individual optimisation processes and Fig. 4 depicts the locations of the signal lights in the map where the fitness difference is represented by the area of the circle denoting the fitness gain for the specific signal lights. Its observed that few signal lights such as 68, 66, 455, 298, 299, 191 and 148 have significantly higher fitness difference implying significantly higher impact on the optimisation process than the rest of the signal lights. The majority of the lights appear to have medium impact falling into the medium range while less than a quarter such as 67, 441, 18, 187 and 233 falls into the low impact level. This figure illustrates two key aspects of the optimisation and problem,

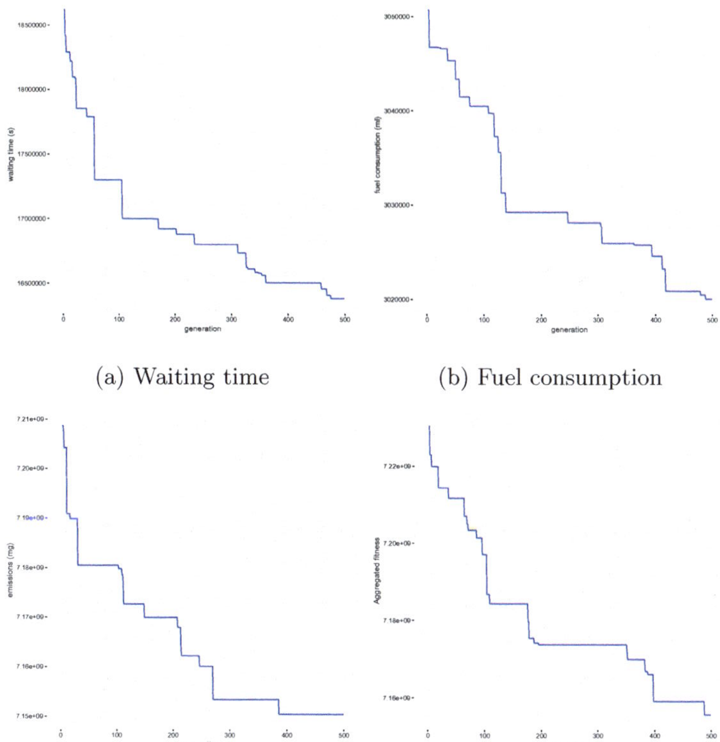

(a) Waiting time

(b) Fuel consumption

(c) Carbon emissions

(d) Overall network performance

Fig. 2. Fitness over generations

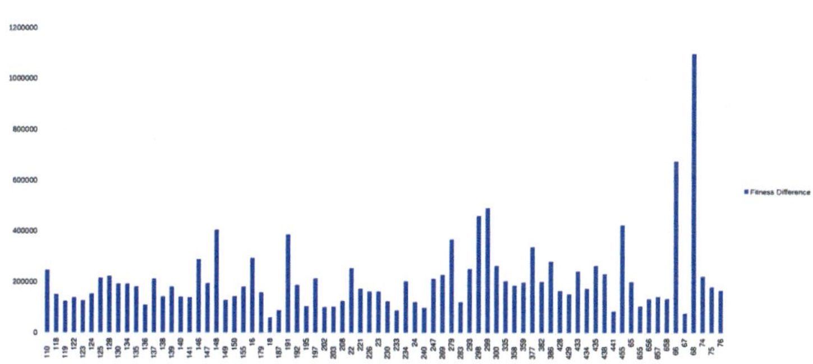

Fig. 3. Fitness difference for the considered 76 lights

firstly it allows the user to highlight those areas of the network that would most benefit from changes to their signal timings which in this instance correspond with major junctions as expected, and secondly can be used as a mechanism for the evolutionary algorithm to focus its effort on those junctions that deliver most benefit e.g. through a differential mutation rate for these variables. This view 'under the hood' of the algorithm, provides an evolutionary algorithm's eye view of the optimisation problem and is useful in communicating algorithm decisions to the user.

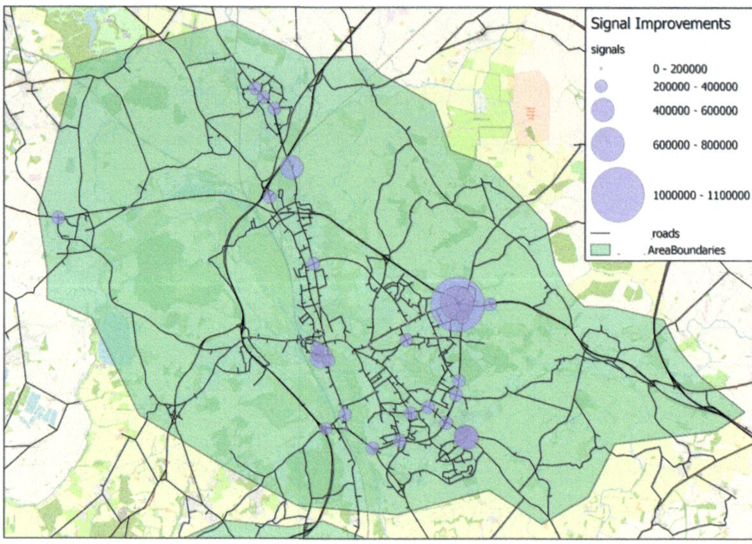

Fig. 4. Fitness difference for the considered 76 lights

5.3 Bench-Marking with Existing Controllers

The performance of the evolutionary algorithm is bench-marked against the Webster [27] and Green Wave methods, which are traditionally traffic signal optimisation techniques in the literature. For these experiments the GA settings are similar to Sect. 5.1 except we consider a larger population here, with a parent population size of $\mu = 10$ and offspring population size of $\lambda = 10$.

Webster. The Webster method is a classic signal timing method, which is based on minimizing traffic delay to calculate the timing plan [27]. For a given intersection, the Webster method calculates the optimal cycle time from the following equation:

$$c_0 = \frac{1.5L + 5}{1 - Y}$$

where Y is the sum of the y values and refers to the intersection as a whole and L is the total lost time per cycle in seconds. The y values indicate the flow to saturation flow ratios for different lanes of the intersections.

Hybrid Greenwave-Webster. Green wave control is another classical technique to regulate traffic signal of urban artery. The control effect is obvious, and the realization is simple. The core of control is to make vehicles successively come across intersections on the artery as many as possible, which can decrease the average number of stops and average delay time of vehicles. Based on the SUMO instructions, the green wave method can only be applicable to the traffic lights with the same cycle times. On the other hand, the researchers have reported that the network delay time is not significantly increased by changing the cycle times obtained by the Webster method within the interval $(0.75C_{opt}, 1.5C_{opt})$ [26], where C_{opt} is the optimum cycle length for a given traffic light. In order to apply the green wave method, we have changed the cycle times of all traffic lights yielded by the Webster method within the mentioned interval and it is assumed that the cycle time of all traffic lights are equal to 27 s.

Results. To initialise the GA we use some Webster solutions, where we use a subset of candidate solutions/individuals randomly generated from a uniform distribution within feasibility bounds similar to previous experiments and the rest of solutions optimised by Webster. It is observed that the GA optimises the solutions over the algorithm run and that the waiting time is improved by 11% with the GA compared to Webster (see Fig. 5). With Webster, local optimisation of each single junction is conducted and this does not consider the interconnections of the nodes of the network. The results suggest that for optimisation of signal timings, the inter-connectivity of the nodes in road network seems to play an important role.

These results further highlight the importance of a good initial population. The suboptimal solutions of Webster provides a good starting point to GA. This is evident when comparing the fitness from the GA with random initialisation versus GA with heuristic initialisation from Webster in Figs. 2a and Fig. 5 respectively. It shows that during 500 generations, the GA with Webster initialisation could achieve fitness over 300% better than the GA with random initialisation. Nevertheless, its theoretically possible for a population based GA to achieve global optimum with high probability given a very large running time. Thus a good starting point can only help in reducing the time a GA takes to reach the optimum.

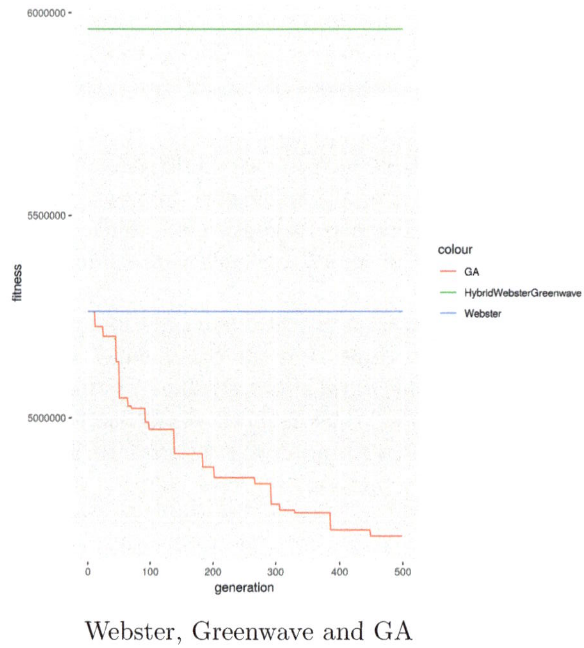

Fig. 5. Waiting time over generations for Webster, Greenwave and GA

6 Conclusions and Future Work

The results of modelling and optimisation of a real world road/city network using a hybrid Webster-GA are presented. Several objectives related to the performance of a road network are considered individually and in combination. Experimental results show that there is an effect from signal timings on road network performance from several different aspects namely, waiting timings, fuel consumption and carbon emissions. Our approach based on GA was benchmarked with classical methods and results show that the GA improved performance by 11%. This is because of the global optimisation of GA considering the network as a whole compared to the other methods considering the nodes separately. The experimental results on relative importance of the nodes further show that certain areas within the road network are more crucial than the others when determining the overall performance of the network. Future work will concentrate on further investigations on the relative importance and inter-dependencies among the different nodes/node clusters of a road network as well as relative importance of the different objectives through multi-objective optimisation.

Another consideration is on how traffic reacts and reroutes according to the improvements made, behaviour changes of the vehicles have the potential to reduce the level of benefit seen. Traffic assignment is computationally costly, but co-evolving traffic routing alongside improvement of will yield solutions that account for changes to traffic routing.

References

1. Specification for traffic signal controller. Technical report, Highways Agency (2005)
2. Armas, R., Aguirre, H., Daolio, F., Tanaka, K.: Evolutionary design optimization of traffic signals applied to Quito city. PLoS ONE **12**(12), 1–37 (2017)
3. Costa, B.C., Leal, S.S., Almeida, P.E., Carrano, E.G.: Fixed-time traffic signal optimization using a multi-objective evolutionary algorithm and microsimulation of urban networks. Trans. Inst. Meas. Control **40**(4), 1092–1101 (2018)
4. Dorigo, M., Birattari, M.: Ant Colony Optimization. Springer, Heidelberg (2010)
5. Eberhart, R., Kennedy, J.: A new optimizer using particle swarm theory. In: MHS 1995. Proceedings of the Sixth International Symposium on Micro Machine and Human Science, pp. 39–43. IEEE (1995)
6. Eiben, A.E., Smith, J.E.: Introduction to Evolutionary Computing. Springer, Heidelberg (2003)
7. Geem, Z.W., Kim, J.H., Loganathan, G.V.: A new heuristic optimization algorithm: harmony search. Simulation **76**(2), 60–68 (2001)
8. Ghanim, M.S., Abu-Lebdeh, G.: Real-time dynamic transit signal priority optimization for coordinated traffic networks using genetic algorithms and artificial neural networks. J. Intell. Transp. Syst. **19**(4), 327–338 (2015)
9. He, J., Hou, Z.: Ant colony algorithm for traffic signal timing optimization. Adv. Eng. Softw. **43**(1), 14–18 (2012)
10. Holland, J.H.: Genetic algorithms. Sci. Am. **267**(1), 66–73 (1992)
11. Humeau, J., Liefooghe, A., Talbi, E.-G., Verel, S.: ParadisEO-MO: From fitness landscape analysis to efficient local search algorithms (RR-7871) (2013)
12. Kai, Z., Gong, Y.J., Zhang, J.: Real-time traffic signal control with dynamic evolutionary computation. In: 2014 IIAI 3rd International Conference on Advanced Applied Informatics, pp. 493–498. IEEE (2014)
13. Kesur, K.B.: Advances in genetic algorithm optimization of traffic signals. J. Transp. Eng. **135**(4), 160–173 (2009)
14. Kesur, K.B.: Multiobjective optimization of delay and stops in traffic signal networks. In: Metaheuristics in Water Geotechnical and Transport Engineering, pp. 385–416 (2013)
15. Kou, W., Chen, X., Yu, L., Gong, H.: Multiobjective optimization model of intersection signal timing considering emissions based on field data: a case study of beijing. J. Air Waste Manag. Assoc. **68**(8), 836–848 (2018)
16. Kwak, J., Park, B., Lee, J.: Evaluating the impacts of urban corridor traffic signal optimization on vehicle emissions and fuel consumption. Transp. Plan. Technol. **35**(2), 145–160 (2012)
17. Li, X., Sun, J.-Q.: Signal multiobjective optimization for urban traffic network. IEEE Trans. Intell. Transp. Syst. **19**(11), 3529–3537 (2018)
18. Li, Z., Schonfeld, P.: Hybrid simulated annealing and genetic algorithm for optimizing arterial signal timings under oversaturated traffic conditions. J. Adv. Transp. **49**(1), 153–170 (2015)
19. Olivera, A.C., García-Nieto, J.M., Alba, E.: Reducing vehicle emissions and fuel consumption in the city by using particle swarm optimization. Appl. Intell. **42**(3), 389–405 (2015)
20. Roshandeh, A.M., Li, Z., Zhang, S., Levinson, H.S., Lu, X.: Vehicle and pedestrian safety impacts of signal timing optimization in a dense urban street network. J. Traff. Transp. Eng. (Engl. Ed.) **3**(1), 16–27 (2016)

21. Schroten, A., van Wijngaarden, L., Brambilla, M., Gatto, M., Maffii, S., Trosky, F., Kramer, H., Monden, R., Bertschmann, D., Killer, M., Lambla, V., Beyrouty, K.E., Amaral, S.: Overview of transport infrastructure expenditures and costs. Technical report, Directorate-General for Mobility and Transport, European Commission (2019)
22. Stevanovic, A., Stevanovic, J., Kergaye, C.: Optimizing signal timings to improve safety of signalized arterials. In: 3rd International Conference on Road Safety and Simulation, Indianapolis, USA, vol. 14 (2011)
23. Stevanović, A., Stevanovic, J., Zhang, K.X., Batterman, S.: Optimizing traffic control to reduce fuel consumption and vehicular emissions: integrated approach with VISSIM, CMEM, and VISGAOST (2009)
24. Storn, R., Price, K.: Differential evolution-a simple and efficient heuristic for global optimization over continuous spaces. J. Global Optim. **11**(4), 341–359 (1997)
25. Tan, M.K., Chuo, H.S.E., Chin, R.K.Y., Yeo, K.B., Teo, K.T.K.: Optimization of traffic network signal timing using decentralized genetic algorithm. In: 2017 IEEE 2nd International Conference on Automatic Control and Intelligent Systems (I2CACIS), pp. 62–67. IEEE (2017)
26. Wang, C.: Analysis and visualization of traffic signal performances. Master's thesis (2016)
27. Webster, F.V.: Traffic signal settings. Technical report (1958)
28. Wu, B., Wang, D.: Traffic signal networks control optimize with PSO algorithm. In: 2016 12th International Conference on Natural Computation, Fuzzy Systems and Knowledge Discovery (ICNC-FSKD), pp. 230–234. IEEE (2016)
29. Zhou, Z., Cai, M.: Intersection signal control multi-objective optimization based on genetic algorithm. J. Traff. Transp. Eng. (Engl. Ed.) **1**(2), 153–158 (2014)

Gender Patterns of Human Mobility in Colombia: Reexamining Ravenstein's Laws of Migration

Mariana Macedo[1(✉)], Laura Lotero[2], Alessio Cardillo[3,4,5], Hugo Barbosa[1], and Ronaldo Menezes[1]

[1] BioComplex Lab, Department of Computer Science, University of Exeter, Exeter, UK
mg615@exeter.ac.uk
[2] Faculty of Industrial Engineering, Universidad Pontificia Bolivariana, Medellín, Colombia
[3] Department of Engineering Mathematics, University of Bristol, Bristol, UK
[4] Department of Computer Science and Mathematics, University Rovira i Virgili, Tarragona, Spain
[5] GOTHAM Lab, Institute for Biocomputation and Physics of Complex Systems, University of Zaragoza, Zaragoza, Spain

Abstract. Public stakeholders implement several policies and regulations to tackle gender gaps, fostering the change in the cultural constructs associated with gender. One way to quantify if such changes elicit gender equality is by studying mobility. In this work, we study the daily mobility patterns of women and men occurring in Medellín (Colombia) in two years: 2005 and 2017. Specifically, we focus on the spatiotemporal differences in the travels and find that *purpose of travel* and *occupation* characterise each gender differently. We show that women tend to make shorter trips, corroborating Ravenstein's Laws of Migration. Our results indicate that urban mobility in Colombia seems to behave in agreement with the "archetypal" case studied by Ravenstein.

Keywords: Gender gap · Ravenstein's laws of migration · Urban mobility · Networks

1 Introduction

Our daily lives are shaped by the convolution of a broad range of individual and social-level demands (*e.g.*, eat, sleep, work, pay bills), and mobility is essential for their fulfilment. Hence, the betterment of our lives passes through the study of how people move [2,17]. Some models on mobility assume that all travellers are – more or less – the same, disregarding the wealth of attributes discriminating one social group from another. Conversely, other studies demonstrate that social demographic attributes like socio-economic status do play a role in

H. Barbosa et al. (Eds.): *Complex Networks XI*, SPCOM, pp. 269–281, 2020.
https://doi.org/10.1007/978-3-030-40943-2_23

mobility [9]. Amidst the plethora of attributes available, gender is a key one because men and women can emerge alternative behaviours [11]. Recently, the gap existing between men and women has become the focus of many studies (*e.g.,* on urban mobility and academic performance) [5,6,13,16]. Despite the efforts made to improve gender equality [4], both gender's routines remain being affected differently [6]. Understanding such differences is crucial to build a better "environment" for everyone [5,8,18], and design interventions on transportation aimed at reducing gender gaps to offer the same mobility opportunities [12,16].

In 1885, Ravenstein published a paper entitled "*The Laws of Migration*" [14] where he highlighted differences between women's and men's mobility arguing that, on average, women migrate more than men. According to Ravenstein, women were more likely to visit areas nearby their "homes," mainly with the purpose of seeking for jobs. Accordingly, women mainly migrated to residential and job rich areas (*e.g.,* industrial). However, society has undergone deep transformation since Ravenstein's study.

In this work, we study the gender-based spatiotemporal differences in urban mobility taking place within the Medellín's metropolitan area (known as Aburrá Valley) in Colombia in two distinct years, 2005 and 2017. Using an approach similar to the one used in [9,10], we find that despite more than one century has passed since Ravenstein's study, women still make shorter trips – remaining closer to their "homes," – and that employment continues to play a considerable role in mobility. Furthermore, our results are in agreement with those of a recent study based on Chilean mobile phone data by Gauvin *et al.* [6] which found, among other things, that men tend to visit more diverse places than women.

2 Dataset

We consider the data collected by two surveys on urban mobility made within the Aburrá Valley's metropolitan area surrounding the city of Medellín (Colombia). Each survey accounts for a distinct year, namely 2005 and 2017. Each interviewed householder is asked about the travels she/he had the day before the interview, providing with the origin and destination zones, the departure and arrival times, the transportation mode used, and the purpose of each travel. The surveys collect only information about travels associated with people's routines. Thus, travels related to a specific time of year are not included. In addition, householders are characterised by their socio-demographic characteristics (age, gender, and occupation) which define their *socioeconomic status* (SES). Each *travel* is divided into one or more *trips* each corresponding to a displacement made with a specific transportation mode. For example, if one traveller went from zone A to zone B walking during the first part and then took a bus, the corresponding travel is made of two trips. Table 1 summarises the main features of each survey such as: the number of travellers (N_P), the number of travels (N_T), and their composition in terms of gender.

Leveraging the meta information available, we can divide the dataset into several subsets based, *e.g.,* on individual occupation or travel's purpose. In Fig. 1, we display the percentage of travels made by each gender grouped

Table 1. Datasets' main properties. For each year, we report the number of zones N_Z, the total area covered, the number of travellers N_P, the fraction of men (women) travellers f^M (f^W), the number of travels N_T, and the fraction of travels made by men (women) f_T^M (f_T^W).

Year	N_Z	Total surface (km^2)	N_P	f^M	f^W	N_T	f_T^M	f_T^W
2005	403	1,043	55,681	0.5143	0.4854	126,164	0.5163	0.4833
2017	521	1,174	30,107	0.5418	0.4582	64,837	0.5434	0.4566

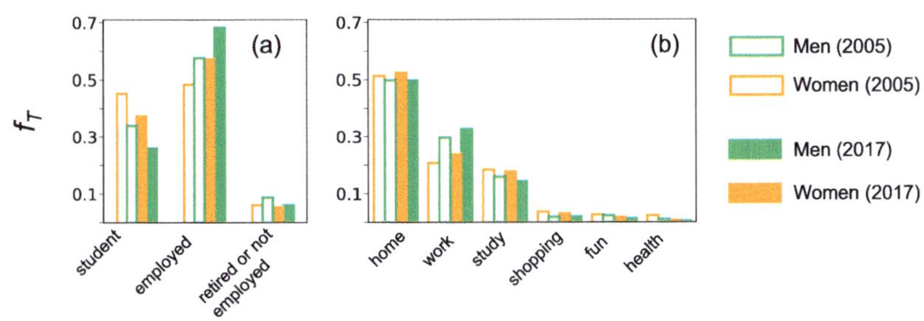

Fig. 1. Fraction of travels made (f_T) by each gender grouped either by occupation (panel (a)), or travel's purpose (panel (b)). The empty bars accounts for data coming from 2005's survey, whereas the filled bars accounts the 2017 case, instead.

either according to the traveller's occupation (panel a), or to travel's purpose (panel b). It is worth noting that the categories displayed in Fig. 1 correspond to 100% of 2005's data, and to 90.36% for occupations and to 92.10% for purpose, in 2017, instead. In 2017, the dataset contains other occupations and trips' purpose such as `housewives` and `give a ride to someone` which are absent in 2005's dataset. According to Fig. 1(a), the majority of our samples is made of students and workers, with more women in the student's group and men in the worker's group (the worker's group showing a rise between 2005 and 2017). The predominance of `students` and `employed` travellers reverberates on the frequency of travel's purposes (Fig. 1(b)), with `work` and `study` being the second and third most frequent classes, preserving the imbalance between genders in both years. Other travel's purposes seem quite underrepresented, probably due to their less periodic nature. However, on average, the fraction of travels made by women due to `other` purposes is higher than the men counterpart. Despite the gender imbalance across occupations and purposes of travels, we show evidences in the following sections that gender can play a role in mobility not because of the abundance of travellers in a class but, rather, because women and men behave differently in our sample (*i.e.*, have distinct routines).

3 Network Description

Travels occur between zones and the Aburrá Valley is divided into N_Z zones. However, there are differences between the zones partitioning in 2005 and 2017, with 2017 displaying a more granular structure due to the growth of the city and its metropolitan area. Since the vast majority of 2017's zones are subsets of 2005 ones, to ensure compatibility between the results, we consider the 2005 zone structure on both surveys. Nevertheless, it is worth noting that the use of 2005 partitioning on 2017 data does not alter significantly the overall behaviour of the distributions of travel's distance, number of locations visited, and number of transport modes used per travel.

The mobility data can be mapped into a weighted spatial network where zones are the nodes, and travels between zones represent the edges [3]. Each edge can be associated with two attributes (weights): the distance between the origin and destination zones, and the number of travels made between those two zones. For each year, we consider initially three different networks: one accounting for the whole mobility, and other two accounting for men and women travels only. In Fig. 2, we present an example of such networks where we display the flow of travels made during certain hours of the day.

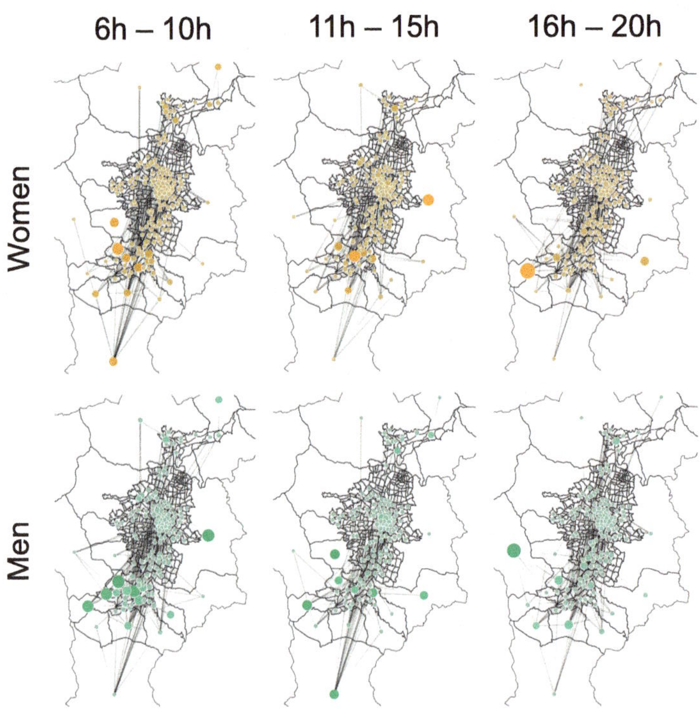

Fig. 2. Network representation of gender mobility *flows* (*i.e.*, number of travels) occurring during the morning (left column), midday (central column), and afternoon (right column). The nodes' size accounts for the in-strength of a zone (*i.e.*, the sum of the weights of all edges entering in a zone), while the edge thickness and colour accounts for the number of travels made between two zones. Data refers to the 2017's survey.

The men's flow is higher than its women counterpart in 69.96% of the zones. The set of departure and arrival zones visited by men and women are statistically different according to the Kolmogorov Smirnov (KS) test with a confidence level of 95% (computed on the set of fractions of travels made to each zone) and p-values of $6.07 \cdot 10^{-14}$ and $7.87 \cdot 10^{-14}$, respectively. If we differentiate travels according to their purpose, the differences between the arrival zones of men and women are higher for employed people and for `work` purpose (KS with p-values of $1.53 \cdot 10^{-90}$ and $8.55 \cdot 10^{-23}$, respectively). The differences between the departure zones are even higher when we consider the `home` travel's purpose (KS with p-value of $2.32 \cdot 10^{-12}$).

4 Spatial Characterisation of Travels

4.1 Analysis of Travel's Distance

One of Ravenstein migration's laws postulates that women tend to make short travels. Here, we check whether this is the case also for Medellín's urban area. We compute the length, x, of each travel as the distance between the centroids of its origin and destination zones; only distances greater than 100 m are considered to avoid underestimated displacements. Moreover, we do not account for travels made within the same zone because we cannot estimate their displacements. After that, we compute the complementary cumulative distribution function (CCDF) quantifying the probability that a travel has length longer than x, $P_{>}(x)$ (Fig. 3). Looking at the CCDF, we do not observe any remarkable differences between them (confirmed also by a KS test). However, the averages of the travel's distances, $\langle x \rangle$, displayed in Table 2, indicate that – on average – men tend to perform longer travels than women. We validate such claim with a t-test which returns p-values of $1.19 \cdot 10^{-8}$ for 2005 and $3.48 \cdot 10^{-11}$ for 2017, instead. When

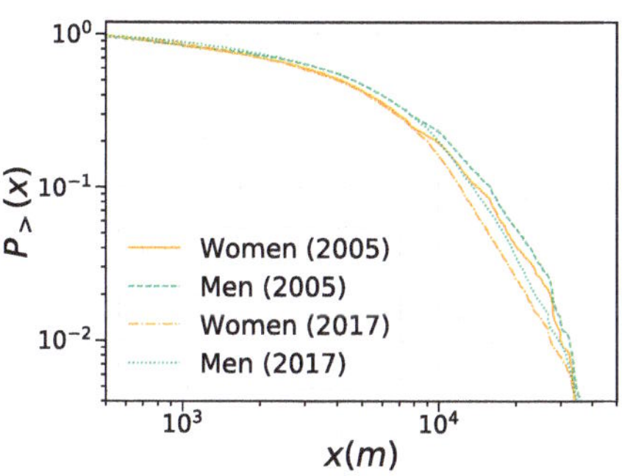

Fig. 3. Complementary cumulative distribution function, $P_{>}(x)$, of the probability of making a travel with a distance between the origin and destination zones equal to x.

Table 2. Average values of travel's distance, $\langle x \rangle$ (m), for travels made by men (M) and women (W) in each year.

Year	Gender	$\langle x \rangle$ (m)
2005	M	6711.42
	W	6031.75
2017	M	6319.32
	W	5542.41

we consider also travels of less than 100 m, the differences between the genders are amplified.

The gender asymmetry between short and long range travels is preserved also when we discriminate them according to either travellers' occupation or travel's purpose (results not shown). If we further divide mobility according to departure time, we find that men are more prone to make short travels between 23 h and 4 h. Such difference might be related with the fact that women in Colombia feel more insecure to move at late night and early morning [7].

4.2 Spatial Coverage

Another relevant aspect is how travels sprawl in space. Overall, we observe that men tend to visit more distinct (unique) zones than women. Such differences can be quantified by computing the Shannon entropy, S^X, of the sequence of zones visited by each person with gender $X = \{M, W\}$ [15] which, up to a multiplicative factor, reads as:

$$S^X = -\sum_{i=1}^{N_Z} p^X(i)\,\log_2 p^X(i). \tag{1}$$

where $p^X(i)$ is the probability that zone i is visited by a travel made by travellers with gender X, which is:

$$p^X(i) = \frac{{N_T}^X(i)}{{N_T}^X}, \tag{2}$$

where ${N_T}^X$ is the total number of travels made by gender X, and ${N_T}^X(i)$ is the number of those that visit zone i. For each year, we compute the entropy of all travels made regardless of traveller's gender (S), of travels made by men (S^M), and by women (S^W). Then, we compute the entropy difference $\Delta S^X = S^X - S$ and study its sign. Table 3 displays the values of S, S^X, and ΔS^X for both genders and years.

Table 3. Entropy of travels made by all travellers, S, or by men, S^M and women, S^W, only. We display also the differences between the entropy of gender X and of the whole population, ΔS^X. Each row accounts for a different survey (year).

Year	S	S^M	S^W	ΔS^M	ΔS^W
2005	7.7761	7.7657	7.7736	−0.0024	−0.0103
2017	6.7359	6.7444	6.7209	0.0085	−0.0149

Given the values displayed in Table 3, in 2017, men displacements appears to be slightly more "explorative" (*i.e.*, more entropic) than women's ones. In fact, women tend to return 2–4% more frequently to zones closer to their origins.

The women's tendency to return to their origin zones is stronger in 2017 than in 2005. In agreement with such trend, if we account also for the travels made within the same zone, we observe 5% more self loops in the women's network than in the men's one. In 2005, instead, men tend to be slightly less entropic than women, albeit such difference is not very high. The analysis of entropy alone is not entirely conclusive but, based on other evidences, we argue that in our case study men's mobility seems to be more explorative than women's one.

4.3 Transportation Multimodality

There is a difference between men and women in the usage of certain transportation modes. For example, men tend to use the car twice much than women, and men reach *directly* their destination more often than women. However, to reach its destination, one might need to use more than a single transportation mode. Here, we quantify the transportation multimodality of each gender. Figure 4 reports the histograms of the fraction of travels made of n trips for occupations and travel's purposes. A quick glance at the histograms reveals that the bulk of travels ($\gtrsim$60%) is made by a single trip, regardless of traveller's occupation, purpose, and gender. Another feature highlighted by the histograms is that men are more inclined to reach their destinations using a single transportation mode, whereas women are more likely to use between 3 and 5 transportation modes than men. For example, in 2017, 69% of `employed` men used one transportation mode (mainly, motorcycle), while only 63% of women did the same (mainly, walking).

Occupation wise, `students` and `retired/unemployed` people reach their destinations mostly directly, whereas `employed` tend to use more than one transportation mode. Travel's purposes display trends similar to those observed for occupations, with `shopping`, `fun`, and – to some extent – `study` appearing more "direct", while the others (especially `work` and `health`) tend to involve multiple trips. Finally, the comparison between travels made in 2005 and 2017 highlights the presence of "longer" travels – in terms of number of trips, – in 2017, suggesting that people have become more multimodal in their displacements. Also, 2017 data display higher gender asymmetry both in terms of number of travels (Table 1) and trips. The asymmetry is bigger for travels made of three trips regardless of whether we select them according to occupation/purpose or consider their aggregate form. Finally, the 2017 survey contemplates four additional travel's purposes: `lunch`, `bureaucratic activities`, `accompany someone`, and `give a ride to someone`. In general, men have lunch more at home than women, and women have a higher amount of travels to perform bureaucratic activities and accompany/pick up someone. However, in the next section, we show that there are also temporal differences in the mobility patterns of men and women.

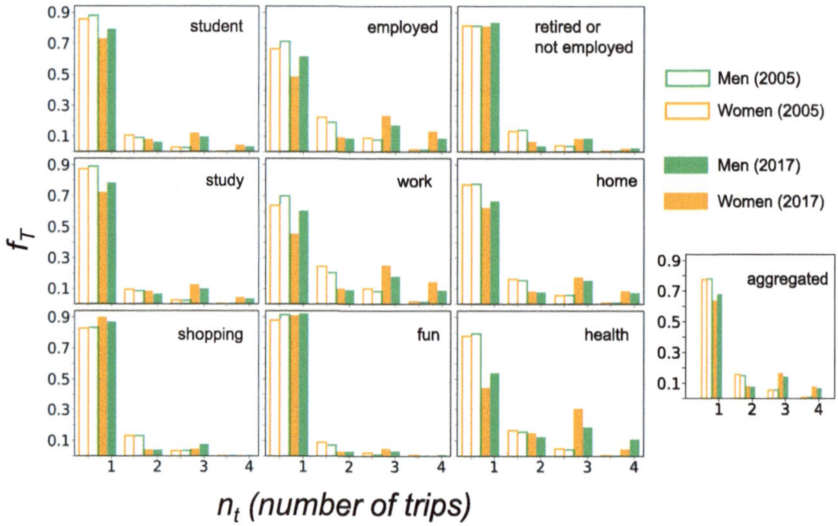

Fig. 4. Histograms of the fraction of travels made (f_T) by n_t trips for several occupations and travel's purposes. The rightmost histogram (i.e., **aggregated**) accounts for the the data aggregated altogether. Empty and filled bars refer to data from the 2005's and 2017's survey, respectively. We use travels made of at most four trips corresponding to the 97.32% and 99.92% of the whole dataset for 2005 and 2017, respectively.

5 Temporal Characterisation of Travels

Travels take place in space but also in time, and their bursty, synchronised, nature is responsible for the emergence of phenomena like traffic jams [7]. After studying the spatial properties of urban mobility, we focus on the temporal perspective. In particular, we study how women/men travel's purposes reverberate on the "rythms" with which travels take place. Figure 5 portrays the evolution in time of travels made with different purposes. For each purpose, we plot the fraction of travels departing at a given time (line plots), as well as the difference between the fraction of travels made by men and women (bar plots). In this way, we capture both the evolution across time of travels (and eventual longitudinal shifts between genders), as well as any eventual gender-based difference.

Eyeballing at Fig. 5 reveals that each travel purpose has a distinct temporal footprint, with its gender components displaying further differences. Some purposes (home, work, and study) exhibit one – or more – clear peak of "activity" along the day. For example, returning home occurs mainly around lunchtime and at the end of the afternoon regardless of the gender. Women tend to leave to go to work about one hour later than men, instead. Other less routinely purposes (shopping, fun, and health), instead, display a temporal profile more diluted along the day. Another feature is that for home, work, and shopping purposes,

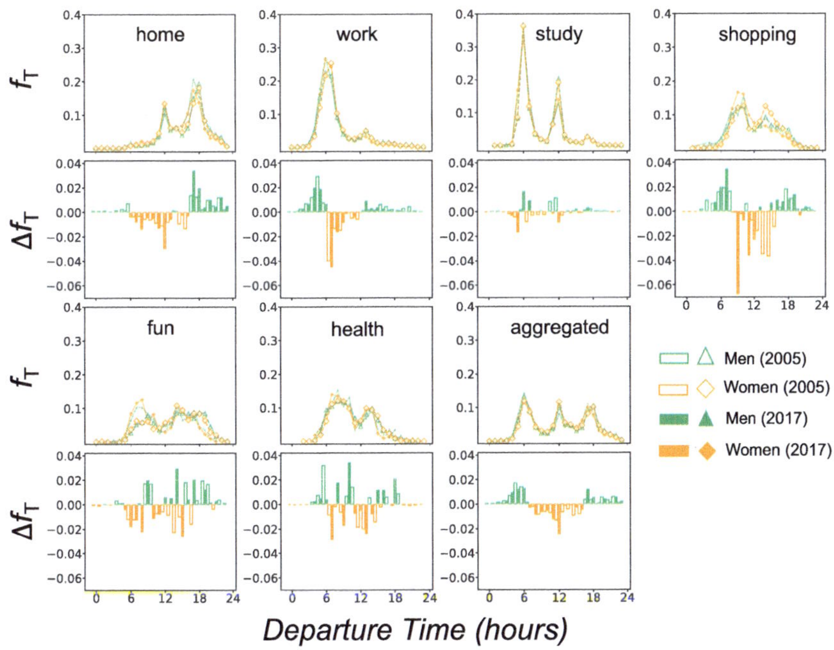

Fig. 5. Fraction of travels made f_T (line plots), and their gender differences Δf_T (bar plots), of several travel's purposes and their `aggregated` case with respect to the departure time (in hours). Empty (filled) symbols refer to data from year 2005 (2017).

the difference plots highlight the predominance (in proportion) of men's travels during the early morning/late evening, whereas women tend to travel more than men during the middle of the day. Curiously, we observe that the predominance of female travels associated with `shopping` occurring during the middle part of the day shifts backward from the second half of the day in 2005, to the first half in 2017. The clearly split gender pattern observed along the day is not present for purposes like `study`, `fun`, and `health`, where the middle part of the day is characterised by the lack of prevalence among genders, albeit the `health` case displays its own unique pattern. `Study` appears to be the most gender balanced purpose, as denoted by the small values of fractions differences.

The aggregated data display, instead, three peaks located in the morning, midday, and late afternoon exhibiting also a synchronisation of both genders at midday. Men still tend to travel more (in proportion) than women during the morning/evening while women's travels are more prominent during the middle part of the day. Occupation wise, unemployed and retired people distribute their travels along the whole day, with women travelling more often in the interval 12 h–18 h, and men in the interval 06 h–12 h, instead. However, the convolution of the temporal curves into the aggregated one smooths out many of the temporal footprints observed, suggesting that gender plays a more prominent role when mobility is studied in terms of people's need/demands. We conclude by noting that we could also make a characterisation in terms of the arrival times. However, such analysis does not highlight any additional feature of mobility.

Finally, we take a step further by presenting a spatiotemporal characterisation of mobility according to two specific travel's purposes recorded exclusively during the 2017's survey: `accompany someone` and `give a ride to someone`. Such purposes constitutes benchmarks to highlight gender differences because they presume the use of transportation modes capable to carry people to their destination (like cars and motorcycles). As we have said, men have more access than women to such modes. Moreover, the act of accompany/give a ride to someone implies a social relationship between the carrier and the carried.

Figure 6 panels (a) and (d) display the evolution in time of the fraction of travels associated with the aforementioned purposes, according to the departure and arrival time of each travel, with men and women curves displaying a longitudinal shift only for the `accompany someone` case. In panels (b) and (e), instead, we report the gender-based differences between the curves displayed in panels a and d. We notice that, in the surroundings of lunchtime women dominate (in proportion) for `giving a ride` purpose. However, in general, the temporal footprints do not display any feature remarkably different from those appearing in Fig. 5. In panels (c) and (f), instead, we display the average of the fraction of all the travels made with a given scope arriving at a given time. The average is computed over the arrival zones. We observe that, for both purposes, men perform more travels (in proportion) than women regardless of the arrival time.

If we perform a spatio-temporal analysis similar to the panels (c) and (f) of Fig. 6 for all the travel's purposes displayed in Fig. 5, we observe that – in both years – women perform (in proportion) more trips from/to `work` than men.

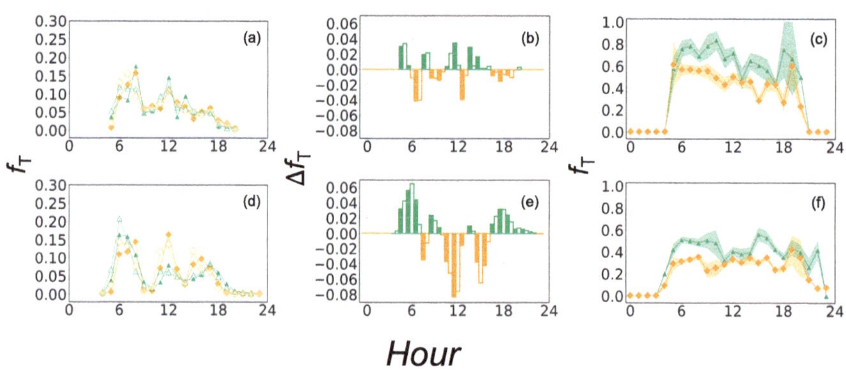

Fig. 6. Spatiotemporal features of the travels made in 2017's survey to `accompany someone` (top row) or `give a ride to someone` (bottom row). Panels (a) and (d) display the fraction of travels (f_T) departing (empty symbols) and arriving (filled symbols) at a given hour. Panels (b) and (e) report the difference (Δf_T) between the men and women curves displayed in panels (a) and (d). Panels (c) and (f) report the average fraction of travels arriving at a given time averaged over all the available zones (the shaded area denotes the standard deviation over different arrival zones).

On the other hand, men tend to perform more trips over the day to the same zones on the purposes of travels: `health` and `shopping`. The majority of women return `home` (at same zone) from 14–16 h in both years. In general, men tend to move (within zones) for `fun` purpose at late night, while women tend to move for the same purpose from 10–17 h, instead. `Study` remains the most gender neutral travel's purpose.

6 Conclusions

In this study, we have analysed the data collected by two surveys on mobility within the Aburrá Valley's metropolitan area surrounding the city of Medellín (Colombia) to quantify how gender reverberates on urban mobility. Using the network paradigm, we have built networks of travels occurring between distinct zones. By leveraging the wealth of meta-information coming along with the data, we have been able to disentangle not only the contributions associated with each gender, but also how distinct job occupation and demands mold the spatio-temporal features that mobility exhibits.

By analysing the spatial properties of travels, we have found that – on average – women prefer to move within or in the proximity of their "home" zones, display a more recurrent mobility pattern, and use more transportation modes (mainly walk and bus) than men. Men tend to display a more exploratory behaviour, with a higher number of unique locations visited, more direct travels (mainly made by car or motorbike), and a mobility predominantly characterised by work-related demands (being employed or go to work). The characterisation of the travel's temporal features, instead, unveiled unexpected features induced by gender. The analysis of the percentage of travels made along the day, grouped according to either occupation or purpose, highlights that each class of travels displays a specific temporal pattern, which is different from its aggregated counterpart [9]. Such phenomenon constitutes one of the hallmarks of complex systems, which is usually encapsulated by the statement "*more is different*" [1]. Specifically, we observe that – on average, – men tend to travel more during the morning/evening, while women, instead, move more during the middle of the day. Such behaviour could be due, on the one hand, to the women's perception of insecurity towards going out during the very early morning or at late night. On the other hand, it could also be due to the higher amount of non-work related travels taking place during the day. Additionally, the curves of travels associated with the home/work mobility display a temporal shift, with men leaving for work one hour earlier, and women leaving to get back home earlier than men. Study's purpose appears to be the most gender neutral activity, as denoted by the small differences between the amount of travels made by each gender. Finally, when we analyse the percentage of travels made with a specific purpose averaged over all the departure zones, we notice that men tend to make more travels than women regardless of the departure time. This is in contrast with the same quantities computed without taking into account the zone of departure.

Summing up, we have seen how gender – in combination with occupation, and demands – molds urban mobility. Our observations are in agreement with Ravenstein's migration's law [14] and recent studies on gender gaps based on mobile phones data [6]. Nevertheless, the availability of detailed meta-information on travels/travellers allowed us to split mobility into distinct components, enabling a more fine grained analysis of the overall phenomenology observed both in space and time. Of course, a better comprehension of gender effects cannot neglect the influence of other factors like age, education, and socio-economic status. Finally, the availability of further studies concerning different cities/countries, as well as, cultures would improve our understanding of the gender effects on mobility by separating them from the spatial environment under scrutiny.

Acknowledgements. AC acknowledges the support of the Spanish Ministerio de Ciencia e Innovacion (MICINN) through Grant IJCI-2017-34300.

References

1. Anderson, P.W.: More is different. Science **177**, 393–396 (1972)
2. Barbosa, H., et al.: Human mobility: models and applications. Phys. Rep. **734**, 1–74 (2018)
3. Barthélemy, M.: Spatial networks. Phys. Rep. **499**, 1–101 (2011)
4. CARE: Gender, power and justice primer. https://www.care.org/our-impact/gender-in-practice
5. Evertsen, K.F., Van der Geest, K.: Gender, environment and migration in Bangladesh. Clim. Dev. **12**, 1–11 (2019)
6. Gauvin, L., et al.: Gender gaps in urban mobility. arXiv preprint arXiv:1906.09092 (2019)
7. Heinrichs, D., Bernet, J.S.: Public transport and accessibility in informal settlements: aerial cable cars in Medellín, Colombia. Transp. Res. Procedia **4**, 55–67 (2014)
8. Levy, C.: Transport, diversity and the socially just city: The significance of gender relations. UCL & Universidad Nacional de Colombia, DPU (2013)
9. Lotero, L., Cardillo, A., Hurtado, R., Gómez-Gardeñes, J.: Several multiplexes in the same city: the role of socioeconomic differences in urban mobility. In: Interconnected Networks, pp. 149–164. Springer (2016)
10. Lotero, L., Hurtado, R.G., Floría, L.M., Gómez-Gardeñes, J.: Rich do not rise early: spatio-temporal patterns in the mobility networks of different socio-economic classes. R. Soc. Open Sci. **3**(10), 150654 (2016)
11. Maharani, A., Sulaiman, H., Aminah, N., Rosita, C.: Analyzing the student's cognitive abilities through the thinking levels of geometry van hiele reviewed from gender perspective. In: Journal of Physics: Conference Series, vol. 1188 (2019)
12. Milan, B.F., Creutzig, F.: Lifting peripheral fortunes: upgrading transit improves spatial, income and gender equity in Medellin. Cities **70**, 122–134 (2017)
13. O'Dea, R., Lagisz, M., Jennions, M., Nakagawa, S.: Gender differences in individual variation in academic grades fail to fit expected patterns for stem. Nat. Commun. **9**(1), 3777 (2018)
14. Ravenstein, E.G.: The laws of migration. J. Stat. Soc. Lond. **48**(2), 167–235 (1885)

15. Shannon, C.E.: A mathematical theory of communication. Bell Syst. Tech. J. **27**(3), 379–423 (1948)
16. Thynell, M.: The quest for gender-sensitive and inclusive transport policies in growing Asian cities. Soc. Incl. **4**(3), 72–82 (2016)
17. Toch, E., Lerner, B., Ben-Zion, E., Ben-Gal, I.: Analyzing large-scale human mobility data: a survey of machine learning methods and applications. Knowl. Inf. Syst. **58**(3), 501–523 (2019)
18. Turner, J.: Urban mass transit, gender planning protocols and social sustainability-the case of Jakarta. Res. Transp. Econ. **34**(1), 48–53 (2012)

Dynamic Network of United States Air Transportation at Multiple Levels

Batyr Charyyev[1], Mustafa Solmaz[2], and Mehmet Hadi Gunes[1](✉)

[1] School of Systems and Enterprises, Stevens Institute of Technology, Hoboken, NJ 07030, USA
{bcharyye,mgunes}@stevens.edu

[2] Computer Science and Engineering, University of Nevada Reno, Reno, NV 89557, USA
msolmaz@nevada.unr.edu

Abstract. United States air transportation contributes to the economy of the country by facilitating the mobility of people and goods. Air transportation is one of the essential systems in the US and the world. In this study, we investigate the dynamics of US air transportation from three angles (i.e., number of flights, number of passengers, and the amount of freight carried) at three levels of airport, city, and state. While there are unique dynamics at each, there is a strong fluctuation in the activity of the links, indicating a highly dynamic system. Backbone analyses of the networks also show that there are specific periods (such as 2007 for flights, 2008 for freight, and 2012 for passengers) in which network changes drastically.

Keywords: Complex networks · Network dynamics · Temporal network

1 Introduction

With globalization, the transfer of goods and people has increased across geographic regions using different methods of transportation systems such as railway, highway, maritime, and airway. Understanding these systems is imperative for the industry and government as they impact the economy and public health. For instance, analysis of the air transportation dynamics is vital for airlines to manage route schedules and increase their market shares effectively. Understanding of air transportation is also essential for public health as air transport enables infected travelers to spread diseases over different regions within a short time.

Air transportation can be abstracted as a network where nodes represent different levels of abstractions and edges represent different quantities of flights.

B. Charyyev and M. Solmaz—Equal contributing authors.

H. Barbosa et al. (Eds.): *Complex Networks XI*, SPCOM, pp. 282–293, 2020.
https://doi.org/10.1007/978-3-030-40943-2_24

Prior complex network analysis of air transportation has shown small-worlds with power-law degree distribution [1]. Complex network analysis provides an understanding of the overall structure of a system, its evolution, and its dynamics. It also enables us to quantify the centrality of individual nodes in a network in terms of different metrics.

Dynamic network analyses have been applied to various complex networks such as Internet flows [3], cattle trade movement [2], migration of people within the US [8,11], air transportation [10], cancer mechanisms [19] and disintegration of terrorist networks [4]. Dynamic network analysis enables us to investigate dynamics in complex systems during different periods. Previous studies focus on the activity of nodes and links; how the weight of links change; formation of communities; and the existence of a backbone; and how the network evolves over time.

In this study, we analyze the United States air transportation network from 2007 to 2016, focusing on the number of flights performed, the number of passengers, and the amount of freight carried between airports, cities, or states. Our analyses focus on link dynamics, such as how long links stay active or inactive in a network or how the weight of the link changes once they become active. We also analyze the backbone for each of the networks and analyze how they evolve with time.

There exist prior work on complex network analyses of air transportation for China [16], India [1], some of the European countries [12], and United States [9,10,15,23]. Prior studies on air transportation networks are based on decade-old data or focus on static behavior of the network. Different from existing studies our study focuses on dynamic properties of the network, in terms of different metrics such as the number of flights, passengers and amount of freight.

In the rest of the paper, Sect. 2 presents related works, Sect. 3 summarizes data collection and filtering along with the methodologies used in this study. Section 4 presents a dynamic network analysis of the US air transportation network using different abstraction levels. Finally, Sect. 5 concludes the paper.

2 Related Work

In this section, we present studies analyzing air transportation systems as complex networks.

Guimera et al. analyzes the global structure of the worldwide air transportation network and show that it is a scale-free small-world network [13]. Authors analyze the network in terms of the centrality and communities. They found that the most connected cities are not necessarily the most central due to the multi-community structure of the network. This was also later confirmed by Tao et al. in analyses of the United States air transportation system [14]. Zanin et al. review the application of the complex networks theory to the air transport system [24]. Authors discuss possible network representation of the air transportation systems, topological and dynamic network analyses, and review the resilience of the air transport network with internal and external disturbances.

Chinese Air Transportation Network (CATN) has been analyzed as a complex network as well. Zhang et al. show a linear correlation between the traffic volume (i.e., passenger and cargo) and the GDP of China [25]. They indicate that there is no correlation between the Chinese GDP and the number of airlines and airports. Wang et al. show that the degree distribution of CATN follows an exponential function with the top 20% of cities accounting for 65% of the routes [22]. From the k-degree analysis, they observed that 4-degree nodes showed the highest average neighbor degree, and it was followed by 5-, 6- and 2-degree nodes. Kai-Quan et al. show that the topology structure of CATN is homogeneous; however, the distribution of the flight flow on CATN is rather heterogeneous [16]. Lin et al. considers the distance between cities and analyzes China's aviation system with a gravitational model [17].

Burghouwta investigates deregulation in the European aviation network and its effect on the structure of the network [6]. Hub and spoke approach of airlines often lead to more routes and a decrease in the operating costs. The author observed that it is harder to impose a hub and spoke system in the European aviation system compared to the US. Cardillo et al. analyses the European air transport network in terms of the resilience of the system against random flight failures in the passenger's rescheduling [7]. Sun et al. investigate the temporal evolution of centrality measures in the European air transportation system [21]. They focus on two layers of the air navigation route network and the airport network to quantify seasonal and weekly variation patterns.

Xu et al. investigates the United States air transportation network between periods of 2002 and 2005, using three different metrics as weight (i.e., the distance between cities, the average daily passengers traveling between cities, and the average one-way fare paid by passengers) [23]. Authors observed that the US air transportation network is a small-world, scale-free network. They also observed the rich-club phenomenon in which large degree cities tend to form interconnections among each other to sustain high volumes of traffic. In a prior work, we analyzed the US air transportation system in terms of topological structure and resilience using July 2011 data [9]. We observed that randomly removing 30% of the airports decreases the size of the giant component by half. The same results were obtained by targeting 10% of airports with the highest degree or 5% of nodes with the highest betweenness. We observed that Denver had the highest number of incoming and outgoing flights followed by Chicago, Atlanta, and Dallas. Tao et al. also analyze the air transportation network of the US between 1990 and 2010 [15]. Authors observed a substantial increase in the number of cities and routes in 2002, which they explained as a result of the 9/11 terrorist attacks leading to change in the air transportation industry. They also observed that in 2002, the efficiency of the air transportation network decreased considerably in terms of shortest path length.

3 Methodology

In this section, we present the data collection and filtering as well as network construction steps.

3.1 Data Collection and Filtering

Flight data for a decade is collected from the United States Department of Transportation website [5]. Data is between the period of 2007–2016 and contains the number of passengers, the total freight carried, and the number of flights between airports for about 375,000 flights. For each flight, there are airport, city, and state codes of the source and destination, which enables us to aggregate the data at the airport, city, and state levels. In the dataset, there exist flights to three regions of the U.S., i.e., Puerto Rico, U.S. Pacific Trust Territories & Possessions, and the U.S. Virgin Islands, which we consider as three states in our state-level models. We also observed that about a thousand flights had the same source and destination. An in-depth analysis of these indicated that some were sightseeing tours such as Grand Canyon. Also, we learned that gate returns are recorded in the data-set with the same source and destination airport from the Department of Transportation. Thus, we excluded such flights from our analyses.

3.2 Network Construction

The collected data is cleaned and analyzed to construct air transportation networks between airports, cities, and states. In a network, nodes represent airports, cities, or states. Directed and weighted edges represent the number of passengers, the total amount of carried freight in pounds, or the number of flights. Three different levels of nodes and edge weights provide us with nine different networks. In the rest of the paper, *flight network* indicates the edge weight determined by the number of flights between two nodes. Similarly, *passenger network* and *freight network* represents networks in which edge weight is determined by the number of passengers, and the amount of freight carried over the link, respectively. The level of aggregation is indicated with the *airport*, *city* and *state* levels. For instance, *airport flight network* represents a network in which nodes are airports, and the number of flights between airports determines edge weight. The processed data and resulting networks are provided at https://github.com/netlab-stevens/US-Air-Transportation.

4 Results

In this section, we analyze the network characteristics of the United States air transportation, considering both annual flights as a single network as well as a temporal network over the decade.

4.1 United States Air Transportation Network

First, we analyze air transportation as a static network for each year, where edges indicate the annual number of flights, the number of passengers, and the amount of freight carried while nodes represent airports, cities or states. From three different representations of both nodes and edges, we obtain nine different abstractions of air transportation and then create networks for each year. Table 1 summarizes the average of the network characteristics for each abstraction across the ten years.

Table 1. Network characteristics of the United States air transportation.

	Nodes	Edges	Average degree	Average weighted degree	Diameter	Average path length	Density	Clustering coefficient	Assortativity
Airport flight network									
Avr	1,224	2.2×10^4	18.07	7,541	8.4	3.25	1.47×10^{-2}	0.44	0.41
Var	660	9×10^5	0.41	2.11×10^5	0.26	4.82×10^{-4}	4.56×10^{-7}	1.8×10^{-4}	9.87×10^{-5}
City flight network									
Avr	1,071	1.82×10^4	16.96	8,620	7.2	3.1	1.58×10^{-2}	0.49	0.32
Var	391	6.23×10^5	0.51	3.22×10^5	0.18	5.31×10^{-4}	8.44×10^{-7}	1.4×10^{-4}	7.06×10^{-4}
State flight network									
Avr	53	2,063	38.93	1.74×10^5	3	1.27	0.74	0.85	0.90
Var	0	1,023	0.36	1.33×10^8	0	1.29×10^{-4}	1.36×10^{-4}	4.2×10^{-5}	6.15×10^{-5}
Airport passenger network									
Avr	1,224	1.96×10^4	15.98	5.5×10^5	8.5	3.33	1.31×10^{-2}	0.4	0.41
Var	660	5.05×10^5	0.23	5.03×10^8	0.5	4.6×10^{-4}	1×10^{-7}	2.30×10^{-4}	1.65×10^{-4}
City passenger network									
Avr	1,071	1.62×10^4	15.12	6.29×10^5	7.4	3.15	1.42×10^{-2}	0.44	0.32
Var	391	3.71×10^5	0.33	6.35×10^8	0.27	1.2×10^{-3}	6.22×10^{-7}	1.88×10^{-4}	8.55×10^{-4}
State passenger network									
Avr	53	1,970	37.19	1.27×10^7	3	1.31	0.72	0.84	0.89
Var	0	1,572	0.56	3.18×10^{11}	0	2.17×10^{-4}	2.08×10^{-4}	3.05×10^{-5}	6.79×10^{-5}
Airport freight network									
Avr	1,224	1.08×10^4	8.77	2.02×10^7	8.8	3.33	7.2×10^{-3}	0.36	0.33
Var	660	4.87×10^5	0.26	1.7×10^{12}	0.62	1.11×10^{-3}	1.78×10^{-7}	2.6×10^{-4}	1.74×10^{-4}
City freight network									
Avr	1,071	8,820	8.26	2.31×10^7	7.6	3.23	7.7×10^{-3}	0.39	0.26
Var	391	3.44×10^5	0.32	2.47×10^{12}	0.49	2.07×10^{-4}	6.78×10^{-7}	3.01×10^{-4}	1.69×10^{-4}
State freight network									
Avr	53	1,664	31.4	4.68×10^8	3	1.42	0.6	0.79	0.8
Var	0	1,498	0.53	1.25×10^{15}	0	2.02×10^{-4}	2.01×10^{-4}	8.07×10^{-5}	1.26×10^{-4}

The yearly analysis shows that the number of nodes and edges sharply decreases in 2008 and 2013 for all networks at the airport and city levels. Additionally, the overall number of edges reduce by 15% to 17% for air and city levels and by 5% to 6% for state level networks of flights, passengers, and freight over the decade. The average weighted degree of the flight networks decreases with time as well. On the other hand, the average weighted degree shows an increasing pattern for the passenger and freight levels since 2008, when the financial crisis peaked. As the average weighted degree for passenger network increases, the average weighted degree for the flight network decreases. This indicates that the smaller flights are being discarded and larger flights are preferred to increase the number of passengers. While the freight network has similar trends at a slower rate of change, it recovers after 2012.

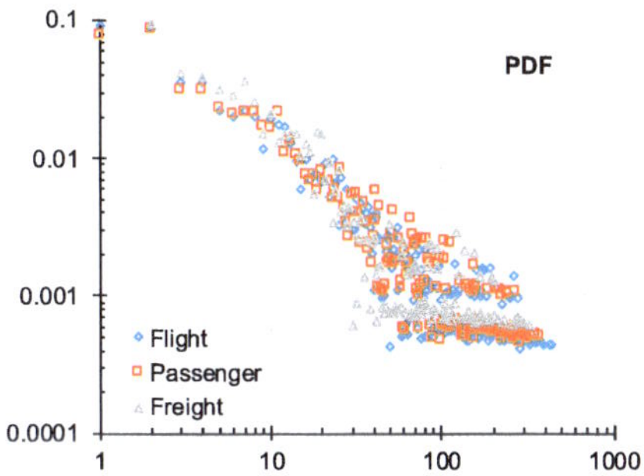

(a) PDF

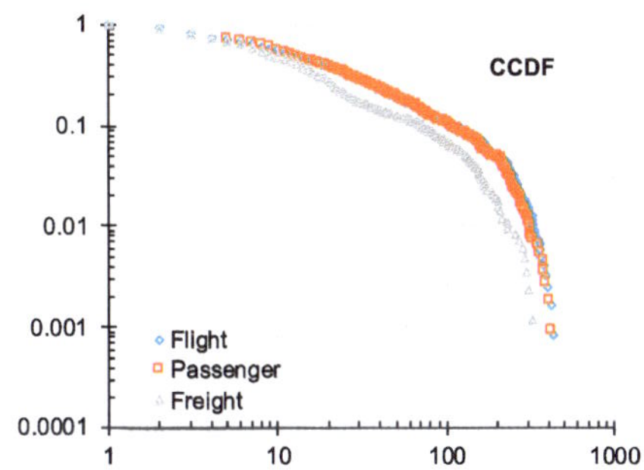

(b) CCDF

Fig. 1. Degree distribution of air transportation network in 2016 at airport level.

Additionally, results indicate that air transportation has a small world characteristic across each abstraction, where networks have a high clustering coefficient and a low average path length. Furthermore, we observe that assortativity decrease for the flight and passenger networks over the years and gets closer to 0, i.e., being non-assortative. Also, both density and clustering have a downward trend at flight and passenger networks, while having an upward trend at freight network at airport and city levels. These results indicate that the passenger network at airport level became more centralized over time. This shows that the airport network of the US has been transforming into a more efficient one in the last decade.

The degree distribution is an important metric summarizing the characteristics of a network. Figure 1 presents the degree distribution of the US air transportation for 2016 at the airport level considering the probability distribution function (PDF) and the complementary cumulative distribution function (CCDF). Overall, we observe that the shape of the degree is an exponential distribution and has similar distributions with different abstractions. Other years show very similar behavior, and hence not shown. For the flight network, airports with the highest degree are DEN (Denver Colorado), ATL (Atlanta, Georgia); for the passenger network, they are ORD (Chicago, Illinois) and ATL (Atlanta, Georgia); and for the freight network, they are MEM (Memphis, Tennessee) and DFW (Dallas/Fort Worth Texas).

Overall, we find exponential degree distributions and small-world networks similar to our prior study where we analyzed a month of data in 2011 [9]. We observe an average path length value between 3.1–3.3 for airport and city levels while it is around 1.3 for state level considering the three edge abstractions.

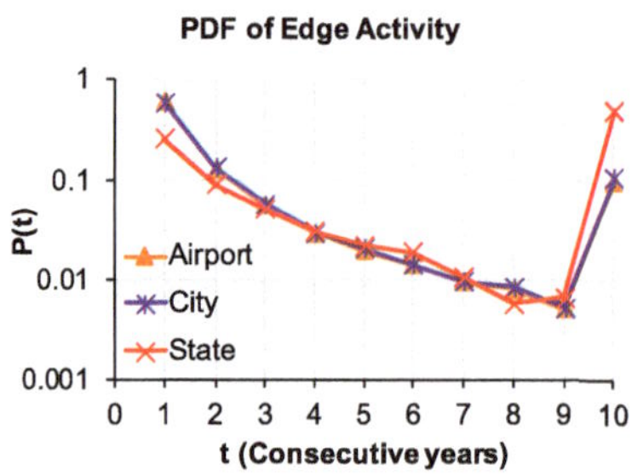

(a) Activity of Edges

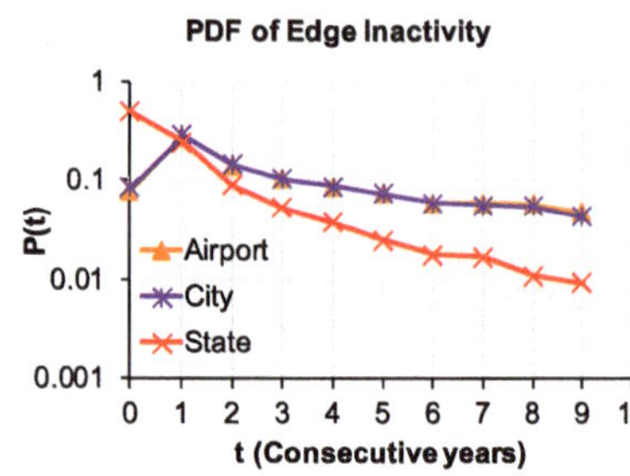

(b) Inactivity of Edges

Fig. 2. The probability of the duration of edge activity and inactivity for the flight networks.

4.2 Network Dynamics

Analysis of the activity and inactivity of the nodes and edges enables us to characterize the network dynamics. Activity (or inactivity) is defined with the *number of consecutive periods in which node or edge remains active (or inactive)*. While nodes in each abstraction of the network usually remain stable, edges show dynamic behavior. Figure 2 shows activity and inactivity of edges for flight networks. Passenger and freight networks have similar results as flight networks; thus, they are not shown.

In analyses of activity and inactivity of edges, airport and city levels show similar behavior, whereas the state level is different. This is mainly because most cities have a primary airport which contributes to the air transportation of that city the most. Overall, the probability of the link to be continuously active or inactive decreases as the number of consecutive years increases. Most of the edges are active for one year with a probability of 0.62, 0.61, 0.26, or inactive with a probability of 0.28, 0.28, 0.24 for airport, city, and state levels, respectively. However, there exist links which are continuously active for ten years with a probability of 0.09 for airport, 0.10 for the city, and 0.49 for the state level. These links are between large cities and major airports.

We observed that most of the edges are active or inactive for short periods, thus we wanted to analyze the appearance and disappearance of the links. We evaluate the fraction of appearing f^a and disappearing f^d links, as a function of their weight. The mechanism is based on [10]. The fraction of appearance and disappearance of the links reveals if there is a correlation between the stability of links and the weight of the links. Fraction of appearing links is $f^a(w) = E^a(w|t)/E(w|t)$, where $E(w|t)$ is the number of links with weight w at time t and $E^a(w|t)$ is the number of such links that were not active at time $t-1$ and thus appearing at time t. Similar logic is applied for fraction of disappearance f^d, where $E^d(w|t)$ is the number of links of weight w active at time $t-1$ but not active at t and $E(w|t)$ is similar as for f^a.

Figure 3 shows the fraction of appearance (f^a) and disappearance (f^d) of links in the airport and state levels. The city level results are similar to the airport

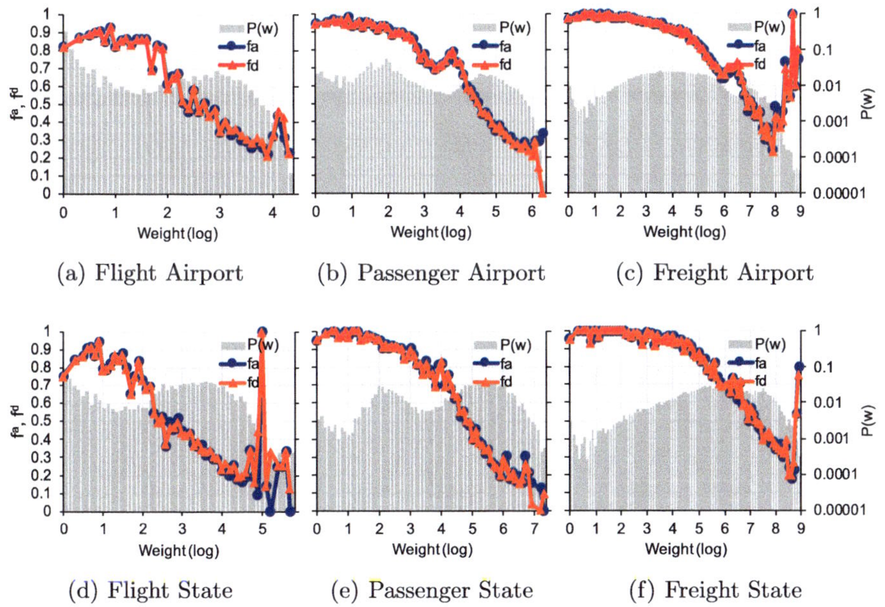

Fig. 3. Fraction of appearance (f^a) and disappearance (f^d) of the links. $P(w)$ is probability of observing weight w.

level results; thus, we did not include them. The fraction of appearance mostly shows similar behavior to the fraction of disappearance, similar to a previous study on migration [8]. When link weight is represented with the number of flights, f^a and f^d shows a fluctuating behavior as shown in Fig. 3(a, d). This is reasonable because the number of flights may change based on the needs and is impacted by economic factors. Overall f^a and f^d decreases with link weight, meaning that stability of links between two nodes increases as the number of flights between two airports or states increases.

When weight is represented with the number of passengers, f^a and f^d decreases, indicating increasing link stability. This is because flights carrying many passengers are economically convenient and have higher stability; thus, they do not appear or disappear frequently. Similar results were obtained from previous analyses of air transportation network on data between periods of 1990–2000 [10]. For freight network in Fig. 3(c, f), there is a positive correlation between links' stability and weight, similar to when weight is represented with the number of passengers. However, links with the largest weights show unstable behavior similar to the migration network [8]. This means that source and destination node for large amounts of freight changes with the time.

Previously, we analyzed links dynamics in terms of how links can switch on and off their activity. Here, we study the evolution of links' weight once they are active. We used the growth rate function $r_{ij}(t) = log(w_{ij}(t+1)/w_{ij}(t))$ of the links which was proposed to analyze firm growth rate [20].

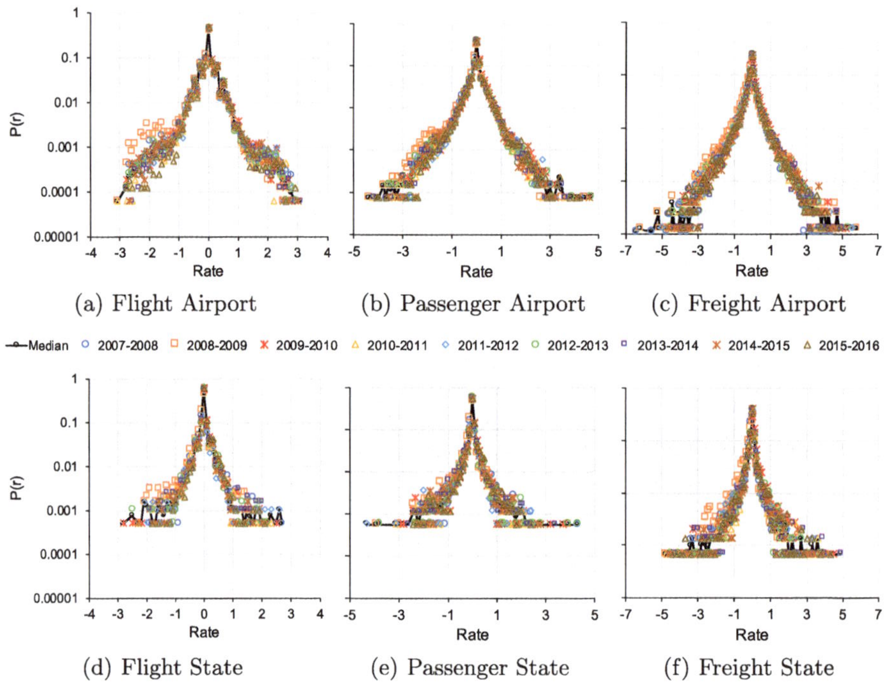

Fig. 4. Growth rate distribution of the link weights for airport and state levels. Each symbol corresponds to one snapshot (i.e., year).

Figure 4 shows the growth rate distribution of the link weights for airport and state levels. We excluded the city level as it is similar to the airport level. Note that the distributions can be superimposed, meaning r_{ij} is independent of time. We observe that most of the weight increments are small, but a sudden large increase or decrease of weights can be observed with a non-negligible probability. Results for flight, passenger, and freight networks show similar behaviors.

Link activity and inactivity showed that most of the links are continuously active or inactive only for a short period of time. However, we observed that there exists a considerable amount of links that are continuously active between big cities. Overall appearance and disappearance of links showed that links are more stable as weight increases. However, for freight network, we observed that links with very large weight show unstable behavior with high f^a and f^d values. Furthermore, we found that in general, link increments are small, but a sudden large increase or decrease of weights can be observed with a non-negligible probability.

	2007	2008	2009	2010	2011	2012	2013	2014	2015	2016
2016	58	62	62	68	70	81	84	86	88	100
2015	61	62	64	68	71	85	86	92	100	88
2014	56	60	63	68	69	84	89	100	92	86
2013	61	64	67	72	75	88	100	89	86	84
2012	62	67	71	78	81	100	88	84	85	81
2011	70	77	80	86	100	81	75	69	71	70
2010	75	79	84	100	86	78	72	68	68	68
2009	76	81	100	84	80	71	67	63	64	62
2008	79	100	81	79	77	67	64	60	62	62
2007	100	79	76	75	70	62	61	56	61	58

(a) Flight State

2007	2008	2009	2010	2011	2012	2013	2014	2015	2016
60	65	63	64	66	66	72	81	84	100
59	63	62	63	65	64	84	86	100	84
62	69	69	68	73	73	83	100	86	81
67	71	71	70	74	71	100	83	84	72
74	84	86	82	90	100	71	73	64	66
76	82	84	84	100	90	74	73	65	66
79	80	86	100	84	82	70	68	63	64
77	84	100	86	84	86	71	69	62	63
82	100	84	80	82	84	71	69	63	65
100	82	77	79	76	74	67	62	59	60

(b) Passenger State

2007	2008	2009	2010	2011	2012	2013	2014	2015	2016
62	64	74	75	75	76	84	86	87	100
60	62	72	75	76	76	88	91	100	87
62	65	78	80	80	82	93	100	91	86
63	67	77	78	79	83	100	93	88	84
65	69	78	86	83	100	83	82	76	76
64	65	77	84	100	83	79	80	76	75
67	71	85	100	84	86	78	80	75	75
71	75	100	85	77	78	77	78	72	74
87	100	75	71	65	69	67	65	62	64
100	87	71	67	64	65	63	62	60	62

(c) Freight State

Fig. 5. Annual evolution of the backbones.

4.3 Network Backbone

In this section, we identify if there is a backbone of the analyzed air transportation networks. Utilizing global thresholding of the links may give misleading results because it may dismiss links with small weights, which are locally important [2]. Hence, we used a disparity filter introduced by Serrano et al. [18], which operates at all scales defined by the weight distribution of links. Using the disparity filter, we generated the backbone of each network for each year. Then, we computed the overlap between all pairs of the backbones. Overlap of two networks is calculated as the ratio of the intersection to the union of edges, $|E1 \cap E2|/|E1 \cup E2|$.

The significance level of disparity filter can be configured with the α parameter. For small values of α the disparity filter is more restrictive. We repeated the analyses for different values of α (i.e., 0.001, 0.1, and 0.5) and present results for the lowest value. Note that as α value increases, the overlap of backbones also increases. Even for small α value, the overlap of two consecutive backbones is around 65% at the airport level, and it reaches up to 90% at the state level.

Figure 5 shows the overlap of backbones as a color-coded matrix for each type of network at the state level. We can see that the color-coded matrix is split into two distinct partitions, where backbones in each partition have higher similarity compared to the backbone from another partition. This behavior is observed in all three networks of flight, passenger, and freight. For instance, in the flight network shown in Fig. 5(a), backbones between periods 2007–2011 have a higher overlap within each other compared to backbones between periods of 2012–2016. These may be due to significant and enduring infrastructure change in 2011, which impacted the backbone of the network. The least similarity across consecutive years is observed between 2007 and 2008. Similar partitions were observed in passenger network shown in Fig. 5(b) in 2012 and freight network shown in Fig. 5(c) in 2008.

5 Conclusion

United States air transportation is a complex system that has a vast contribution to the economy. We analyzed the US air transportation system in terms of the number of flights, the number of passengers, and the amount of freight carried. We perform these analyses at the airport, city, and state levels. Results showed that the air transportation network is a small-world and has an exponential degree distribution. We observed that most of the links are continuously active or inactive for short periods of time, the overall stability of links increase as their weight increase, and once links are active weight increments are small. We also observed that there exist distinct periods in which the backbone of the network in each abstraction level changes drastically, which requires further investigation. In the future, we are planning to further investigate the cause and impact of the backbone changes in the United States and global air transportation networks.

References

1. Bagler, G.: Analysis of the airport network of India as a complex weighted network. Physica A **387**(12), 2972–2980 (2008)
2. Bajardi, P., Barrat, A., Natale, F., Savini, L., Colizza, V.: Dynamical patterns of cattle trade movements. PLoS ONE **6**(5), e19869 (2011)
3. Bakhshaliyev, K., Canbaz, M.A., Gunes, M.H.: Investigating characteristics of internet paths. ACM Trans. Model. Perform. Eval. Comput. Syst. (TOMPECS) **4**(3), 16 (2019)
4. Behzadan, V., Nourmohammadi, A., Gunes, M., Yuksel, M.: On fighting fire with fire: strategic destabilization of terrorist networks. In: Proceedings of the 2017 IEEE/ACM International Conference on Advances in Social Networks Analysis and Mining, 31 July 2017, pp. 1120–1127. ACM (2017)
5. Bureau of Transport Statistics: Air Carrier Statistics. https://www.transtats.bts.gov/Tables.asp?DB_ID=111
6. Burghouwt, G., Hakfoort, J.: The evolution of the European aviation network, 1990–1998. J. Air Transp. Manag. **7**(5), 311–318 (2001)
7. Cardillo, A., Zanin, M., Gómez-Gardenes, J., Romance, M., del Amo, A.J., Boccaletti, S.: Modeling the multi-layer nature of the European Air Transport Network: resilience and passengers re-scheduling under random failures. Eur. Phys. J. Spec. Top. **215**(1), 23–33 (2013)
8. Charyyev, B., Gunes, M.H.: Complex network of United States migration. Comput. Soc. Netw. **6**(1), 1 (2019)
9. Cheung, D.P., Gunes, M.H.: A complex network analysis of the United States air transportation. In: Proceedings of the 2012 International Conference on Advances in Social Networks Analysis and Mining (ASONAM 2012), 26 August 2012, pp. 699–701. IEEE Computer Society (2012)
10. Gautreau, A., Barrat, A., Barthélemy, M.: Microdynamics in stationary complex networks. Proc. Natl. Acad. Sci. **106**(22), 8847–8852 (2009)
11. Goldade, T., Charyyev, B., Gunes, M.H.: Network analysis of migration patterns in the united states. In: International Conference on Complex Networks and their Applications, vol. 29, pp. 770–783. Springer, Cham (2017)
12. Guida, M., Maria, F.: Topology of the Italian airport network: a scale-free small-world network with a fractal structure? Chaos, Solitons Fractals **31**(3), 527–536 (2007)
13. Guimera, R., Mossa, S., Turtschi, A., Amaral, L.N.: The worldwide air transportation network: anomalous centrality, community structure, and cities' global roles. Proc. Nat. Acad. Sci. **102**(22), 7794–7799 (2005)

14. Jia, T., Jiang, B.: Building and analyzing the US airport network based on en-route location information. Physica A **391**(15), 4031–4042 (2012)
15. Jia, T., Qin, K., Shan, J.: An exploratory analysis on the evolution of the US airport network. Physica A **1**(413), 266–279 (2014)
16. Kai-Quan, C., Jun, Z., Wen-Bo, D., Xian-Bin, C.: Analysis of the Chinese air route network as a complex network. Chin. Phys. B **21**(2), 028903 (2012)
17. Lin, J.: Network analysis of China's aviation system, statistical and spatial structure. J. Transp. Geogr. **1**(22), 109–117 (2012)
18. Serrano, M.Á., Boguná, M., Vespignani, A.: Extracting the multiscale backbone of complex weighted networks. Proc. Nat. Acad. Sci. **106**(16), 6483–6488 (2009)
19. Solmaz, M., Lane, A., Gonen, B., Akmamedova, O., Gunes, M.H., Komurov, K.: Graphical data mining of cancer mechanisms with SEMA. Bioinformatics **35**(21), 4413–4418 (2019)
20. Stanley, M.H., Amaral, L.A., Buldyrev, S.V., Havlin, S., Leschhorn, H., Maass, P., Salinger, M.A., Stanley, H.E.: Scaling behaviour in the growth of companies. Nature **379**(6568), 804 (1996)
21. Sun, X., Wandelt, S., Linke, F.: Temporal evolution analysis of the European air transportation system: air navigation route network and airport network. Transp. B: Transp. Dyn. **3**(2), 153–168 (2015)
22. Wang, J., Mo, H., Wang, F., Jin, F.: Exploring the network structure and nodal centrality of China's air transport network: a complex network approach. J. Transp. Geogr. **19**(4), 712–721 (2011)
23. Xu, Z., Harriss, R.: Exploring the structure of the US intercity passenger air transportation network: a weighted complex network approach. GeoJournal **73**(2), 87 (2008)
24. Zanin, M., Lillo, F.: Modelling the air transport with complex networks: a short review. Eur. Phys. J. Spec. Top. **215**(1), 5–21 (2013)
25. Zhang, J., Cao, X.B., Du, W.B., Cai, K.Q.: Evolution of Chinese airport network. Physica A **389**(18), 3922–3931 (2010)

Economical Networks

Mining the Automotive Industry: A Network Analysis of Corporate Positioning and Technological Trends

Niklas Stoehr[1], Fabian Braesemann[2(✉)], Michael Frommelt[1], and Shi Zhou[3]

[1] IBM, AI Core, GER, New York, USA
[2] Saïd Business School & Oxford Internet Institute, University of Oxford, Oxford, UK
f.braesemann@sbs.ox.ac.uk
[3] Department of Computer Science, University College London, London, UK

Abstract. The digital transformation is driving revolutionary innovations and new market entrants threaten established sectors of the economy such as the automotive industry. Following the need for monitoring shifting industries, we present a network-centred analysis of car manufacturer web pages. Solely exploiting publicly-available information, we construct large networks from web pages and hyperlinks. The network properties disclose the internal corporate positioning of the three largest automotive manufacturers, Toyota, Volkswagen and Hyundai with respect to innovative trends and their international outlook. We tag web pages concerned with topics like e-mobility & environment or autonomous driving, and investigate their relevance in the network. Sentiment analysis on individual web pages uncovers a relationship between page linking and use of positive language, particularly with respect to innovative trends. Web pages of the same country domain form clusters of different size in the network that reveal strong correlations with sales market orientation. Our approach maintains the web content's hierarchical structure imposed by the web page networks. It, thus, presents a method to reveal hierarchical structures of unstructured text content obtained from web scraping. It is highly transparent, reproducible and data driven, and could be used to gain complementary insights into innovative strategies of firms and competitive landscapes, which would not be detectable by the analysis of web content alone.

Keywords: Automotive industry · Network analysis · Complex networks · Digitisation · Web page mining · Competition

Code and Data: www.github.com/Braesemann/MiningAutomotive-Industry

1 Introduction

1.1 Motivation

Environmental change and the ongoing digitisation cause large-scale transformations in the economy, as boundaries of production, distribution and consumption

H. Barbosa et al. (Eds.): *Complex Networks XI*, SPCOM, pp. 297–308, 2020.
https://doi.org/10.1007/978-3-030-40943-2_25

are reshaped [1–3]. Industries are impacted differently depending on factors like contribution to greenhouse gas emissions, automation capabilities, customer proximity, and labour complexity [2]. One of the biggest, yet most strongly affected industries is the automotive industry [3–7].

According to estimates of the International Organization of Motor Vehicle Manufacturers (OICA), more than 5% of the world's total manufacturing employment is directly involved in the production of vehicles and parts [8]. However, the economic importance of the automotive industry reaches far beyond that. Many of the most groundbreaking innovations of the 20th-century, mass production, just-in-time and multi-divisional business organisation originated in automotive companies and left a recognisable geographical footprint [5,9].

The need for sustainable mobility solutions will drive further innovation in fields relevant to the automotive industry such as e-mobility, connectivity and autonomous driving [10–12]. These innovations provide unparalleled opportunities for value creation, but also major risks as new market participants might introduce services, which could threaten the established ecosystem [5,6]. Such radical shifts are creating the need for analytic feedback and status reports. Not only executive boards, policy-makers and investors, but more importantly, millions of employees are interested in a successful and sustainable transformation and therefore rely on objective assessments of car manufacturers' positioning with regards to novel developments and technologies.

1.2 Contributions

In this paper, we provide a new perspective on the possibilities for analysing shifting industries. Using only publicly available data, we present a network-centred approach that allows for conclusions on innovative orientation and international outlook of the world's leading car manufacturers, Toyota, Volkswagen, and Hyundai.[1]

In this study, we enrich a quantitative analysis of web page content with meta-information. This approach surpasses conventional search engine optimisation and web page analysis, which are increasingly important tools for analysing markets and competitive landscapes, as for example, the recent introduction of the *Google Market Finder* tool shows. Our method exploits the web page structure of the firms to obtain complex networks, where each web page is considered a node and each hyperlink referring to another web page is considered a link. This novel approach of exploiting the interplay between content and network analysis allows mapping a company's corporate positioning based on publicly available information. The approach is thus highly transparent and can be reproduced to monitor and contrast company web pages to gain insights into their innovation and market strategies.

[1] These are the three biggest automotive manufacturers by production numbers in 2016: Toyota (10.2 mio. vehicles), the Volkswagen Group (10.1 mio cars), and Hyundai with 7.9 mio. units [13].

The study exemplifies the value of applying network analysis as a complementary tool to text mining and content analysis. In times of abundant non-structured text information that can easily be obtained from different online sources via web-scraping, the combination of network analysis and content analysis allows for imposing structure on unstructured big data, as the hierarchy of information, which is implicitly stored in the web-page network topology, is maintained.

2 Related Work

This work is related to two strands of literature: studies on the transformation of the automotive industry and network analyses of web pages.

Change in the automotive industry has been intensively investigated in past studies [9,14–17], focusing on changing geographies and challenges of the technological transformation, and on technical, environmental, and management implications. Moreover, network analysis has been applied in the field, mainly to understand Supply-Chain systems in the automotive industry [18,19].

While web page networks of car manufacturers have, to our knowledge, not been investigated previously, similar approaches were applied in tourism and e-commerce. For instance, Baggio et al. investigate the links between tourism destination websites to find the statistical characteristics of the underlying graph [20], and Wang et al. establish principles of an improved link structure for a hypothetical e-supermarket website [21].

Additionally, many pioneering studies in network science have focused on the world wide web: in 2000, Broder et al. discovered the bow tie structure of the world wide web [22], and Meusel et al. aimed to understand it's growth mechanism in constructing a giant network of large parts of the world wide web [23]. Barabási and Albert used web data to demonstrate their theory on the emergence of scaling in random networks [24], and the PageRank algorithm was introduced for ranking web pages in a directed graph of the world wide web [25].

The content and structure of web pages has previously been investigated from the perspective of search engine optimisation [26], data retrieval [28], website design [29], and web navigation [30].

Our approach builds on the previously mentioned studies, but assumes a different perspective. Instead of analysing content and structure of web pages to optimise properties of individual websites, we explore the potential of applying network analysis of web pages to gain insights into business models and innovative orientation of companies and industries.

3 Research Hypotheses

We analyse the properties of web page networks of leading car manufacturers and tag individual web pages that mention relevant keywords to observe their position in the network.[2] If car manufacturers aimed to associate themselves with innovative topics, which are transforming the automotive industry, it could be expected that this is reflected in the positioning of innovative keywords and their associated network centrality. This leads to our first research hypothesis:

> *H1: Car manufacturer web pages dealing with innovative topics tend to be higher ranked than other web content.*

Furthermore, firms eager to display their innovative efforts in a good light should be more likely to describe important content with positive language as a means of customer communication. Accordingly, we hypothesise:

> *H2A: Well connected pages are more likely to show a positive sentiment than peripheral ones.*
> *H2B: Pages focusing on innovative topics are characterised by a positive sentiment.*

To test this, we perform sentiment analysis on the content of each web page in the network. Additionally, the web page hierarchy should allow to gain insights into the international orientation of the firms. This leads to our last hypothesis:

> *H3: The size of car manufacturer country domain networks corresponds to the country market size.*

This hypothesis is examined via analysing the prevalence of different country domains and target markets throughout the network.[3]

4 Methodology

Our approach consists of four steps. First, we apply web-crawling on the company web pages in order to obtain the network data. Secondly, we visualise the networks and derive properties and centrality measures from the data. Thirdly, we search for relevant keywords on innovative topics and apply sentiment analysis to the web pages. Lastly, we compare the manufacturers with respect to their international orientation in 25 national markets.

After retrieving the network data, we turn towards a comprehensive analysis of link structure and node meta-information, using a number of tools. We use the graph visualisation tools *Gephi* [31] and *Graphviz* [32]. The Python package *NetworkX* [33] is used for obtaining the network and node properties. The search engine optimisation software *Screaming Frog* and *SEO Powersuite* are employed to get additional node information.

---

[2] For this part of the analysis, which hinges on the identification of specific keywords, we analyse the US domains of the company websites (www.toyota.com, www.vw.com, www.hyundaiusa.com), as the United States is the second largest global car market behind China, and the largest English-language car market.

[3] For the international comparison, we use the manufacturers' international web pages (www.toyota-global.com, www.volkswagen.com, www.hyundai.com/worldwide), as a starting point for the data collection.

To obtain the network data, we implement a crawler able to retrieve an extensive number of pages from the web. For this purpose, we use the *Python* package *Beautiful Soup* for HTML and XML parsing. Essentially, the crawler starts at a single initial web page, where it retrieves the web page including all it's hyperlinks. Following the breadth-first paradigm, it then visits all the web pages that the start page links to and stores them as nodes connected via the hyperlinks (links). The crawler then repeats this process on the second level of web pages before it goes on to the next level. We terminate the crawler after the hyperlinks on the 6th level have been added to the network (Fig. 1A). The data was collected in December 2018.

Based on the key themes identified in the literature [3–7], we establish three major innovative trends affecting the automotive industry: (1) 'e-mobility and environment', (2) 'autonomous driving and artificial intelligence', and (3) 'connectivity and shared mobility'. These trends are identified in the textual components[4] of each web page by counting relevant keywords.[5] For the sentiment analysis, we use the natural language processing toolbox *TextBlob* [34] and analyse the textual components of each individual web page with regards to their polarity on a continuous, symmetric sentiment scale ranging from −1 (negative) to +1 (positive) [27].

To capture the international orientation of the car manufacturers, we examine the peculiarity and prevalence of web pages of different country domains, starting from the international landing page of the three firms. The domain affiliation is derived from the country tag, e.g '.ca' for Canada, '.uk' for United Kingdom etc. The number of web pages per country domain are then compared with the size of the national market in terms of car sales and registrations [35].

At this point, we want to emphasise that this study is exploratory: the purpose is to investigate the value of complementing content analysis with network centrality measures on the example of car manufacturer web pages. Further research is needed to test the generalisability of the findings beyond the case presented here (for example in comparing different industries, contrasting the web page findings with patent data, and tracking developments over time).

---

[4] The textual components are all HTML tags (predominantly "title" and "body") of a web page. Specifically excluded are the tags "script", "style", "head", "[document]". This way, we only include textual components visible to the user.

[5] The list of keywords has been created prior to looking at any company website, based only on the qualitative definitions in the literature. The respective keywords are:

E-mobility & environment: e-mobility, battery, environment, biological, eco, ecological, electric, hybrid, environment, environmental-friendly;

Connectivity & shared mobility: connectivity, shared, mobility, sharing, interconnectedness, cloud, cloud computing, wifi, 5G;

Autonomous driving & artificial intelligence: autonomous, self-driving, ai, machine learning, artificial intelligence, intelligent, neural network, algorithm.

5 Results

5.1 Network Structure

The obtained web page networks of the US domains of Hyundai, Toyota, and Volkswagen show distinct structures (Fig. 1B). In the Toyota and Volkswagen networks, several navigational pages connect the major elements of the websites, while the web pages of Hyundai fan out into only two major components.

The networks also differ with regards to the positioning of innovative topics.[6] In contrast to Volkswagen, Hyundai and Toyota refer to 'e-mobility & environment' in large parts of their networks; also in the sections 'model-selection' and 'configurator'. This result most likely reflects their focus on hybrid vehicles. All manufacturers refer to 'connectivity & sharing' only in peripheral categories ('offers' and 'operations'). The topic 'autonomous driving & artificial intelligence' is discussed in designated sections on Hyundai's 'newthinking' and Volkswagen's 'media' website; in contrast, it is essentially absent from Toyota's web page. These results provide support for hypothesis $H1$: the manufacturers tend to display content on innovative topics at prominent places in their web page networks.

Despite these differences, the networks show a similar degree distribution and network properties (Fig. 1C and Fig. 1D). According to the high nodal average degree of 93 and a network density of 0.07, the web page network of Toyota with its 2,654 nodes and 353,327 links is the largest and most densely connected one.

5.2 Sentiment and Centrality

To understand the role of positive language on the manufacturer web pages ($H2$), we display the Volkswagen network coloured according to sentiment (Fig. 2A). Qualitatively, we note a positive correlation between node degree and sentiment: more central nodes appear to have a more positive sentiment. As the inset shows, node degree centrality and sentiment are actually characterised by an inverse U-shaped relation. In contrast to our initial hypothesis $H2A$, not the most central pages, but intermediately ranked pages reveal the most positive sentiment. These pages are most likely content-driven (e. g. pages that are linked to in the 'media' or 'newsroom' sections of the website). Less central nodes might rather describe specific topics such as warranty issues or model specifications, and the most central nodes appear to fulfil navigational purposes, characterised by a more neutral sentiment. Awareness of this arguably unintentional but yet noticeable use of language can provide a competitive edge, when firms consistently associate their content on innovative topics with positive sentiment.

In all three manufacturer web page networks, sites dealing with innovative content tend to be more central than other web pages (Fig. 2B, left panel); providing quantitative support for $H1$. The figure displays the distribution of

[6] In the network visualisations, the nodes are coloured according to the keyword-category that appears most often in a web page. If none of the keywords occurs, the node is coloured in grey.

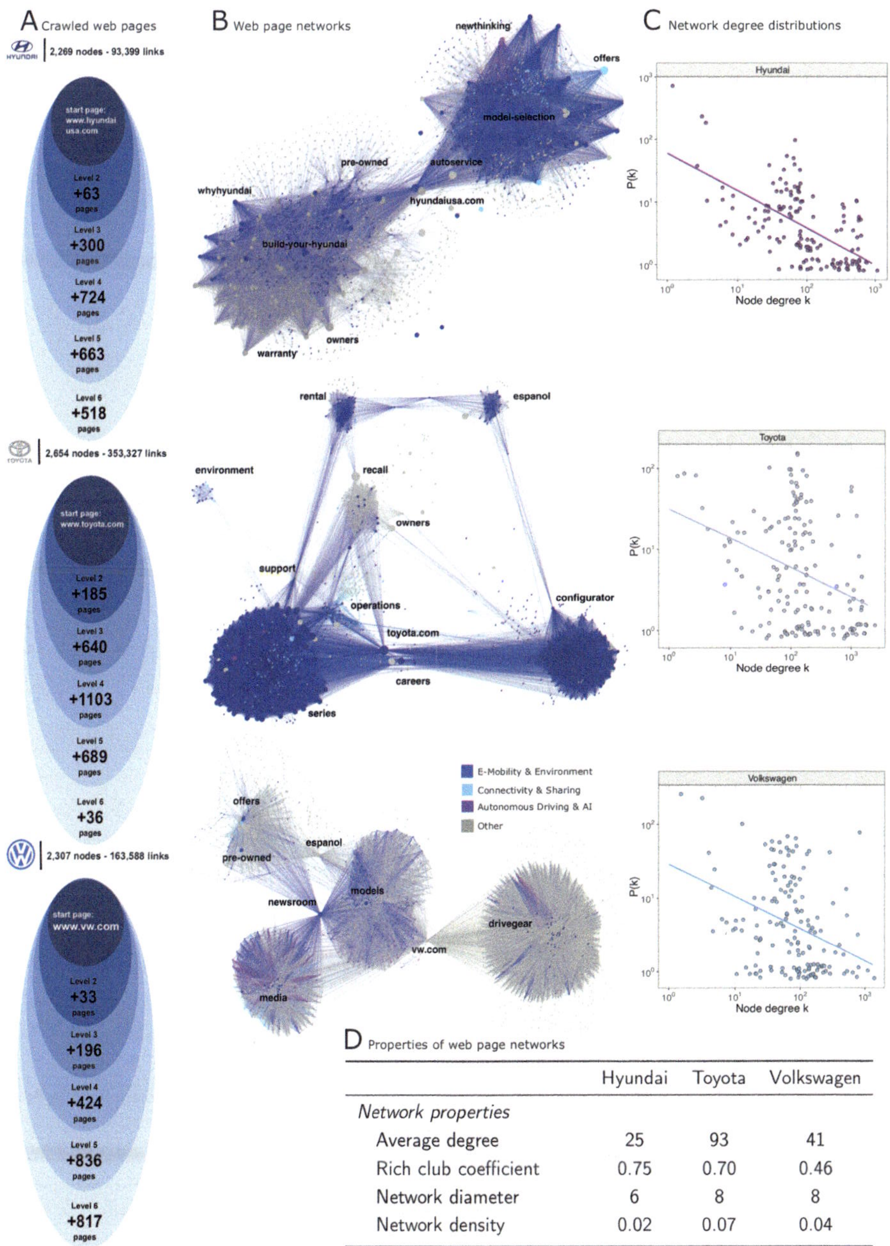

	Hyundai	Toyota	Volkswagen
Network properties			
Average degree	25	93	41
Rich club coefficient	0.75	0.70	0.46
Network diameter	6	8	8
Network density	0.02	0.07	0.04

Fig. 1. **(A)** Crawled web pages of three car manufacturers: crawlers collected all sub-pages (nodes) and hyperlinks (links: in- and out-links are combined as undirected links) to a depth of 6 levels in a breadth-first manner. **(B)** Resulting web page networks: pages mentioning keywords on innovative trends are highlighted. The networks differ in their structure and positioning of innovative contents. **(C)** Degree distribution of the networks **(D)** Network Properties: average node degree, average rich club coefficient, network diameter, and network density vary between the three manufacturer web pages.

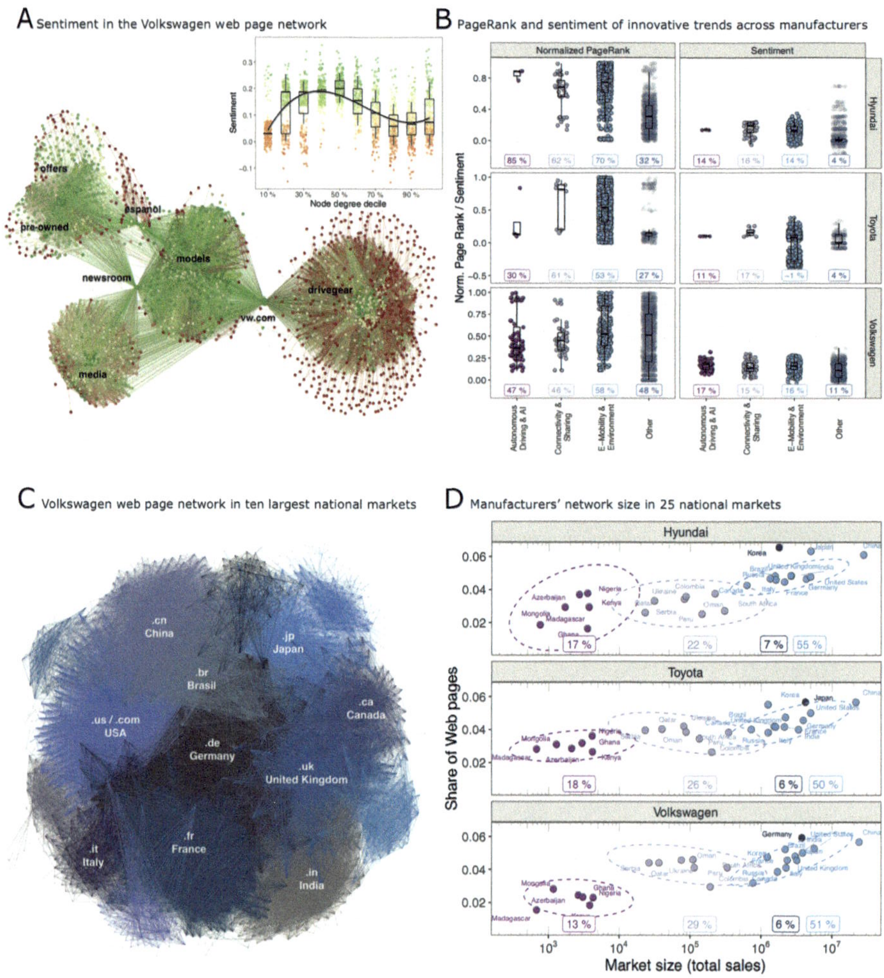

Fig. 2. (**A**) Distribution of sentiment in the Volkswagen web page network: nodes with intermediate degree centrality show, on average, a more positive sentiment (inset). (**B**) Distribution of PageRank centrality and sentiment per topic and manufacturer: pages on innovative topics (in particular e-mobility and environment) tend to be higher ranked and to have more positive sentiment than other sub-pages. (**C**) International Volkswagen web page network (crawled from international landing page) in ten largest car markets: Volkswagen's German home market network forms a central cluster, densely connected with other key markets in Europe and North America. (**D**) Global sales (log-scale) and share of web pages in market groups: the three manufacturers similarly focus their web presence on a cluster of large car markets, with their home market web page network being disproportionately extensive.

the normalised PageRank per page (PageRank divided by the highest ranked page's value) and the average rank per category (labels below the box plots). Pages with the theme 'e-mobility & environment' are, on average, more central than 'other' web pages with a normalised PageRank of 0.7 vs. 0.32 (Hyundai), 0.53 vs. 0.27 (Toyota), and 0.57 vs. 0.48 (Volkswagen).

With regards to hypothesis $H2B$, sites on innovative themes use more positive language (Fig. 2B, right panel). An exception is Toyota's content on e-mobility with a slightly negative average sentiment of −0.01. We could not find statistical significant relations between page centrality and sentiment per topic category. Nonetheless, the results suggest that car manufacturers tend to display content on innovative trends at more prominent positions in their networks. These findings emphasise the importance of linking content analysis with structural properties that maintain the hierarchical structure of the analysed content. Considering the web page network allows for identifying the manufacturers focus on certain topics. Without considering the network, the firms prioritisation of these topics would have not been detectable, given the relative low count of pages with innovative content (in particular on 'AI' and 'Connectivity' related topics).

5.3 Country Domain Analysis

For analysing the international orientation of the firms, we crawl web page network data from the firms international landing pages. Figure 2C shows the Volkswagen network of the sub-domains referring to the ten largest national markets in terms of car sales and registrations. The country domains agglomerate in densely connected clusters of different size, indicating the importance of the respective target markets. Volkswagen's German web page cluster is disproportionately large and connected with most other markets. Hence, it is at the core of the international network, reflecting Volkswagen's focus on it's home market.

This finding does not only hold for Volkswagen, but also for the other two car manufacturers (Fig. 2D). The figure shows the size of the sub-domain networks (number of web pages) per manufacturer in 25 national markets for which we could obtain the number of passenger car sales [35]. Applying a k-means clustering algorithm on the number of sales (log-scale) yields three distinct groups of national markets: small markets of African and Asian countries (less than 1,000 cars sold in 2017), a second cluster of Global South countries with an intermediate market size (1,000–1,000,000 cars), and group of twelve large markets (more than 1 million cars), consisting of OECD and Newly Industrialised countries.

As hypothesised ($H3$), the manufacturers show a similar pattern in terms of their web presence in these groups: 13–18% of the firms' web pages target the group of small markets, while 22–29% relate to the medium-sized markets. With 50–55% of all their web pages, the firms' online presence clearly focuses on the group of large national markets with more than a million annually sold cars. The home market's web page network of each company is disproportionately large with 6–7% of all web pages. This finding provides insights into the competitive strategies of the three firms, as they clearly focus their online marketing activities and product range to a limited number of particularly important markets.

6 Discussion

6.1 Summary

In the digital era, the automotive industry remains one of the cornerstones of global manufacturing, not only in terms of employment and trade, but more importantly for its role in introducing new technologies [9]. This work presents a network-centred approach to gain insights into the innovative focus of three large car manufacturers by analysing the firms' web pages. In crawling publicly available content and structural meta-information from the websites of Toyota, Hyundai, and Volkswagen, we construct complex networks and analyse their properties. The interplay of content and hierarchical meta-information reveals meaningful patterns hidden from conventional web page content analysis.

The analysis of the web page networks reveals that the three firms centrally present content on the topic 'e-mobility & environment'; however, the topics 'connectivity & sharing' and 'autonomous driving & artificial intelligence' are dealt with only in more specialised sections. The companies tend to make use of more positive language on pages on innovative trends and emphasise such content at prominent places of the networks. Our analysis of national sub-domains shows that the manufacturers concentrate their online marketing efforts on a limited number of target markets, with a particular focus on their home market. Our approach exploits only publicly available information and is easily reproducible. It promotes transparency and could be used as a complementary tool to monitor shifting industries and competitive landscapes, as it helps to discriminate unstructured textual web data by relevance, which is implicitly captured by the hierarchical web page network structure.

6.2 Limitations and Future Research

Since this is an exploratory study, it has some limitations and future research is needed to establish generalizability and applicability of our findings beyond the case presented. Web-crawling in an iterative manner can yield instable results, depending on the starting page of the crawler. Thus, data collection should comprise an ensemble of methods conducted from several starting pages. Moreover, websites change frequently; hence, our approach is time-sensitive. To leverage it's potential, the websites should be monitored over longer periods of time, so that evolving structures and topics can be identified [36].[7]

Moreover, mentioning digital trends does not necessarily accompany putting the digital transformation into practice, but it demonstrates awareness and prioritisation. If many highly ranked pages are concerned with digital trends, the website is more likely to be presented in a search engine's result when a user searches for "e-mobility", "autonomous driving", or other innovative trends.

[7] Alternatively, the web page structures of those countries could be compared, in which the manufacturers apply different marketing strategies (there might be markets where the adoption rates of digital technologies in the automotive sector are higher).

References

1. Langley, M., World Economic Forum: Digital transformation: understanding the impact of digitalization on society. World Economic Forum (2017)
2. European Economic and Social Committee: Impact of digitalisation and the on-demand economy on labour markets and the consequences for employment and industrial relations. European Economic and Social Committee (2017)
3. Weinelt, B., World Economic Forum: Digital transformation: digital transformation of the automotive industry. World Economic Forum (2016)
4. OECD: Recent developments in the automobile industry. Economics Department Policy Notes, no. 7 (2011)
5. Knoedler, D., Wollschlaeger, D., Stanley, B.: Automotive 2030: racing toward a digital future. IBM Institute for Business Value (2019)
6. Mohr, D., Gao, P., Kaasm, H.W., Wee, D.: Disruptive trends that will transform the auto industry. McKinsey & Company (2016)
7. Kuhnert, F.: Five trends transforming the automotive industry. PwC (2018)
8. OICA: Total manufacturing employment 2018. http://www.oica.net/category/production-statistics/2017-statistics/. Accessed 16 Dec 2018
9. Ferrazzi, M., Goldstein, A.: The new geography of automotive manufacturing. International Economics, Chatham House (2011)
10. Braesemann, F., Stoehr, N., Graham, M.: Global networks in collaborative programming. Reg. Stud. Reg. Sci. **6**(1), 371–373 (2019)
11. Stephany, F., Braesemann, F.: Coding together - coding alone: the role of trust in collaborative programming. SocArxiv preprint https://doi.org/10.31235/osf.io/8rf2h (2019)
12. Stephany, F., Braesemann, F.: An exploration of Wikipedia data as a measure of regional knowledge distribution. In: International Conference on Social Informatics, vol. 10540, pp. 31–40 (2017)
13. OICA: World motor vehicle production 2016. http://www.oica.net/wp-content/uploads/World-Ranking-of-Manufacturers.pdf. Accessed 16 Dec 2018
14. Traub, M., Voegel, H., Sax, E., Streichert, T., Haerri, J.: Digitalization in automotive and industrial systems. In: Design, Automation & Test in Europe Conference & Exhibition (DATE), 1203–1204 (2018)
15. Fridman, L., et al.: MIT autonomous vehicle technology study: large-scale deep learning based analysis of driver behavior and interaction with automation, arXiv preprint arXiv:1711.06976 (2017)
16. Günther, H.-O., Kannegiesser, M., Autenrieb, N.: The role of electric vehicles for supply chain sustainability in the automotive industry. J. Clean. Prod. **90**, 220–233 (2015)
17. Thoben, K.D., Wiesner, S., Wuest, T.: "Industrie 4.0" and smart manufacturing a review of research issues and application examples. Int. J. Autom. Technol. **11**(1), 4–19 (2017)
18. Kito, T., Brintrup, A., New, S., Reed-Tsochas, F.: The structure of the Toyota supply network: an empirical analysis. Said Business School WP 3 (2014)
19. Swaminathan, A., Hoetker, G., Mitchell, W.: Network structure and business survival: the case of US automobile component suppliers. University of Illinois at Urbana-Champaign (2002)
20. Baggio, R., Corigliano, M.-A., Tallinucci, V.: The websites of a tourism destination: a network analysis. In: Information and Communications Technologies in Tourism, pp. 279–288 (2007)

21. Wang, Y., Wang, D., Ip, W.H.: Optimal design of link structure for e-supermarket website. IEEE Trans. Syst. Man Cybern. Part A: Syst. Hum. **36**(2), 338–355 (2006)
22. Broder, A., et al.: Graph structure in the web. Comput. Netw. **33**(1–6), 309–320 (2000)
23. Meusel, R., Vigna, S., Lehmberg, O., Bizer, C.: Graph structure in the web—revisited: a trick of the heavy tail. In: Proceedings of the 23rd International Conference on World Wide Web, pp. 427–432 (2014)
24. Barabási, A.-L., Albert, R.: Emergence of scaling in random networks. Science **286**(5439), 509–512 (1999)
25. Page, L., Brin, S., Motwani, R., Winograd, T.: The pagerank citation ranking: bringing order to the web. Stanford InfoLab (1999)
26. Ortiz-Cordova, A., Jansen, B.J.: Classifying web search queries to identify high revenue generating customers. J. Am. Soc. Inf. Sci. Technol. **63**(7), 1426–1441 (2012)
27. Falck, F., Marstaller, J., Stoehr, N., et al.: Measuring proximity between newspapers and political parties: the sentiment political compass. Policy Internet (2019). https://sci-hub.tw/https://onlinelibrary.wiley.com/doi/pdf/10.1002/poi3.222
28. Gowda, T., Mattmann, C.A.: Clustering web pages based on structure and style similarity (application paper). In: IEEE 17th International Conference on Information Reuse and Integration (IRI) (2016)
29. Wan, H.A., Chung, C.-W.: Web page design and network analysis. Internet Res. **8**(2), 115–122 (1998)
30. Sahebi, S., Oroumchian, F., Khosravi, R.: An enhanced similarity measure for utilizing site structure in web personalization systems. In: 2008 IEEE/WIC/ACM International Conference on Web Intelligence and Intelligent Agent Technology, vol. 3, pp. 82–85 (2008)
31. Bastian, M., Heymann, S.., Jacomy, M.: Gephi: an open source software for exploring and manipulating networks. In: Third International AAAI Conference on Weblogs and Social Media (2009)
32. Ellson, J., Gansner, E.R., Koutsofios, E., North, S.C., Woodhull, G.: Graph drawing software: Graphviz and dynagraph - static and dynamic graph drawing tools, pp. 127–148. Springer (2003)
33. Schult, S.A., Hagberg, A.A., Swart, P.J.: Exploring network structure, dynamics, and function using NetworkX. In: Proceedings of the 7th Python in Science Conference (SciPy2008) (2008)
34. Loria, S.: TextBlob: Simplified text processing (2018). https://textblob.readthedocs.io/en/dev/index.html. Accessed 16 Dec 2018
35. OICA: Personal car registrations and sales 2017. http://www.oica.net/wp-content/uploads/Sales-Passenger-cars-2017.pdf. Accessed 16 Dec 2018
36. Stoehr, N., Yilmaz, E., Brockschmidt, M., Stuehmer, J.: Disentangling interpretable generative parameters of random and real-world graphs. arXiv:1910.05639 cs.LG, (NeurIPS, Graph Representation Learning) (2019)

Finding the Worldwide Industrial Transfer Pattern Under the Perspective of Econophysics

Lizhi Xing[1,2](✉) and Yu Han[1]

[1] Beijing University of Technology, Beijing 100124, China
itwasa@163.com
[2] Indiana University, Bloomington, IN 47408, USA

Abstract. Industrial transfer is the inevitable trend of economic development. The traditional industrial transfer theory tends to adopt partial data and methodologies from reductionism, and thus can't tackle with the highly non-linear systematic problems like the mechanism and evolution path of international, regional, and domestic industrial transfer. With the properties of structural complexity, dynamic evolution and multiple linkages, complex networks can better reflect the interdependent and mutually restricted relation between different levels and components of the industrial structure, pinpoint the optimization and control nodes. Currently, there are only a few available researches on such weighted, directed and dense networks reflecting the topological complexity of global value chain, with the results being unsystematic and impractical. This paper utilizes the available ICIO data to build the Binary GISRN model in accordance with crucial flows of materials, energy, and information among industrial sectors all over the world. Also, methods of defining and measuring the networks' redundancies are devised to figure out the trigger of worldwide industrial transfer pattern according to the link prediction method, thus blazing a new trail for the evolutionary economics.

Keywords: Global Value Chain · Inter-Country Input-Output table · Complex network · Network pruning · Link prediction · Industry transfer pattern

1 Introduction

Industrial transfer refers to the phenomenon that in market economy, some enterprises in developed regions adapt to the changes in regional comparative advantages and then relocate part of their manufacturing capacity to the developing regions through cross-regional direct investment, resulting in the shift of the spatial distribution of industrial sectors on the ***Global Value Chain (GVC)*** from developed regions to developing regions. Take the global manufacturing sectors as an example. From the first industrial revolution to the

H. Barbosa et al. (Eds.): *Complex Networks XI*, SPCOM, pp. 309–321, 2020.
https://doi.org/10.1007/978-3-030-40943-2_26

present, three waves of industrial transfers have taken place, each reshaping the global economic structure. The first industrial transfer was in the early 20th century when the United Kingdom transferred "excessive industries" to the United States, which now turns into the most powerful country in the world. The second wave occurred in the 1950s. The United States' traditional industries, such as textiles and iron were relocated to the defeated countries–Japan and Germany, who then transferred low-value labor-intensive industries to Taiwan, Hong Kong, South Korea, and Singapore, now known as the Four Asian Tigers. The third wave started with the reform and opening up of China in the 1980s. At that time, China, as the core of the third industrial transfer, became the destination of various industries from all over the world.

Described by a complex network model, the problem of industrial transfer can be solved by network evolution, which is reflected by new nodes introduced into the network, new edges between nodes, the variation of edge weight, to name a few. In this case, analyzing the rules and mechanisms of network evolution from the perspective of econophysics can solve the industrial economic problems, thus better applying complex network theory to the study of industrial economics.

In our opinion, research on the mechanism and path choice of regional industrial transfer based on ***Link Prediction*** [1] is at the leading edge of researches on the regional industrial transfer from the weighted to the dense network at home and abroad, featuring prospects, innovativeness, and high application value. At present, the most urgent task is to figure out how to combine the regional industrial system with the complex network theory and establish a new concept, new theory and new method based on link prediction method, and then provide a more complete theoretical decision model and the corresponding algorithms for the decision-making in regional industrial transfer.

2 Modeling

2.1 Modeling Data

Industrial transfer also refers to the enterprise-driven transfer of a country's industrial sectors due to changes in resource supplies or product demands, which is a dynamic process in both time and space. As an important way for countries/regions to readjust and upgrade their industrial structure, it is the economic behavior that enterprises or industrial sectors at large relocate from one region to another following the flow of production factors. If the research objective were the industrial transfer phenomenon on the GVC, the ***Inter-Country Input-Output (ICIO)*** data would have been needed to construct the network model. ICIO data have proven itself to be a reliable source for analyzing economic globalization. Thanks to it, sectors all over the world can form a sophisticated GVC, bringing the advantages of simultaneous study on international and domestic economies in detail as a holistic network.

2.2 Network Modeling

To establish an industrial complex network, a sector within a region is considered as a node and the inter-industry ***Input-Output (IO)*** relation as a tie, and its weight represents the sale and purchase relations between producers and consumers. Thus, a graph $G = (V, E, W)$ containing n nodes is drawn to represent sectors within a nation or region and denote a node set V. Pairs of nodes are linked by ties to reflect their interdependencies, thereby forming an asymmetric edge set E. However, in valued graphs, a set E can be replaced by weight set W, which can be extracted from the region of inter-country and inter-industry use and supply in ICIO table.

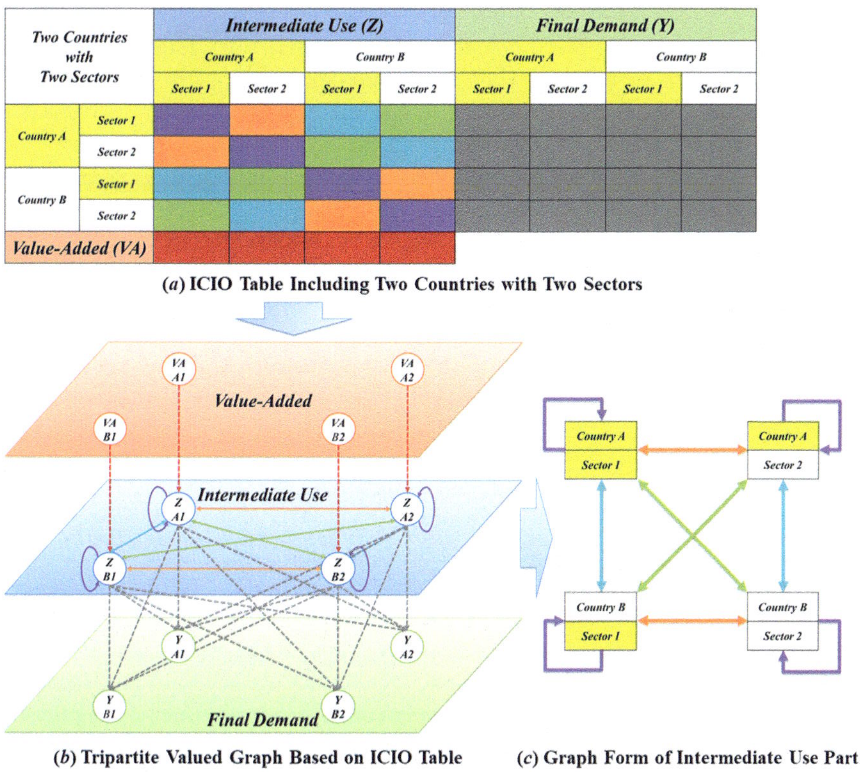

Fig. 1. The relationship between ICIO table and GVC network.

Note that, typical IO or ICIO table includes three different areas, namely ***Value-Added, Intermediate Use and Final Demand***. It is possible that the whole global economic system can be abstracted to a ***Multi-Layer Network*** as shown in Fig. 1(b). Also, the intermediate use layer can construct multi-layer network too if every country/region is taken as a single layer [2].

In this paper, the table of intermediate inputs in ***Eora Multi-Region Input-Output Table Database (MRIOV199.82***, and the simplified version with 26-sector harmonized classification is named ***Eora26)*** [3] is straightforwardly adopted to build graph G for it covers the largest number of countries and the longest period among all the ICIO databases, which is the aggregation of every single ***Industrial Value Chain (IVC)*** in the major economic entities. We name this single layer ICIO network model ***Global Industrial Value Chain Network (GIVCN)*** [4] since its purpose is to reflect how economic shocks propagate and expand along the GVC, as well as to what extent the industrial impact generates on the national level.

2.3 Network Pruning

There is an important premise lying ahead of link prediction. That is, the GIVCN model is too dense to embody the main topological structure of an economic system (the number of edges, including self-loops, is almost equal to the square number of nodes), which means useful forecasting results with existing methods are unavailable. No matter how we plan to do, the first mission is network pruning.

By comparison of the numerical matrix of ***Strongest Relevance Path Length (SRPL)*** [4] and edge weight, we notice that some elements in the same place of them are identical. That is to say, a part of SRPLs between any two sectors (could be the same one) is directly equal to their IO values. This phenomenon means there exists both the strongest and the most immediate industrial relevance in the IO network. Thus, these same elements could be extracted to form a new matrix, and the equations are:

$$\tilde{w}_{ij} = \begin{cases} w_{ij}, & w_{ij} = SRPL_{ij}^{(N)} \\ 0, & otherwise \end{cases} \tag{1}$$

Taking $\tilde{W} = \{\tilde{w}_{ij}\}$ as an adjacency matrix, we establish ***Global Industrial Strongest Relevant Network (GISRN)*** to abstract the optimized value chains within the scope of the global production system, which could be called the ***"Backbone of GVC"***. Notice that, a given node's self-loop may disappear in this newly refined network, which depends on whether its inner-node relevance path length is a SRPL or not.

Since the GISRN model has retained the crucial linkages between industrial sectors, there is no need to achieve this purpose through the heterogeneity of network, which is based on the edge weights. Besides, the self-loops, standing for the intra-industry consumption, also become useless to predict the inter-industry relations. Link prediction will, therefore, be carried out in six binary GISRN-Eora26 models (build a model every four years between 1990 and 2015). The topological structure of binary GISRN-Eora26-2015, which contains 4,914 industrial sectors and 18,085 inter-industry IO relations in 189 countries/regions (much less than the square number of industrial sectors), is as shown in Fig. 2.

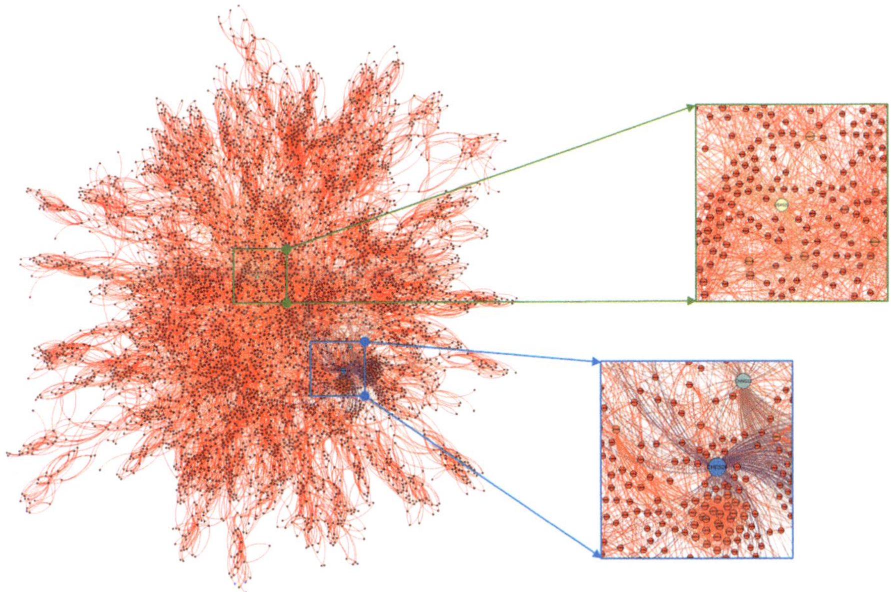

Fig. 2. Binary GISRN-Eora26-2015.

After a large-scale industrial transfer, the relationship between the relevant industry sector (whether it is transferred or transferred) and its neighboring ones will change significantly. In fact, despite of the various political factors, all kinds of the possibility at the economic level have long been embedded in the topological structure of GVC. Therefore, based on the Binary GISRN model, an attempt has been made to restore the dynamic mechanism of industrial transfer at the global level through the link prediction algorithms, which is also the evolutionary mechanism of GVC.

3 Methodology

We selected from three types of existing structural similarity indices and used those that fit Binary GISRN model's features to carry out the link prediction.

3.1 Similarity Index Based on Local Information

Common Neighbors. The simplest similarity index based on local information is the ***Common Neighbors (CN)*** index. The basic assumption of the application of CN index in the link prediction is that two of the non-connected nodes tend to be connected if they have a lot of common neighbors. Therefore, it is defined as: for the nodes i in the network, its neighbor set is $\Gamma(i)$, and then the similarity of nodes i and j will be defined as the number of their common neighbors. That is:

$$S_{ij} = |\Gamma(i) \cap \Gamma(j)| \tag{2}$$

where the right side of the equation represents the potency of set. Obviously, the number of their common neighbors is equal to the number of two-order paths between nodes i and j, which is $S_{ij} = (A^2)_{ij}$.

Adamic-Adar. If we consider the information about the degree of two nodes' common neighbors, the ***Adamic-Adar (AA)*** index will be a good choice. The idea of AA index is: the common neighbor nodes with a small degree will contribute more than those with a larger degree. AA index gives, according to the degree of the CN nodes, each node a weight, which is equal to the reciprocal of the log of the node's degree. In this manner, the AA index is defined as:

$$S_{ij} = \sum_{k \in |\Gamma(i) \cap \Gamma(j)|} \frac{1}{logK(k)} \tag{3}$$

Resource Allocation. The definition of ***Resource Allocation (RA)*** index is that, given that there is no directly connected nodes i and j in the network, some resources can be allocated from node i to j, during which their common neighbors will become the transmission medium. If each medium has a unit of resources and allocates them to its neighbors, the number of resources accessible to node j can be defined as the similarity between nodes i and j, i.e.:

$$S_{ij} = \sum_{k \in |\Gamma(i) \cap \Gamma(j)|} \frac{1}{K(k)} \tag{4}$$

Preferential Attachment. The scale-free network structure can be generated with the method of ***Preferential Attachment (PA)***, in which the probability of a new edge connected to a node i is proportional to $K(i)$. This mechanism has also been applied to the network without consideration of growth, for instance, a link is removed at the first step, with another link added. The probability of the new edge connecting nodes i and j is directly proportional to the product of the degree of two nodes. From this, we can define the PA index as:

$$S_{ij} = K(i)K(j) \tag{5}$$

3.2 Similarity Index Based on Path

Considering the potential influence of the three-order path range based on CN index, we can define the similarity index based on ***Local Path (LP)*** as:

$$S = A^2 + \alpha A^3 \tag{6}$$

where α is an adjustable parameter, A represents the adjacency matrix of the network, and $(A^3)_{ij}$ represents the number of paths whose length between nodes i and j is 3. When $\alpha = 0$, the *LP* index will be degraded to *CN* index. In other words, *CN* index can be regarded as an index based on the path in nature, but it only considers the number of two-order paths. *LP* index can be extended to cases of higher-order, i.e. the case of n-order:

$$S = A^2 + \alpha A^3 + \alpha^2 A^3 + \cdots + \alpha^{n-2} A^n \tag{7}$$

As n increases, the computational complexity of the *LP* index will be increasing. In general, the computational complexity of n-order path is $O(N \langle K \rangle^n)$. But when $n \to \infty$, the *LP* index is equivalent to the *Katz* index concerning all the paths of the network.

3.3 Similarity Index Based on Random Walk

Many kinds of similarity indices are defined based on the random walk process, including ***Average Commute Time (ACT), Cos+, Random Walk with Restart, SimRank,*** etc. Considering the binary GISRN-Eora26 model is a small-scale network, the ACT index is selected as the representative of this sort of index.

From the view of the whole network, the ***Mean First Passage Time (MFPT),*** denoted by $E(i, j)$, is the expected number of steps when a random walk starts at source node i needs to reach sink node j for the first time [4]. Then the ACT between nodes i and j is:

$$n(i, j) = E(i, j) + E(j, i) \tag{8}$$

The smaller the ACT of the two nodes is, the closer the two nodes will be. Then we can define the similarity based on the ACT:

$$S_{ij}^{ACT} = \frac{1}{(E(i, j) + E(j, i)} \tag{9}$$

In directed but unweighted network, $E(i, j)$ probably not equals to $E(j, i)$, which means ACT index depends on bidirectional MFPT.

3.4 Econophysics Background

In sum, from the performance of link prediction algorithms, we can divide the 6 indices into four categories: the first one includes the *LP* index and the *CN* index, the prediction accuracy of which are both proportional to the importance of common neighbors of given nodes, and the latter is the special case of the former (i.e. the adjustable parameter of *LP* index is zero); the second one that is based on the opposite assumption of the first one includes the *RA* index and the *AA* index, and the difference between them is that the *RA* index is more sensitive to the heterogeneity of network and thus prior to the *AA* index in the case of higher average degree centrality; the third one is the ACT index, which measures the distance of inter-node information transfer path from two directions; the fourth category is the PA index, which is correlated to certain importance of the nodes themselves, not to the path between them.

According to the law of gravity, we can further explain the econophysics significance of the ACT index and the PA index. The former only relates to the inter-node distance (the expected number of steps) is a concept in the sense of probability during the random walk, which is probably not equal in the forward and backward directions. Therefore, the premise of ACT index is that inter-industry closeness is inversely proportional to their integrated round-trip distance on the GVC. The latter, on the other hand, ignores the influence of distance and is only related to certain importance of a node (just like the mass of matter), which means the bigger the local influence of two industrial sectors, the closer the relationship between them. In other words, they respectively concern about the denominator and numerator of the formula for gravity, and their econophysics hypothesis are as follow: the influence of industrial sector in the backbone network of GVC can promote industrial transfer, and the length of inter-industry IVC is one of the factors hindering industrial transfer.

4 Experimental Results

4.1 Overall Statistics

Considering the relative stability of the industrial structure in a short period, the proportion of E^P is set as 10%. In the simulation, the LP index sets the weight calculated in the second step as 0.5. That is to say, the indirect influence caused by the adjacent nodes is attenuated to 50%. During the accuracy calculation of ***Area under the Receiver Operating Characteristic Curve (AUC)***, the more the number of samplings, the closer the result is to the accurate value, so each algorithm sets the sampling frequency to 10,000 times in this step. Finally, the link prediction simulation results from 24 binary GISRN-Eora26 models are shown in Table 1.

Table 1. Accuracy of Link Prediction Results in Binary GISRN-Eora26 Models.

Year	CN	AA	RA	PA	LP	ACT
1990	0.717	0.689	0.711	0.792	0.827	0.746
1995	0.712	0.689	0.700	0.802	0.826	0.788
2000	0.715	0.681	0.711	0.775	0.822	0.791
2005	0.708	0.684	0.693	0.775	0.816	0.760
2010	0.719	0.662	0.725	0.797	0.801	0.777
2015	0.708	0.682	0.708	0.786	0.826	0.749

In Table 1, the *LP* index represents the highest accuracy, and the *AA* index the lowest. From the results, it can be concluded that there is a discrepancy shown in the conditions and tendencies of industrial transfer on the GVC and within regions. In the following parts, we explicate what characteristics each index can reflect on the industrial transfer at a global level.

4.2 Industrial Convergence

The *LP* index, derived from the *CN* index, takes into consideration the relationship between industrial sectors and global upstream and downstream sectors, as well as direct trading with other countries/regions. Moreover, it expands the study on industrial transfer possibility, along the value chain and the supply chain, to a broader GVC level. However, higher-order local path algorithms (for example, the *Katz* index when $n \rightarrow \infty$) do not apply to the binary GISRN-Eora26 model. This is because most of industrial sectors on the GVC only have strong connection to their neighboring sectors (they may be the different sectors in the same country/region, or the similar sectors in other countries/regions), and the internal relationship of the industrial sector will be compromised if the assessment of industrial transfer trends is expanded to the GVC level. This is just like predicting the distribution of sectors close to the production end through changes in close-to-market-end sectors (the retail industry), not to mention that industrial sectors in binary GISRN-Eora26 models spread all over the global economic system.

According to the results, the *LP* index is better than the *CN* index. This shows that industrial transfer on the GVC, driven by economic globalization, gradually develops toward the goal of ***Industrial Convergence***. From the perspective of the information and communication industry, industrial convergence means that the industrial boundaries blur based on technical and digital convergence. From the perspective of its causes and process, industrial convergence is regarded as a process that gradually completes technical convergence, product and business convergence, market convergence, and finally industrial convergence. From the perspective of product services and industrial organization structure, as the function of a product changes, the boundaries of the institutes or corporate organizations start to blur. From the perspective of industrial innovation and development, it refers to the dynamic development process wherein different or the same industries, based on technological and regulatory innovation, interpenetrate, interweave and in the end blend into one, gradually acquiring new industrial forms.

4.3 Mega-Merger Tendency

The *PA* Index, with the second-best predictive effect in binary GIRSN-Eora26 models, gives rise to a crucial economic problem: industrial production specialization on the GVC brings greater influence to some industrial sectors, and between the two, the close IO relations are more likely to be established, thus creating the so-called ***Mega-Merger Tendency*** in the global economic system.

According to Ohlin's Factor Endowment Theory, presuming that two countries are at the same technological level of making a product, the discrepancy of prices would be due to different costs which arises from different prices of production factors. Further, prices of production factors depend on a country's relative abundance of factors, referred to as the endowment differences, and the price difference that consequently follows results in international trade and international division of labor [5]. The theory assumes that factors are homogeneous, having no difference, and can be transferred. Nevertheless, factor endowments of countries/regions are different in both quantity and quality, so it is difficult to plausibly explain the emergence of strong industrial sectors in a certain region if quality differences of factors are ignored.

Nowadays, vertical specialization can be found in every country/region worldwide. Differences in technological and capital factors have promoted industrial transfer on a global scale. At the same time, various multilateral trade agreements have removed barriers to market entry in many countries, and the technology diffusion effect is constantly increasing overall. As a result, competitive industrial sectors tentacles extend from domestic to international. According to the gravity model, this process can be described like that the breadth of the industrial sectors' impact (measured by the degree of corresponding nodes in the network model) may undermine the impact of location factors on the possibility of establishing relevance between sectors. However, in recent years, political games (e.g., British Brexit) and trade friction (e.g., China-US Trade War) between some countries/regions have impeded the flow of production factors from one country to another.

4.4 Industrial Agglomeration

As defined in the ACT index, if the adjacency matrix is asymmetric, the prediction accuracy of this index is inversely proportional to the sum of MFPTs in both directions. In the Binary GISRN model, this means if two industrial sectors are both upstream and downstream sector to one another with few medium sectors connecting them, the possibility of establishing direct linkages between them will be higher. In other words, industrial sectors will possibly form a kind of symbiotic relationship in the economic sense when that does happen, eventually resulting in the phenomenon of ***Industrial Agglomeration***. In the high-tech parks with a high degree of openness, thousands of domestic and foreign enterprises on the same IVCs stay together, significantly reducing the transaction costs and promoting the flow of various production materials and innovation elements. It is therefore reasonable to assume that the ACT index has the capability of interpreting the development mechanism of regional economic system from the perspective of industrial economics.

However, the ACT index still needs to be improved due to its application only in the unweighted network in this paper. If the common neighbors of two given nodes incorporate important hub nodes in the network, the possibility of shortcuts between them will notably increase, which also works for LP index and ACT index. Their real difference lies in whether the transfer efficiency after the first step will go down or not. Obviously, since the prediction accuracy of the ACT index is lower than the LP index in the Binary GISRN model, one should not ignore the attenuation of value transfer efficiency, which will even be aggravated with the extension of IVCs. In the follow-up study, the ACT index will be applied to the weighted and directed GIVCN model, because each IO relation embodies the local heterogeneity of GVC and constitutes the source of information asymmetry.

4.5 Niche Advantage

Both RA index and AA index are used to find pairs of nodes with weaker common neighbors and give them a higher chance of being connected. In the Binary GISRN model, this means if there are some structural holes on the GVC, i.e., the IO relations between some industrial sectors and their upstream and downstream ones appear to be weak, their peripheral sectors will establish direct supply chain beyond them. According to the analysis on the LP index, as the industrial boundaries become increasingly blurred, some medium sectors face the result of being marginalized or even eliminated. As a result, there is the inevitability that the industrial classification for the ICIO database be updated to better reflect GVC. In addition, the simulation result that the prediction accuracy of the RA index is higher than the AA index again justifies that the backbone of GVC is unbalanced, even though the average degree centrality $\langle K \rangle$ of Binary GISRN models are merely between 3 and 4.

By comparison, the prediction accuracy of these two indices turns out to be the lowest, because the IO structure reflects the relatively stable counter-balance between the industrial sectors after the economic system has evolved over a long period of time, and it is contrary to the general law of the value-added process that certain industrial sectors bypass the weak production links to directly connect each other. But some industrial transfer phenomena do occur between two countries/regions of considerable geographical distance or weak direct and indirect industrial relevance. Part of this is due to policy changes at the national macro-level (e.g., the Marshall Plan after World War II), and more importantly, the transferee of industrial transfer often has ***Niche Advantage*** that the transferor lacks (e.g., China' large amounts of cheap labor and the huge consumer market in the 1980s). In the follow-up study, research ideas from spatial econometrics will be drawn on, and relevant data of a country/region's niche advantage beyond ICIO data collected.

5 Discussion and Conclusions

After a year-by-year discovery on the GVC, how to predict the evolution trend of the industrial structure still captures the interest of scholars and policymakers. The reason for such enduring fascination is that when a wide consensus was finally reached on the importance of geographical proximity, agglomeration and local spillovers, the industrial globalization from breadth to depth had already made the GVC evolve into a complex economic system with mobile boundaries. It is the transformation that undermines the traditional perspectives and forces us to critically think over why clusters and districts exist, extend, exhaust or expand finally. From the econophysics angle, we try to simulate this process by link prediction, and hence give economic meanings to the most practical results. Some conclusions based on link prediction in binary GISRN-Eora26 models are as follows:

(1) The industrial transfer is a branch of regional economics, but its research methods should not be limited to that of economics. The regional industrial structure is an explicitly defined complex system, in which the internal relationships can be described in detail by IO data, and the tools and methods developed in statistical mechanics and theoretical physics can be used effectively to model and analyze the complex system of regional industries. This method reflects the research prospects, for it spans multiple disciplines including economics, management, physics, and statistics.
(2) Traditional theories and methods can't fulfill the needs of the regional industrial transfer. The IO-based system study on the regional industrial complex network and industrial transfer is a holistic research theory, and the application of link prediction and accuracy analysis to this field is a breakthrough in this research method. The integration of new theories and new methods fundamentally guarantees the integrity and systematization of the research.
(3) The application of link prediction obtains index accuracy of different levels, because the government agencies of countries/regions will promote the upgrading of industrial structure through macro-control means, while the global economic system lacks division of labor by overall planning. And complicated factors in international trade aggravate the development imbalance between countries/regions.

However, there is still a lot of room for improvement in the research content covered. The Binary GISRN model ignores the edge weight of network and thus loses a lot of information that can reflect its heterogeneity. To solve this problem, each link prediction algorithm must be further refined to be applied to a weighted and directed GISRN network or even more dense GIVCN model. In the optimization, we can use the dynamic mechanism behind many economic phenomena as the basis for designing the link prediction algorithm, such as the dissipative structure theory and the law of gravity.

Acknowledgment. The authors acknowledge support from National Natural Science Foundation of China (Grant No. 71971006), Beijing Natural Science Foundation (Grant No. 9194024) and Study Abroad Foundation of China Scholarship Council (CSC No. 201806545030).

References

1. Lü, L.Y., Zhou, T.: Link prediction in complex networks: a survey. Phys. A: Stat. Mech. Appl. **390**(6), 1150–1170 (2011)
2. Alves, A., Luiz, G., Mangioni, G., Rodrigues, F., Panzarasa, P., Moreno, Y.: Unfolding the complexity of the global value chain: strength and entropy in the single-Layer, multiplex, and multi-layer international trade networks. Entropy **20**(12), 909 (2018)
3. Lenzen, M., Moran, D., Kanemoto, K., Geschke, A.: Building Eora: a global multi-region input-output database at high country and sector resolution. Econ. Syst. Res. **25**(1), 20–49 (2013)
4. Xing, L.Z.: Topological Complexity and Evolutionary Mechanism of Global Value Chain. Science Press, Beijing (2020)
5. Ohlin, B.G.: Interregional and International Trade. Harvard University Press, Cambridge (1935)

Similarity Analysis in Multilayer Temporal Food Trade Network

Natalia Meshcheryakova[1,2](✉)

[1] National Research University Higher School of Economics, Myasnitskaya Str. 20, 101000 Moscow, Russia
natamesc@gmail.com

[2] V.A. Trapeznikov Institute of Control Sciences of Russian Academy of Science, Profsoyuznaya Str. 65, 117342 Moscow, Russia
https://www.hse.ru/en/staff/natamesc

Abstract. We analyze export/import food trade network that contains several layers. Each layer accounts for a particular commodity that countries trade with. The network has directed weighted edges. We look at statistical and topological similarity of layers in order to detect dependencies between different products trade. The measures include the estimation of out-degree correlation as well as the analysis of communities. We apply a normalization technique to the initial graphs that takes into account individual attributes of nodes and the possibility of groups formation. The most important elements of the networks are considered in order to compare different layers. Additionally, we analyze the network in time and detect the most similar periods of trade. The analysis of trade in dynamics gives the opportunity to track changes in export/import patterns. The results may have a significant contribution to the further analysis of food security of countries and the development of trade processes.

Keywords: Graph similarity · Multilayer network · Temporal network · Food trade network

1 Introduction

A large number of processes in our world can be represented as networks. A network is an advanced tool to analyze data, where relations between different objects exist. Moreover, these relations can have a complex structure and evolve in time. Thus, network analysis is not an elementary process that requires multi-stage approach and deep investigation. For instance, if a network has a multilayer structure [1] it requires the analysis of each layer separately as well as the whole graph should be studied entirely. On the other hand, if a network changes over time (temporal networks [2]) we need to investigate each time slot and study the dependencies of periods. Some networks can have both multilayer and temporal structure that makes the analysis even more complex.

One of the examples of networks that have both multilayer and temporal structure is an international food trade network. This process includes regular

H. Barbosa et al. (Eds.): *Complex Networks XI*, SPCOM, pp. 322–333, 2020.
https://doi.org/10.1007/978-3-030-40943-2_27

trade by different types of product categories that can be represented as layers in a network. Export and import processes play significant role in both country's economics as well as in human lives [3]. People and nations depend on quality and quantity of food they obtain. Additionally, they also depend on stability of nutrition and variety of products they can afford and access. Thus, this type of relations should be studied extensively. This may include the detection of unstable and influential elements, the analysis of products substitutability, temporal development, etc. International food networks is considered in [4] in the context of layers reducibility. Structural properties of trade network are investigated in [5–9]. Multiplex cascading failure model on interantional trade network is studied in [10]. Influence of countries and their connections in food export/import networks is analyzed in [11,12].

Still, there are few works that consider specifically food trade network and take into account the complex structure of this type of relations. Our main goal in this work is the analysis of a food trade network with respect to its peculiar properties. In particular, we address to the problem of similarity estimation within different layers and the detection of resemble periods of food trade through 20 years period. The analysis of layers correlation can show us the possible substitutability of commodities or strong dependencies between different products, while temporal analysis can help us to discover key changes in networks through the period in question and to detect resemble periods of food trade.

Different techniques can be applied to the networks in order to estimate their similarity. Some well-known nodes similarity measures [13–15] can be adopted to compare the same node in two different graphs that further can be aggregated into network similarity measures. Another approach includes adjacency matrices correlation that is highly unstable for small changes in one of the graphs. The idea of assortative mixing approach based on degree correlations are discussed in [16–18] and can also be adapted to the graphs comparison. The approaches that are based on common subgraphs and graphlet kernels are studied in [19–22]. Simulation-based approach for rooted graphs is described in [23]. In [24] graph similarity measure based on spectrum comparison is proposed. Nodes neighborhood is studied in [25] in order to compare nodes in two graphs. Von Neumann entropy approach is proposed in [4] in order to reduce the number of layers in a network. Multi-criterial approach for the network comparison that includes topological and structural properties is considered in [26].

Despite the fact that there exist the numerous of measures for the graph comparison most of them do not consider the possible complex structure of a network as directions and weights of edges, graph connectivity, groups formation, structure of communities, etc. Moreover, elements of a network (nodes and edges) are treated equally while in real world participants may have different magnitude and contribution to the network system. Therefore, we propose an adaptation of some classical approaches to the comparison of complex network structures as well as develop new models of graph similarity measures based on community analysis.

This paper is organized as follows. First, we provide a description of data source and our network. Second, we describe modifications of some network similarity measures and propose new approach. Third, we apply similarity metrics to food trade networks for different layers periods and analyze the results. Finally, we aggregate the obtained results and make a conclusion.

2 Data Description

Trade export/import statistics is very popular for the network analysis as this kind of data represents relationships between countries. We address to the problem of world food trade as this process has a high impact on all people around the world and directly influence on food security of countries.

Our main data source is World Integrated Trade Solution database [27] that was developed and compiled by several data sources. Particularly, the UNSD Commodity Trade database (UN Comtrade), that is a part of WITS database, provides with detailed bilateral trade statistics among hundreds of goods and services. The key advantages of this data are that it is free access, collects data since 1962 and covers almost all countries and territories in the world.

Trade data is represented in various classifications that are used to standardize reported values and make data comparable. The two main classifications for trade data reports are Harmonised System (HS) and Standard International Trade Classification (SITC) are distinguished by goods detalization and periods covering. In this work we focus on SITC data as it provides more suitable products division into categories for our purposes.

Digit codes are used for the division of products into categories and subcategories. The longer a code is, the more detailed a particular category is. SITC classification codes vary between 1 and 5 digits; in this work we choose 3-digit codes and selected 33 types of products, included different products preparations. Each product represents a particular layer of a network.

Another important point is that each flow is represented by two numbers as both partners (exporter and importer) provides with their statistics separately. Ideally, these two numbers should be consistent but in real world not many flows have the identical statistics. In particular, among 1 700 019 flows in our data only 16% has the difference less than 10% and almost 13% of reported statistics differ in 10 times. Hence, the problem that arises here is to choose one value of a flow between two values, that can diverge tremendously. Particularly, we cannot choose exporter or importer data entirely as some countries do not report at all which means that we can lose information about existing flows.

In our work we use two-stage approach in order to choose a value of a flow. First, we select a set of countries that report their statistics in agreement with UN recommendations [28]. Next, we choose a value of a flow of that country that is appeared in the set of good reporters. In case when both countries are good reporters or both countries are not in this set we estimate the concordance of their reported statistics with the statistics of countries from the set of good reporters. And in this case we choose a value of a country that is mostly cohere in reports with good reporters with respect to all flows in a network.

We also choose several periods in order to analyze networks in dynamics. Our data is annual and covers period from 1996 to 2017. The last two years are still poorly represented in database this is why we do not analyze them. As the result, we obtain 726 networks, where each network corresponds to a product group for a particular period.

Next, we describe two approaches to the comparison of layers and temporal changes in networks.

3 Similarity Analysis

3.1 Graph Transformation

In this work we propose two approaches of graphs comparison. The first approach is based on nodes degree matching and the second one is based on communities matching. But before we proceed with these two methods we need to describe an essential stage of graphs preprocessing.

The graphs consist of nodes that are countries and weighted directed edges that represent money flows of export/import in thousand dollars. Clearly that the volumes of trade differ for different types of commodities and it is meaningless to compare networks with absolute weights. For instance, in 2017 the total trade of food preparations is almost 80 times as large as total trade of some sort of flour. Hence, we need to normalize weights and reduce them to one dimension in order to adequately compare layers and networks in dynamic.

There are numerous approaches of data normalizing and a researcher can choose any of them depending on the problem statement. The simplest technique of normalization is the division of all weights to the total sum of weights or to the maximal weight in each graph. However, the diversity in volumes is very huge for all graphs, which means that these normalizations make most of the flows negligible.

We apply to our graphs a normalization of weights based on direct influence estimation. This method is described in [29,30] and the main advantage of this approach is that it takes into account individual attributes of nodes and the possibility of a group influence. Precisely, each node has a threshold value that indicates a critical loss of incoming weights. This threshold can be derived from the external data sources as well as from graph characteristics. In our work we use a threshold of a fixed percent of maximum between total import and total export of a node, i.e. $threshold_i = q\% \cdot \max(import_i, export_i)$ for node i. As for q we analyzed different values and decided on $q = 35\%$. The formula is motivated by the fact if a country imports more that exports to other countries then the loss in import value could not be covered by the reduction of export value. On the other hand, if a country exports more than imports from other countries then the possible loss in import value could be compensated by the reduction of export value.

After we derive thresholds for each node we estimate the power of direct influence of nodes on each other. Obviously, a node might not overcome the threshold value of its partner on its own. However, if nodes unite into a group their total export could be higher than a threshold of their common partner they point to. Hence, the influence of nodes is calculated as a maximal possible contribution of its weight to a group that overcomes a threshold. The influence value varies between 0 and 1 and it equals to 0 when a node has no impact in any group and it equal to 1 when a node overcomes a threshold value on its own. Additionally, we can fix a maximal possible group size; in our work we limit it to 5 nodes.

As the result, we obtain graphs of direct influences, where all values are normalized with respect to the size of nodes and the possibility of group formation. Moreover, some weak connections are eliminated. After we reduce all graphs to one dimension we can compare them both in the context of layers as well as in dynamics.

3.2 Graphs Similarity Based on Out-Degree Correlation

The first approach to the graph comparison is based on nodes degree vectors. The natural presumption is that two graphs are considered to be similar if the same nodes are connected mostly identically in both graphs. If we take into consideration directed weighted networks it leads to the fact that we should also pay attention to the directions and magnitude of arrows.

As described in the previous section we firstly normalize all weights in all graphs and derive new weighted graphs, where a weight indicates the intensity of direct influence. Consequently, we are able to compare weights in different graphs and layers as they are homogeneous.

The following idea is straightforward. In general, we have two graphs (or two layers) we want to compare, G_1 and G_2, with corresponding adjacency matrices, $A(G_1)$ and $A(G_2)$, where $a_{ij}(G_k)$ is an influence value of node i on node j in graph $G_k, k \in \{1, 2\}$. Then, for each node i we can take rows i in adjacency matrices of both graphs denoted by $A_{i*}(G_k)$ with the exclusion of the i-th column. Next, we calculate the correlation coefficient of two rows in both graphs $corr(A_{i*}(G_1), A_{i*}(G_2))$. Finally, the similarity measure can be estimated as the normalized sum of correlations between each nodes out-degrees in both graphs, i.e.

$$SIM_{outdeg}(G_1, G_2) = \frac{\sum_{i=1}^{n} corr(A_{i*}(G_1), A_{i*}(G_2))}{n} \quad (1)$$

If the sets of nodes are distinct in two graphs then we can complete these graphs with corresponding isolated nodes in order to make the sets of nodes equivalent.

However, this pure approach has some shortages. Generally, we also take into account unimportant elements of a network, whose values are not stable from network to network, but their occurrence in the formula above can cause erroneous results. Thus, we can select elements we want to compare in accordance with some magnitude of nodes, that can be derived externally or can be estimated from a graph itself. In our research we select top-40 countries by GDP for each particular year. In other words, we compare the position of key participants and their connections with other important elements of a system. Note that the selection of important elements is not related to the normalization of the initial graph to the graph of direct intensities. The scheme of this approach is shown on Fig. 1.

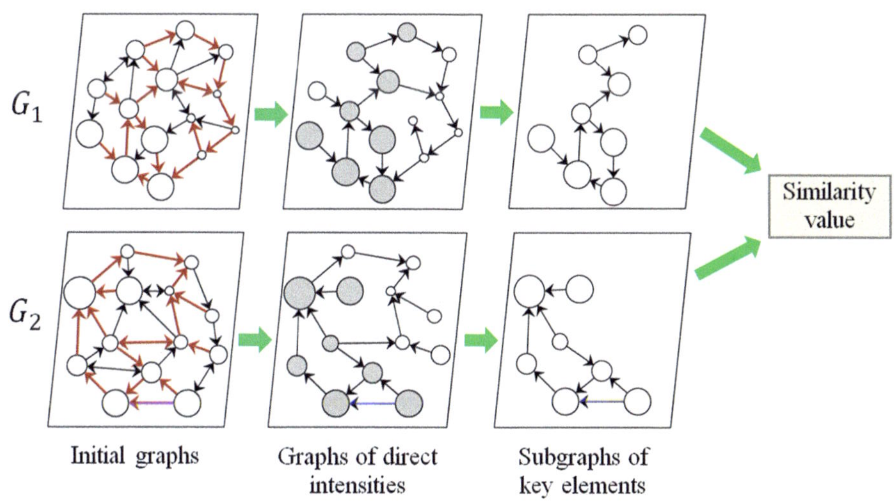

Fig. 1. Preprocessing scheme for the graphs comparison.

In this work we consider only out-degree correlation on nodes as normalized out-going edges reflect the influence distribution of elements in a graph. However, in-degree analysis can be taken into account as well. Moreover, the results obtained by out-degree and in-degree can be combined together.

3.3 Graphs Similarity Based on Communities Neighborhood

In this section we investigate how communities of nodes change over time and in layers space. For that purpose we employ Infomap community detection algorithm [31] that can be applied to weighted directed networks. As in previous section, we run the algorithm on normalized graphs of direct intensities in order to eliminate weak and unimportant connections that can warp the results.

There exist several ways of partition similarity measure in one graph [32] that can be adapted to community comparison in two graphs. One of the most popular measures is Jaccard index [33] that is calculated as the ratio of intersection of two partitions and the union of these partitions. If we talk about the communities in two different networks, we can also compare corresponding communities as the ratio of their intersection and their union. However, it might be not clear which communities to compare as they can be absolutely different in two networks.

In recent research we compare community neighbors of each node in two graphs that afterwards is aggregated to similarity measure of two graphs.

More precisely, we have two graphs, G_1 and G_2, and each node i in graph G_k is associated with some community $C_j(G_k) : i \in C_j(G_k)$. Hence, we apply Jaccard index in order to compare two communities of node i in two graphs. Then, the obtained indices can be aggregated to the similarity measure of two graphs:

$$SIM_{comm}(G_1, G_2) = \frac{1}{n} \cdot \sum_{\substack{i=1 \\ i \in C_j(G_1) \\ i \in C_l(G_2)}}^{n} \frac{|(C_j(G_1) \cap C_l(G_2)) \setminus \{i\}|}{|(C_j(G_1) \cup C_l(G_2)) \setminus \{i\}|}, \tag{2}$$

It is important to point out that we exclude a current node from consideration as it always appears in intersection of two sets and add excessive scores to the sum. On the other hand, the denominator of the ratio can be zero as the union of two sets may include node i solely. It only means that node i is separated from other nodes in both graphs, that definitely cause similarity at this point. Hence, it this case we add 1 to the sum.

As in the previous section it is unnecessarily to compare all the elements of the networks as unimportant nodes may change the neighborhood frequently and diminish the results. For that reason we select 40 countries with the highest GDP in each period and compare communities with respect to these countries. The scheme of this approach in shown on Fig. 2

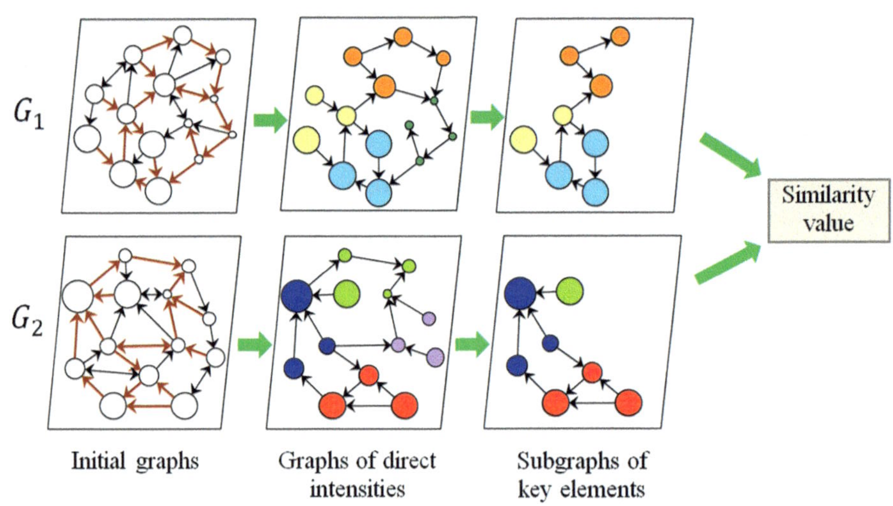

Fig. 2. Preprocessing scheme for the graphs comparison with communities.

4 Similarity Analysis in International Food Trade Network

In this section we apply the proposed measures to the food trade networks. For each considered year we compare all pairs of layers that correspond to different product groups and for each product group we compare networks for various periods. The main results by out-degree correlation are represented in Table 1.

Table 1. Top-5 similar layers by out-degree correlation.

1996	2002	2008	2013	2017
Fresh fruit vs. Fruit prep (0.542)	Cereal prep vs. Food prep (0.620)	Cereal prep vs. Chocolate (0.645)	Cereal prep vs. Chocolate (0.557)	Cereal prep vs. Food prep (0.537)
Cereal prep vs. Food prep (0.512)	Cereal prep vs. Chocolate (0.620)	Cereal prep vs. Food prep (0.579)	Fresh fruit vs. Fresh veg (0.554)	Cereal prep vs. Chocolate (0.522)
Fruit prep vs. Veg prep (0.507)	Fruit prep vs. Veg prep (0.522)	Fresh veg vs. Sugar prep (0.524)	Cereal prep vs. Food prep (0.505)	Fresh fruit vs. Fruit prep (0.506)
Cereal prep vs. Sugar prep (0.433)	Chocolate vs. Food prep (0.493)	Fruit prep vs. Sugar prep (0.517)	Fresh fruit vs. Fruit prep (0.491)	Sugar prep vs. Chocolate (0.499)
Fresh fish vs. Fresh fruit (0.429)	Fresh veg vs. Sugar prep (0.473)	Chocolate vs. Food prep (0.515)	Fruit prep vs. Fresh veg (0.463)	Fresh fruit vs. Fresh veg (0.495)

We can see from Table 1 that the most similar layers occur in homogeneous product groups (fruits vs. vegetables, chocolate vs. sugar, etc.). This might be explained by the fact that close product groups require similar environment. Hence, if a country exports one product then it has opportunities to grow and export the other product. However, some heterogeneous layers are also similar according to proposed measure. For instance, cereals and chocolate trade are very close according to out-degree correlation due to the fact that large economics as Germany, Belgium and USA are major exporters of these products according to FAO statistics [34].

Table 2 represents statistics for community matching measure.

Table 2. Top-5 similar layers by community matching.

1996	2002	2008	2013	2017
Wheat vs. Barley (0.556)	Maize vs. Wheat flour (0.615)	Cereal prep vs. Chocolate (0.565)	Cereal prep vs. Chocolate (0.676)	Dried meat vs. Butter (0.618)
Barley vs. Maize (0.524)	Barley vs. Maize (0.547)	Fresh fruit vs. Fresh veg (0.524)	Butter vs. Food prep (0.591)	Fish prep vs. Dried fruit (0.545)
Cereal prep vs. Food prep (0.518)	Chocolate vs. Margarine (0.542)	Fresh meat vs. Eggs (0.523)	Chocolate vs. Food prep (0.575)	Butter vs. Barley (0.541)
Wheat vs. Maize (0.489)	Cereal prep vs. Food prep (0.527)	Cereal prep vs. Sugar prep (0.516)	Wheat vs. Barley (0.548)	Cereal prep vs. Chocolate (0.536)
Sugar prep vs. Chocolate (0.462)	Fresh meat vs. Milk (0.518)	Maize vs. Other Cereals (0.476)	Cereal prep vs. Food prep (0.542)	Sugar prep vs. Chocolate (0.491)

Comparing to the previous results we can notice that the most similar layers also include grass family (wheat, barley, maize, other cereals) and in livestock products (meat, eggs, milk and milk products). It might be explained that, first of all, these products can be produced all together at one land, and, secondly, these product groups are one of the most important for consumption and major exports as USA, Australia, Canada, Russia, Netherlands, Germany, France, etc. have a stable and well-functioning process of trade by these product groups.

It is also important to emphasize that not all cereals products are similar. For instance, rice layer and other grass products are least similar by the proposed measures. This can be explained by the fact that the major exporters of rice, that are India, Thailand and Vietnam, have more dense trade with other Asian countries than with Western countries.

It is also essential to study the dynamical changes in layers similarity. In this part of a research we compare all years with each other for each commodity. The results are illustrated on Fig. 3.

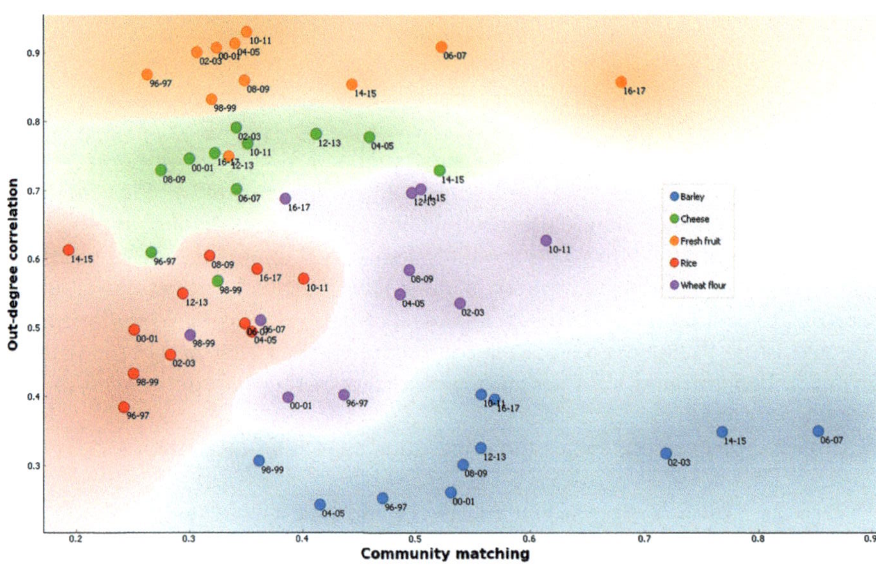

Fig. 3. Periods similarity by two proposed measures.

We select several products and take only adjacent years. Each bubble on Fig. 3 corresponds to a pair "year-year". Axes on Fig. 3 reflect the proposed similarity measures. We can see from Fig. 3 that most of the selected products remain stable by out-degree correlation measure while they are very diverse by community matching. Additionally we can notice that the commodities are well-separated from each other with respect to these measures, which implies that it is essentially important to study international food trade considering each product group rather than aggregated trade.

Figure 4 illustrates the dynamical changes for both measures separately. Again, we can see that out-degree correlation is mainly stable for selected products while community matching is rather fluctuating. Note that out-degree correlation measure reflects the intensities of direct connections that are normalized to [0; 1]. As a consequence, it does not show the volumes of trade that tend to increase in time.

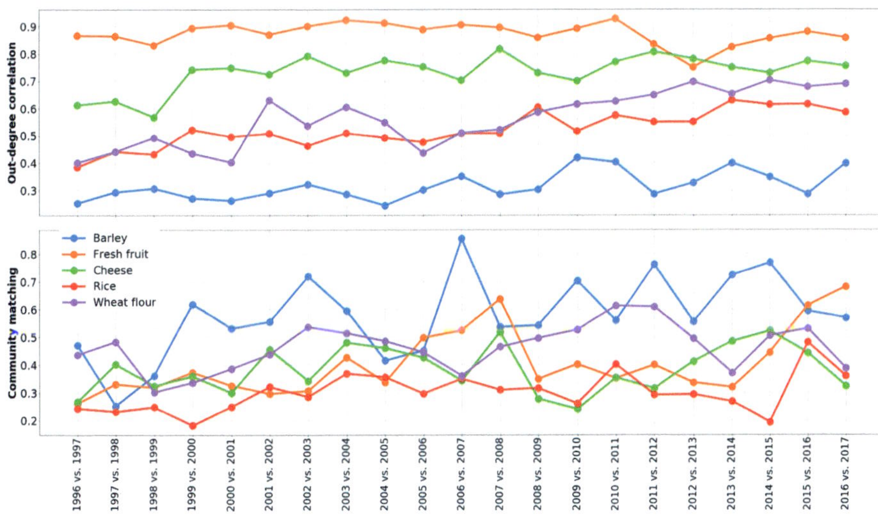

Fig. 4. Dynamics of similarity measures.

5 Conclusion

In current research we propose two approaches of weighted directed graph comparison based on well-known models. The first method is based on out-degree correlation for each node in two different graphs. The second approach compares communities in networks obtained by Infomap algorithm. The novelty of this work resides in the fact that we apply special normalization techniques to the initial graph that takes into account individual attributes of nodes and tends to estimate the intensities of direct influences. As the result, we obtain reconstructed and sparser networks with comparable weights. Another point is that we compare only important elements of a network that keep stability and may influence the whole structure. The choice of the most important elements can be done according to graph characteristics as centrality measures or with the help of external attributes. Finally, the proposed measures can be combined with each other into one similarity index.

We apply the described measures to a real network of international food trade. This network has a complex structure as it includes many layers and evolves over time. We compare layers with each other for every year and analyze the similarity of time slots for each commodity. The obtained results show that the most similar trade takes place in those layers that hit into one product family (grass products with grass products, livestock with livestock, etc.) along with the fact that trading partners remain the same for each major exporter. It can also be seen that for some commodities trading partners form stable trade groups over the period in question while for other products trading partners are less consistent.

The proposed measures can also be applied to other complex structures as they consider various special aspects that can be adjusted to the particular problem statement.

Acknowledgements. The article was prepared within the framework of the Basic Research Program at the National Research University Higher School of Economics (HSE) and supported within the framework of a subsidy by the Russian Academic Excellence Project '5–100'. This work is also supported by the Russian Foundation for Basic Research under grant 18-01-00804a Power of countries in the food security problem.

References

1. Kivela, M., Arenas, A., Barthelemy, M., Gleeson, J.P., Moreno, Y., Porter, M.A.: Multilayer networks. J. Complex Netw. **2**(3), 203–271 (2014)
2. Holme, P., Saramäki, J.: Temporal networks. Phys. Rep. **519**(3), 97–125 (2012)
3. Food and Agriculture Organization of the United Nations. http://www.fao.org/home/en/. Accessed 10 Nov 2019
4. De Domenico, M., Nicosia, V., Arenas, A., Latora, V.: Structural reducibility of multilayer networks. Nat. Commun. **6**(1), 1–9 (2015)
5. Nemeth, R., Smith, D.: International trade and world-system structure: a multiple network analysis. Rev. (Fernand Braudel Center) **8**(4), 517–560 (1985)
6. Smith, D.A., White, D.R.: Structure and dynamics of the global economy: network analysis of international trade 1965–1980. Soc. Forces **70**(4), 857–893 (1992)
7. Barigozzi, M., Fagiolo, G., Garlaschelli, D.: Multinetwork of international trade: a commodity-specific analysis. Phys. Rev. E **81**(4), 046104 (2010)
8. Mastrandrea, R., Squartini, T., Fagiolo, G., Garlaschelli, D.: Reconstructing the world trade multiplex: the role of intensive and extensive biases. Phys. Rev. E **90**(6), 062804 (2014)
9. Alves, A., Luiz, G., Mangioni, G., Rodrigues, F.A., Panzarasa, P., Moreno, Y.: Unfolding the complexity of the global value chain: strength and entropy in the single-layer, multiplex, and multi-layer international trade networks. Entropy **20**(12), 909 (2018)
10. Lee, K.-M., Goh, K.-I.: Strength of weak layers in cascading failures on multiplex networks: case of the international trade network. Sci. Rep. **6**(1), 26346 (2016)
11. Aleskerov, F.T., Meshcheryakova, N.G., Sergeeva, Z., Shvydun, S.V.: Centrality measures and clustering analysis in a retail food network. In: 2017 IEEE 11th International Conference on Application of Information and Communication Technologies, vol. 1, pp. 48–52. Institute of Electrical and Electronics Engineers (2017)
12. Meshcheryakova, N.: The impact of indirect connections: the case of food security problem. In: Complex Networks and Their Applications VII. Studies in Computational Intelligence, vol. 813, pp. 80–90. Springer, Cham (2019)

13. Faust, K.: Comparison of methods for positional analysis: structural and general equivalences. Soc. Netw. **10**(4), 313–341 (1988)
14. Leicht, E.A., Holme, P., Newman, M.E.J.: Vertex similarity in networks. Phys. Rev. E **73**(2), 026120 (2006)
15. Newman, M.E.J.: Networks: An Introduction. Oxford University Press, Oxford (2010)
16. Newman, M.E.J.: Mixing patterns in networks. Phys. Rev. E **67**(2), 026126 (2003)
17. Xulvi-Brunet, R., Sokolov, I.M.: Changing correlations in networks: assortativity and dissortativity. Acta Phys. Pol. B **36**, 1431–1455 (2005)
18. Valdez, L.D., Buono, C., Braunstein, L.A., Macri, P.A.: Effect of degree correlations above the first shell on the percolation transition. EPL (Europhys. Lett.) **96**(3), 38001 (2011)
19. Messmer, B.T., Bunke, H.: Efficient subgraph isomorphism detection: a decomposition approach. IEEE Trans. Knowl. Data Eng. **12**(2), 307–323 (2000)
20. Raymond, J.W.: RASCAL: calculation of graph similarity using maximum common edge subgraphs. Comput. J. **45**(6), 631–644 (2002)
21. Shervashidze, N., Vishwanathan, S.V.N., Petri, T., Mehlhorn, K., Borgwardt, K.M.: Efficient graphlet kernels for large graph comparison. J. Mach. Learn. Res. - Proc. Track **5**, 488–495 (2009)
22. Macindoe, O., Richards, W.: Graph comparison using fine structure analysis. In: 2010 IEEE Second International Conference on Social Computing (2010)
23. Sokolsky, O., Kannan, S., Lee, I.: Simulation-based graph similarity. In: Lecture Notes in Computer Science, pp. 426–440 (2006)
24. Wilson, R.C., Zhu, P.: A study of graph spectra for comparing graphs and trees. Pattern Recogn. **41**(9), 2833–2841 (2008)
25. Zager, L.A., Verghese, G.C.: Graph similarity scoring and matching. Appl. Math. Lett. **21**(1), 86–94 (2008)
26. Aleskerov, F., Shvydun, S.: Stability and similarity in networks based on topology and nodes importance. In: Complex Networks and Their Applications VII, pp. 94–103 (2018)
27. WITS - About WITS. https://wits.worldbank.org/about_wits.html. Accessed 10 Nov 2019
28. United Nations Statistics Division - International Merchandise Trade Statistics. https://unstats.un.org/unsd/tradereport/introduction_MM.asp. Accessed 10 Nov 2019
29. Aleskerov, F., Meshcheryakova, N., Shvydun, S.: Centrality measures in networks based on nodes attributes, long-range interactions and group influence. arXiv preprint arXiv:1610.05892 (2016)
30. Aleskerov, F.T., Meshcheryakova, N.G., Shvydun, S.V.: Power in Network Structures. Models, Algorithms, and Technologies for Network Analysis. Springer Proceedings in Mathematics and Statistics, vol. 197, pp. 79–85 (2017)
31. Rosvall, M., Axelsson, D., Bergstrom, C.T.: The map equation. Eur. Phys. J. Spec. Top. **178**(1), 13–23 (2009)
32. Fortunato, S., Hric, D.: Community detection in networks: a user guide. Phys. Rep. **659**, 1–44 (2016)
33. Jaccard, P.: Etude de la distribution florale dans une portion des Alpes et du Jura. Bull. de la Societe Vaudoise des Sci. Naturelles **37**(142), 547–579 (1901)
34. Ranking of countries by commodity exports. http://www.fao.org/faostat/en/#rankings/countries_by_commodity_exports. Accessed 10 Nov 2019

Transactional Compatible Representations for High Value Client Identification: A Financial Case Study

Irene Unceta[1,2](✉), Jordi Nin[3], and Oriol Pujol[2]

[1] BBVA Data & Analytics, Barcelona, Catalonia, Spain
irene.unceta@bbvadata.com

[2] Department of Mathematics and Computer Science, Universitat de Barcelona, Barcelona, Catalonia, Spain
{irene.unceta,oriol_pujol}@ub.edu

[3] ESADE, Universitat Ramon Llull, Barcelona, Catalonia, Spain
jordi.nin@esade.edu

Abstract. High value client identification is a crucial task to any company. In the banking industry, high value is not solely related to purchasing power, but also to an intensive use of financial services, such as card payments or bank wire transfers. This is why transactional data is a valuable source of information. In this work we propose a method to estimate the net worth of individuals for whom we lack any transactional data, either because they are non-clients or because they conduct their main activity elsewhere. We exploit the representation learned by a value prediction model trained over a signed graph of social financial relationships between BBVA clients to infer a transactional compatible representation of clients outside the graph. As a result, we obtain a new model that can predict value labels for both client and non-client data. Our results show an improvement in prediction accuracy over the previous baseline in a 2 million client database.

Keywords: Financial network · Graph embeddings · Value prediction

1 Introduction

Identifying and nurturing high value clients is crucial to any business strategy. An accurate prediction of client value allows, among other things, an effective allocation of marketing expenditure or the mitigation of exposure to losses. Not in vain do companies devote numerous resources to properly segmenting their client portfolio [13]. In the context of the financial industry, value is related to wealth, *i.e.* those individuals who own large amounts of money are perceived as valuable; but also to a high need. Due to their greater purchasing power, high net worth clients are characterized by an intensive use of financial products and

H. Barbosa et al. (Eds.): *Complex Networks XI*, SPCOM, pp. 334–345, 2020.
https://doi.org/10.1007/978-3-030-40943-2_28

services [17]. Identifying them correctly is therefore a necessary step towards maximizing business objectives.

Traditional approaches to value prediction exploit socio-demographic information [15], including the age, the annual income or the number of active financial products. While all these attributes provide insights into the characteristics of customers, there is little they can tell us about their behavior. And perhaps more importantly, they provide us with no knowledge about how clients are related to each other. The old saying of "rich get richer" states that clients with similar characteristics tend to stay close to each other and, more specifically, that high value clients tend to be closely related [6]. These relations in the real world may be established in terms of friendship, marriage, mutual business or spatial proximity. Irrespective of their form, it seems reasonable to assume that such connections should have a reflection in the financial realm as money transfers between individuals or even as shared financial products in this part of the population. Note that, we can think in two kinds of transactions: professional one, for example a client-supplier payment, or personal relations, for instance an instant money transfer to share the expenses of a dinner.

However, transactional data is scarce. While a bank may posses socio-demographic data of both clients and non-clients, this is not necessarily the case when it comes their transactional information. Mainly because clients tend to conduct their financial activity in more that one institution at the same time. But also because there exists a large number of non-transactional clients. Hence, it is usually the case where transactional data is simply not available for a large percentage of the client portfolio.

Given this context, our work aims at answering the following three questions:

- Is it possible to extract meaningful knowledge for high value estimation from customer connections?
- More specifically, can we exploit knowledge of the structure of financial interactions to obtain a compatible representation of non-transactional client information? and if so, can this knowledge be used to enrich high value client estimation models?
- Finally, is the learned representation general enough to be exported to other regions and contexts?

Recent advances in representation learning in the domains of computer vision, speech recognition and natural language processing have devised a methodology to extract meaningful information from the topology of graphs [4,5,7,11,12]. These advances move away from using *ad hoc* heuristics to extract structural information from graphs, such as degree and neighborhood statistics. In turn, they allow the representation of graph connectivity into a lower dimensional latent space by means of embeddings. These embeddings can then be exploited to enrich previously existing classification or regression models. Notably, they can be used to endow these pre-existing models with richer information in those cases where the transactional data itself is not directly accessible.

In this paper we present a case study with data from BBVA. In particular, we use financial activity during the period 2017–2018 to build a graph representing the interactions among transactional clients. We exploit the topology of this graph to train a model to infer the value label for individuals for whom we lack any transactional data, either because they are non-clients or because they conduct their financial activities elsewhere. We validate our results by first training a baseline model for high value client prediction which builds upon non-transactional data and comparing its results to those obtained when including the added knowledge from client transactions. Our main contributions are summarized as follows:

- A detailed description of a high value prediction system based on client non-transactional data.
- A method to represent a graph structure in a lower dimensional embedding latent space.
- A model that projects non-transactional data onto the learned representation.

The rest of this paper is organized as follows. First, in Sect. 2 we present a summary of related work on representation learning in general and graph embeddings in particular. Section 3 describes our context and presents the use case. We introduce our proposed approach in Sect. 4. In Sect. 5, we carry out a set of experiments to empirically validate our theoretical proposal. Finally, the paper ends with our conclusions and future work in Sect. 6.

2 Background on Graph Embeddings

In machine learning (ML) embeddings are defined as non-linear low dimensional mappings of high-dimensional data into a vector space endowed with certain metric properties. Embeddings have been popularized by their application in recommenders systems [18] and natural language processing. For example, word embeddings are known to produce good numerical representations of words by capturing relevant semantic features [1]. Embeddings provide a method to map data into vector representations that capture meaningful properties for the considered task. This is of particular relevance when dealing with complex data structures, such as graphs. Indeed, the difficulty of directly working with graphs structures, has motivated the increased interest of the community in the search for meaningful representations of these structures.

Graph embeddings have been widely used in recommender systems, where the most successful embedding approaches rely on matrix and tensor factorization. Methods such as singular value decomposition [9], factorization variants in the complex domain [14] or non-negative rank decompositions are some of the proposed techniques in this research line. A substantial number of techniques proposed for finding structure preserving embeddings follow the rationale behind word embeddings, such as word2vec [11]. Here the context of a node is defined by means of random walks, where each generated path plays the equivalent role of a sentence. DeepWalk [12] and Node2Vec [5] are two examples of such kind of embeddings.

Finally, we can also find deep learning solutions to graphs embeddings. One strategy consists of using autoencoders. Following this line of thought we find methods such as structural deep network embeddings [16] or deep neural networks for learning graph representations [3] where the full graph adjacency matrix is used and fully embedded. Another very successful approach is that of graph convolutional networks [7] where a convolution-like operation is applied over the graph. In this scenario an embedding for a node can be obtained by iteratively aggregating neighborhood embeddings.

3 Identifying High Value Customers

Valuable clients constitute a high net worth demographic characterized by its great financial needs and high purchasing power. Usually, such clients correspond to less than 30% of the portfolio. However they are responsible for around the 60% of the profits [10]. Hence, their associated value and the need to profile them correctly.

3.1 A Preliminary Approach

Traditionally, high value is predicted using socio-demographic information, such as the age, the municipality of residence or the ownership of a house. While this information may not be highly predictive, it is readily accessible and thus provides a suitable starting point. Indeed, collecting socio-demographic data of both clients and non-clients is reasonably easy. This data is then used to estimate the net worth of those individuals for which the high value label is unknown, be it because they have no active contracts with the bank or because they mainly operate elsewhere. While models trained in this form perform relatively well, a large amount of information escapes their reach.

3.2 Exploiting Transactional Data

A more comprehensive vision of a client is provided by data related to financial transactions. These include bank transfers, direct debits and card payments, as well as active accounts, mortgages or financial products in general. In the context of value prediction, this information is specially relevant because it can be used to study connections among clients. Unfortunately, there exists no unified protocol to encode financial transactions. Usually, not even within the same corporation. Moreover, even while such transactions may be correctly labeled and registered, individual entities can only access a subset of this information.

In those cases where clients allocate their funds in a single bank, labeling them according to their value is a reasonably easy task. However, due to a tendency to distribute financial assets across different entities, this is not usually the case. Clients tend to conduct financial activities in more than one bank at the same time, so that each individual entity is left with only a partial view of their overall financial situation. This challenge becomes even greater when it comes to predicting the value associated to non-clients. A further complication comes with exploiting transactional data, a task which is, in general, hard.

Traditional techniques to exploiting structural information in graphs rely upon summary graph statistics for node strength and degree, *ad-hoc* features that encode local neighborhood structures [2], or common diffusion models describing spread dynamics [8]. While these metrics and models provide insights into the relationships between nodes, they fail to replicate the full topology of the network. Moreover, it can be very time-consuming to extract them and they are often inflexible to the learning process. Hence, financial relationships are seldom exploited for predictive purposes.

Thanks to recent advances, however, graph representations can be extracted directly in the form of low dimensional embeddings [4,19] that can then be used to train classification models. Moreover, they can also be exploited to enrich the socio-demographic attributes, by forcing non-transactional data into a representation compatible with that learned by the embeddings. In what follows we present a use case where we show the utility of this technique to exploit the BBVA transactional topology in Spain to predict the value of individuals for whom the connectivity data is unknown.

3.3 Case Study: High Value Client Prediction at BBVA

BBVA is a global institution that operates in many different countries across the world. However, it does so under different brands. It therefore lacks a unique operation protocol. Moreover, because different regulatory frameworks apply, data collection, storage and usability vary from market to market. Thus, extracting the data necessary to build the corresponding connectivity graph for each particular context is not always feasible. These difficulties motivate the research in inferring potential transactionality representations that can be exported and exploited when this kind of data is unavailable or scarce.

Additionally, properly building the adjacency matrix of transactions among private customers is not straight forward. Each customer can have more than one active account with the bank and each account can have more than one associated client. When it comes to encoding the transactions themselves, receptors of wired transfers are not necessarily unique, nor are the senders.

Instead of directly training a new model using both transactional and non-transactional data, we devise a solution to infer a transactional compatible representation of socio-demographic data. This approach has the advantage of portability. Once having learned this lower dimensional representation, it is possible to project new socio-demographic data onto this space. This effectively amounts to "imagining" the connectivity of those nodes for whom we lack this information. With this information we can then enrich the more readily available socio-demographic information. In other words, we can export the representation learned for a given transactions graph to other contexts where this data is not available and hence improve the pre-existing client value prediction models, or use it to infer potential transactionality for potential clients.

4 Inferring Transactional Compatible Representations from Non-transactional Data

In this section we explore the hypothesis that non-transactional data encodes topology patterns of the potential transactions that can be exploited to predict the value of clients. This hypothesis leads to the following question: can we create a transaction compatible representation of clients starting from non-transactional data that may help us to identify high value clients? To answer this question we propose a solution composed of the following two stages:

- **Financial graph embedding:** We train a dual loss autoencoder to project the structure of the financial interactions graph onto a low-dimensional embedding that forces a clustered representation. This embedding preserves the topological features as well as the client value labelling for each node.
- **Transactional compatible neural network:** We use the embedded data to train a neural network to predict high value. Due to the compositional nature of this model, its last layers capture high-level knowledge encoded in the embeddings. We freeze these last layers and use them as fixed weights for a second neural network, which uses non-transactional data as input. The initial layers of this network attempt map input data into a transactional compatible representation that can be exploited by the frozen layers to produce an enriched prediction.

A diagram for this pipeline in shown in Fig. 3, where non-transactional data is represented in green and data corresponding to connections between clients is shown in yellow. The shaded yellow nodes represent the graph embeddings

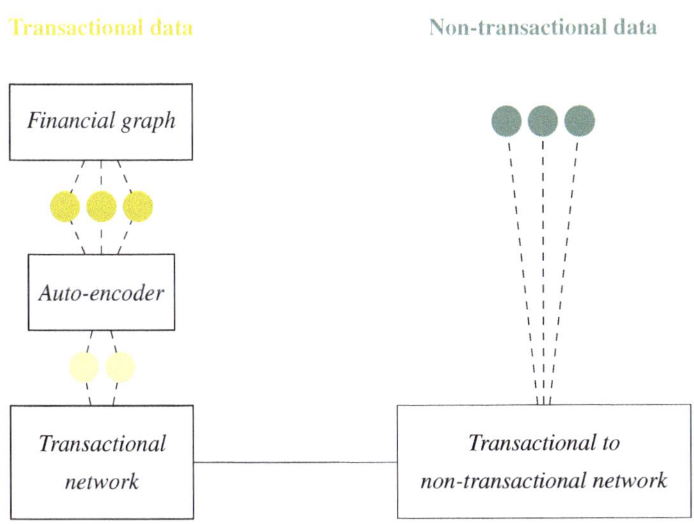

Fig. 1. Diagram for proposed solution.

learned for the transactional information. This data is fed to an initial neural network the last layers of which are frozen and passed on to a second network. This second network projects non-transactional data onto the frozen weights and outputs a value prediction compatible with the original graph structure.

4.1 Financial Client Graph Autoencoder

In this work we use a dual loss autoencoder to obtain graph embeddings. The rationale behind this particular representation is the following: on the one hand, we need a vector representation of the graph that captures its topology with accuracy. For this reason, we use a reconstruction autoencoder that converts the graph labeled adjacency matrix into a representation that captures the topology such that it can be faithfully reconstructed. On the other hand, we want to endow the embedded vector with a metric that clusters together structures corresponding to the same target, to help in the future classification task. Thus, besides the standard reconstruction loss to embed the topology of the graph, we also define a second loss that takes into account the target class label. We use a simple linear layer that forces the learned representation to be directly separable. Figure 2 shows a diagram of the architecture for such a dual loss autoencoder (Fig. 1).

4.2 Transactional Compatible Artificial Neural Network

Once the topological embedding for each node is obtained, the next step consists of training a neural network that hybridizes the non-transactional data with transactional knowledge. we do this in two steps:

- **Step 1:** We train an artificial neural network using the obtained embeddings with the goal of predicting the desired target outcome. We call this model the transactional network, $\mathcal{N}_T$. As it is commonly accepted, we would expect the last hidden layers in this architecture to be able to capture broader and more abstract features of how the topology of the graph serves as a predictive feature for our outcome. The network described in this step is shown in the left hand side of Fig. 3, with the frozen layers depicted in black.
- **Step 2:** We build another artificial neural network using non-transactional data. The last layers of this ANN are exactly the same as the last layers from the $\mathcal{N}_T$ network trained in step 1. The right hand side of Fig. 3 shows an schematic representation of the resulting architecture and the location the frozen layers. This model is referred to as the transactional to non-transactional network, $\mathcal{N}_{TNT}$. Its goal is to create a representation from non-transactional data compatible with the transactional layers. We expect non-transactional features that may help in inferring the expected transactionality of a customer to implicitly emerge in this setting.

As a result of the above, we obtain a new model, the $\mathcal{N}_{TNT}$, that distills the non-transactional data to match the requirements of high-level transactional mappings for high value client prediction.

5 Experiments

In this section we describe our experiments to evaluate our proposal on a subset of the BBVA client portfolio. We first emulate a pre-existing non-transactional model by training a classifier to predict client value from socio-demographic data. We use this classifier as a baseline to compare our results when enriching this predictor with data from financial transactions.

5.1 Socio-Demographic Dataset

We use a private dataset which contains socio-demographic information from the BBVA client portfolio. We discard professional and corporate customers and filter the data to select only private accounts. The dataset consists of descriptive attributes such as age, employment situation or zip code by December 2018. Due to the sensitive nature of banking data, we anonymize and identify all individuals using randomly generated IDs. We remove null and missing values and discard all features which provide no useful information for inference. We convert all categorical attributes to numerical using either one-hot encoders or by directly encoding class labels with integer values. We standardize all features to zero-mean and unit variance. The resulting dataset contains information about 9 variables for 2,018,465 private customers. We label individuals according to whether they have been identified as high value or not. High value clients correspond to 15% of the total population. We perform a stratified split to obtain training and test sets of relative sizes of 0.8 and 0.2, respectively.

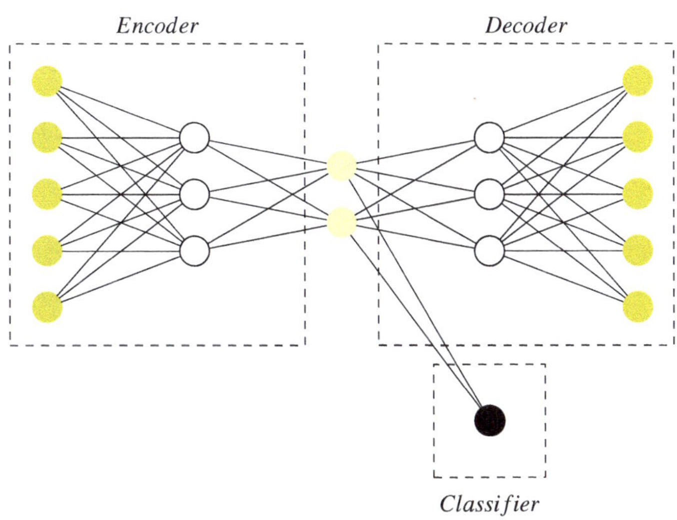

Fig. 2. Dual loss auto-encoder. A low-dimensional embedding is learned by simultaneously training the reconstruction and classification losses.

5.2 Financial Transactions Graph

In addition to socio-demographic data, we also use information about the connections among those clients in the dataset which are transactional. We identify a connection between two clients when there exists a wire transfer between them or when they have shared a common product, such as a bank account, a mortgage or a consumer credit, during the past two years (2017–2018). We use this data to build a network of financial transactions where customers are represented by nodes and linked by an edge if there exists a connection between them. Because we are primarily interested in modeling the relationships between individuals, we assume all edges to be undirected, independently of the direction of money transfer. We weight all edges according to the labels of the arrival node, so that each edge of the graph is represented in the adjacency matrix with a value in $\{+1, -1\}$. Thus, connections with nodes identified as high value have a weight equal to 1 and the others have a weight equal to -1. As a result, we obtain a signed network of interactions.

5.3 Experimental Set up

As previously explained, the full experimental pipeline is as follows: we first train a double-loss autoencoder as a pre-processing step to learn a lower-dimensional representation of the connectivity graph. We encode the full transactional dataset using the embedding learned by the autoencoder and feed it to the transactional network ($\mathcal{N}_T$). We freeze the last layers of this network to train the non-transactional network and compare the obtained results to those of a baseline model ($\mathcal{N}_{NT}$) for the same data.

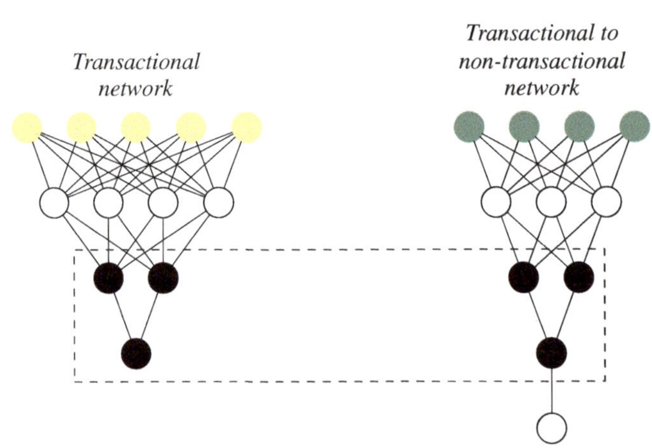

Fig. 3. Dual representation. The last layers (bold) are shared among both networks.

Non-transactional Baseline Network ($\mathcal{N}_{NT}$). We use the socio-demographic data to train two baseline models. We first train an artificial neural network with two hidden layers of sizes 32 and 16. We use selu neurons with a fixed dropout of 0.9 for all layers and an Adam optimizer with a learning rate equal to 10^{-3}. In addition to this model and as a further sanity check, we also train a cross-validated gradient-boosted tree to predict high value labels.

Dual Loss Autoencoder. We build the autoencoder using symmetric encoding and decoding modules, both consisting of three hidden layers with 256, 128 and 32 neurons, respectively. We split training data in two. We use the first half to train the reconstruction weights, while the other half is used to train the classification loss that forces the embedding into a suitable class separable representation. We train the autoencoder imposing an L2 regularizer onto the weights of the last encoding layer and using a least squares classification loss. We use selu neurons, no dropout and a least squares reconstruction loss with a default parameter Adam optimizer.

Transactional Network ($\mathcal{N}_T$). We use the embeddings learned by the autoencoder to train the transactional network. This network consists of five fully connected layers with 512, 256, 128, 32 and 16 selu neurons, no dropout and a sotfmax cross entropy loss with a default parameter Adam optimizer.

Transactional to Non-transactional Network ($\mathcal{N}_{TNT}$). To infer the connectivity of the non-transactional nodes, we force the socio-demographic dataset into a representation compatible with that learned by the transactional network. To do so, we retain the weights of its last 3 layers. As shown in Fig. 3 We build an additional network with five fully connected layers of sizes 512, 256, 128, 32 and 16, for which the last 3 layers share the same weights as the network trained using the graph embeddings. By freezing the final layers of this network, we force the incoming non-transactional data into a suitable representation to reproduce the learned connectivity. The final layer is composed of two neurons that produce the one-hot encoded output that is fed to a cross entropy softmax loss. We train the weights using an Adam optimizer with a fixed learning rate and no dropout. Due to the class imbalance in the dataset, we train all networks using balanced batches, except for the autoencoder, for which we impose class penalty weights onto the classification loss.

5.4 Discussion of Results

Table 1. Description of results

	$\mathcal{N}_{NT}$	$\mathcal{N}_T$	$\mathcal{N}_{TNT}$
Accuracy	0.69 (0.67)	0.65	0.65
Balanced accuracy	0.71 (0.71)	0.60	0.72
Recall (high value)	0.74 (0.78)	0.46	0.82
Agreement with $\mathcal{N}_{NT}$	1.00	0.58	0.91

Our experimental results are detailed in Table 1. We report both the balanced and the standard accuracy for the transactional network and for the transactional to non-transactional network, as well as for the baseline models (results for the gradient-boosted tree are shown in parenthesis). Additionally, we show the recall values for the positive class (high value) for all trained models. Finally, we also report the level of agreement between the predictions output by the $\mathcal{N}_T$ and $\mathcal{N}_{TNT}$ networks and those obtained using the baseline model.

When compared with the baseline model, we obtain an improvement of 8% in the recall of the positive class (4% when compared with the boosted tree), meaning that high value individuals are more accurately identified by the $\mathcal{N}_{TNT}$. This increment in the recall metric can be misleading if the precision or the balanced accuracy are compromised. However, when compared to the baseline model, we obtain an improvement of 1% in balanced accuracy when identifying high value clients. This ratifies that the obtained value is a real neat gain. These improvements are of paramount importance: our method is not only able to identify a larger number of high value customers but also improves the balanced accuracy. While the improvement in this last figure may seems like a little improvement, due to the large amount of profit related to this population, it translates into a high revenue. This is the ultimate goal of this use case, the next natural step of which would be to create a client portfolio including all the possible high value customers to help the marketing department to correctly identify them.

6 Conclusions and Future Work

In this paper we propose and validate a method to identify high value clients by exploiting information about their connections in terms of shared financial transactions and products. We evaluate our approach through a case study using data from BBVA. We build transactional compatible representations for non-transactional customers and use this knowledge to enrich a pre-existing model for value prediction based on socio-demographic data. Empirical results demonstrate the value of transactional data and its potential for predicting high worth labels.

Future developments should focus on selecting only those clients for whom BBVA is their primary bank. In addition to this, we would like to validate the country hypothesis by extracting data from other markets where the bank conducts its business, such as Mexico or Turkey. Finally, we want to improve transactional data including other information sources, such as credit card purchases, debits or housing-related expenses.

Acknowledgment. This work has been partially funded by the Spanish project TIN2016-74946-P (MINECO/FEDER, UE), and by AGAUR of the Generalitat de Catalunya through the Industrial PhD grant 2017-DI-25. We gratefully acknowledge the support of BBVA Data & Analytics for sponsoring the Industrial PhD.

References

1. Bengio, Y., Ducharme, R., Vincent, P., Jauvin, C.: A neural probabilistic language model. J. Mach. Learn. Res. **3**(Feb), 1137–1155 (2003)
2. Benoit, D.F., den Poel, D.V.: Improving customer retention in financial services using kinship network information. Expert Syst. Appl. **39**(13), 11435–11442 (2012)
3. Cao, S., Lu, W., Xu, Q.: Deep neural networks for learning graph representations. In: AAAI, pp. 1145–1152 (2016)
4. Goyal, P., Ferrara, E.: Graph embedding techniques, applications, and performance: a survey. Knowl.-Based Syst. **151**, 78–94 (2018)
5. Grover, A., Leskovec, J.: Node2Vec: scalable feature learning for networks. In: SIGKDD, pp. 855–864 (2016)
6. Hodgson, G.: Banking, finance and income inequality. Positive Money (2013)
7. Kipf, T.N., Welling, M.: Semi-supervised classification with graph convolutional networks. In: ICLR (2018)
8. Kiss, C., Bichler, M.: Identification of influencers - measuring influence in customer networks. Decis. Supp. Syst. **46**(1), 233–253 (2008)
9. Koren, Y., Bell, R., Volinsky, C.: Matrix factorization techniques for recommender systems. IEEE Comput. Mag. **42**(8), 30–37 (2009)
10. Targeting and Rewarding High Value Customers. https://www.kobie.com/2015/03/06/targeting-and-rewarding-high-value-customers/
11. Mikolov, T., Chen, K., Corrado, G., Dean, J.: Efficient estimation of word representations in vector space. arXiv preprint arXiv:1301.3781 (2013)
12. Perozzi, B., Al-Rfou, R., Skiena, S.: DeepWalk: online learning of social representations. In: SIGKDD, pp. 701–710 (2014)
13. Ryals, L., Knox, S.: Cross-functional issues in the implementation of relationship marketing through customer relationship management. Eur. Manag. J. **19**(5), 534–542 (2001)
14. Trouillon, T., Welbl, J., Riedel, S., Gaussier, É., Bouchard, G.: Complex embeddings for simple link prediction. In: ICML, pp. 2071–2080 (2016)
15. Verhoef, P.C., Donkers, B.: Predicting customer potential value an application in the insurance industry. Decis. Supp. Syst. **32**(2), 189–199 (2001)
16. Wang, D., Cui, P., Zhu, W.: Structural deep network embedding. In: SIGKDD, pp. 1225–1234 (2016)
17. Wu, S.Y., Wu, A.: Information asymmetry, bargaining power and customer profitability: an empirical investigation on bank-client relationship. In: AAA 2008 MAS Meeting Paper (2007)
18. Zhang, S., Yao, L., Sun, A., Tay, Y.: Deep learning based recommender system: a survey and new perspectives. ACM Comput. Surv. **52**(1), 5:1–5:38 (2019)
19. Zhou, J., Cui, G., Zhang, Z., Yang, C., Liu, Z., Sun, M.: Graph neural networks: a review of methods and applications. arXiv preprint arXiv:1812.08434 (2018)

Social Problems

A Complex Network Approach to Structural Inequality of Educational Deprivation in a Latin American Country

Harvey Sanchez-Restrepo[1](✉) and Jorge Louça[2]

[1] Faculty of Sciences, University of Lisbon, Cidade Universitária, 1649-004 Lisbon, Portugal
harvey@comunidad.unam.mx

[2] Information Sciences, Technologies and Architecture Research Center, ISCTE-IUL, 1649-026 Lisbon, Portugal
jorge.l@iscte-iul.pt

Abstract. To guarantee the human right to education established by the fourth UNESCO's Sustainable Development Goal, a deep understanding of a big set of non-linear relationships at different scales is need it, as well as to know how they impact on learning outcomes. In doing so, current methods do not provide enough evidence about interactions and, for this reason, some researchers have proposed to model education as a complex system for considering all interactions at individual level, as well as using computer simulation and network analysis to provide a comprehensive look at the educational processes, as well as to predict the outcomes of different public policies.

The highlight of this paper is modeling the structure of the inequality of a national educational system as a complex network from learning outcomes and socio-economic, ethnicity, rurality and type of school funding, for providing a better understanding and measuring of the educational gaps. This new approach might help to integrate insights improving the theoretical framework, as well as to provide valuable information about non-trivial relationships between educational and non-educational variables in order to help policymakers to implement effective solutions for the educational challenge of ensuring inclusive and equitable education.

Keywords: Structural network · Large-scale assessments · Policy informatics

1 Introduction

The 193 countries attached to Unesco promulgated the Sustainable Development Goals (SDG), the fourth goal (SDG-4) establishes that "education is a human right" [1,2] and that the essential axes of quality in education must be learning and equity, since that all human beings have the right to learn, the State is obliged to guarantee the exercise of this right to all citizens equally

H. Barbosa et al. (Eds.): *Complex Networks XI*, SPCOM, pp. 349–358, 2020.
https://doi.org/10.1007/978-3-030-40943-2_29

[3,4]. However, Unesco points out that inequality in education has progressively increased and that the most impoverished bear the worst consequences [5,6]. In Latin America, the results of the last Large-scale Assessment of learning (LSA) show that most of the countries have high percentages of children with low-level learning after several years of schooling [7].

At global level, the lack of quality in education is of such magnitude that Unicef estimates that 250 million children, after having attended school, fail to develop the minimum learning in Literacy and Numeracy, both necessary to continue learning at the following educational levels [8]. For facing this challenge, many governments have promoted multiple reforms to improve the quality of their education systems, however, the modest improvements in learning have been accompanied by huge inequalities between different population groups, raising many questions to the policymakers [9,10].

In the last years, models based on Network science have been proposed as an alternative representation of systems, overall, because many systems can be described by complex interconnected networks as a result of self-organized processes [11]. One of the main advantages of modeling educational systems as a network, or multiple networks, is establishing connections between population characteristics of individuals as random phenomena with probabilities of occurrence given by data. This approach allows the identification of the factors that influence learning based on both topological and statistical parameters for unveiling some hierarchical structures related with inequality and educational deprivation.

For providing a robust model for better understanding inequality gaps, in this research we use concepts from network science as a tool for studying complexity and global and local properties of the structure of inequality in learning outcomes observed in a Latin American country. The analysis is based on statistical properties of the networks related with low-level-of-proficiency students and the Socio-economic Status (SES) of the student's family, Rurality of the area where the school is located (RA), Type of school (TS), and Ethnicity (ET) for analyzing out-of-equilibrium states [12].

1.1 Dataset

For developing the model, a multivariate dataset integrates learning outcomes of every student who has completed the k-12 education process, estimated by the ability's parameter θ^j through a LSA carried out in Ecuador in 2017, using a standardized computer-based test[1] and integrated with a robust dataset with more than 240 variables coming from surveys to student's families and teachers. For building the scores, ability and psychometric parameters were estimated by Item Response Theory as usual, through a 2P-Logistic model [12,13], following Eq. 1:

[1] Full dataset is available in http://www.evaluacion.gob.ec/evaluaciones/descarga-de-datos/, selecting the option Ser Bachiller 2017–2018 and microdato for downloading the full data.

$$P(\theta_j) = \frac{e^{[\alpha_i(\theta_j - \beta_i)]}}{1 + e^{[\alpha_i(\theta_j - \beta_i)]}} \text{ with } \theta_j, \alpha_i, \beta_j \in (-\infty, \infty) \quad (1)$$

After estimation process, raw scores were re-scaled to a standardized Learning index ($LI_j \in [4.0, 10.0]$), a monotonous transformation of θ^j, where higher levels of learning are more likely to have higher scores [12,13]. The scores are on a continuous scale corresponding to four levels of achievement, according with the LSA and national standards, according with a technical description[2], all students are classified in Levels of Achievement (LA) stablished by a Bookmark process carried out by an expert pedagogical group on each subject [12]. The three psychometrical cut points s_1, s_2, s_3 correspond to four LA, where L_0 corresponds to those who have not reached a minimum level of learning, L_1 to the minimum acceptable, L_2 at the level of achievement raised by the system and L_3 corresponds to a performance higher than the standard.

1.2 Deprivation Learning Index

For estimating the Deprivation Learning Index (DLI), we use the family of scores $\{LI_j\}_j = 1, \cdots, N$, of those students with low level of achievement L_0-class where s_1 is the first cut point - the minimum score to be located at level L_1. They are students suffering learning deprivation according with the sociological proposal that states 'there is an irreducible nucleus of needs that are common to every human being', while relative deprivation, estimated by LI_j for each student, becomes from 'needs, thresholds and satisfactions are determined by each society' [14], both established by the Bookmark process.

For the L_0-class, absolute deprivation is given by $H = n(L_0) \sum n(LI_j)$, where $n(LI_j)$ represents the number of students below the first LA, the intensity $\lambda(LI_j)$ is given by the distance to reach the first level L_1, then DLI is given by $\delta_j = H \cdot \lambda(LI_j)$, which represents a measure of the collective learning deficit, which considers the magnitude - the number of students with low performance - and intensity - how much below the minimum performance level are located [14].

As can be seen, histogram in Fig. 1 shows that 22% does not meet the learning minimums at the end of the compulsory cycle.

Scores distribution in Fig. 1 allow the study of equity and it can be deepened by analyzing the levels of deprivation - absolute and relative - experienced by different population groups and their relationship with the socioeconomic status and ethnicity of students.

1.3 Model Specification

In the last years, models based on graph theory have been proposed as a parallel representation of psychometric constructs such as intelligence, leadership or depression [15]. These models have in common that the covariance among the

[2] Technical and pedagogical details about design can be found in http://www.evaluacion.gob.ec/evaluaciones/ser-bachiller/.

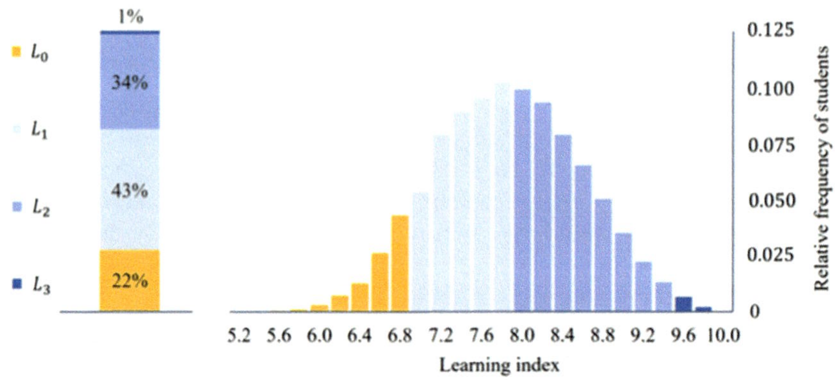

Fig. 1. Distribution of the students among Learning index (scores).

observable variables could be explained from the identification of patterns found among a set of interactions between these variables, measured through a set of informative items. The model for building the network is based on interactions between two nodes, representing the level of learning of each student directed to the set of each factor categories, where the edge is weighted by $\lambda(LI_i)$ [16]. For carrying on the analysis, the model runs in three phases: (1) analyzing the scores for assigning a LA to each student for identifying those located in level L_0, (2) estimating the SES for aggregated levels and subpopulation groups using the cut points for splitting in deciles, and (3) analyzing the associated factors to learning for creating the family of sequences $\left\{\theta^j \rightarrow L_k^j \rightarrow (SES_d^j)\right\} \forall j$ [17,18] for each student, where L_k^j is the LA, and SES_d^j corresponds to the SES decile of the j-th student.

To carry out a refinement of the variables that increase the deficit of basic skills in the population, the SES and the estimated deciles for the general population are preserved during all the stages, all other estimates are made again over the L_0-group. As each student is represented by a node, a set of edges, weighted by $\lambda(LI_j)$, are first directed to one of the SES-decile nodes $\left\{\theta^j \rightarrow L_k^j \rightarrow (SES_d^j)\right\} \forall j$, a process which allows to analyze aggregated inequality at school level, as well as In-degree distribution for SES nodes.

For extending the model and knowledge about social determinants, TS, RA and ET are included in the analysis one by one for analyzing their effects through the sequence $\left\{\theta^j \rightarrow L_k^j \rightarrow (SES_d^j) \rightarrow (TS_{C2}^j, RA_{C1}^j, ET_{C3}^j\right\} \forall j$, where C denotes an index for each subcategory of the factors RA, TS and ET. Network analysis was carried out by Gephi 0.9.2 and statistical estimations and plots with R 3.5.0 and Orange 3.3.8.

2 Socioeconomic Status and Student's Learning Outcomes

To estimate the size of the gap at the macro level, the first network shown in Fig. 2 integrates the Weighted In-degree distribution of directed edges from nodes indicating subpopulation groups to those representing SES deciles, given by $\{SES_d\}$ $d \in \overline{1,10}$, where each edge represents one student in L_0-class. As inequality implies asymmetries, in conditions of total equity - where socioeconomic factors would not produce differences - we might expect equal distribution of L_0-edges over the network for all deciles, but the distribution is not like that. Therefore, the study of equity can be deepened by analyzing the levels of absolute and relative deprivation experienced by different population groups and their relationship with the SES of the students.

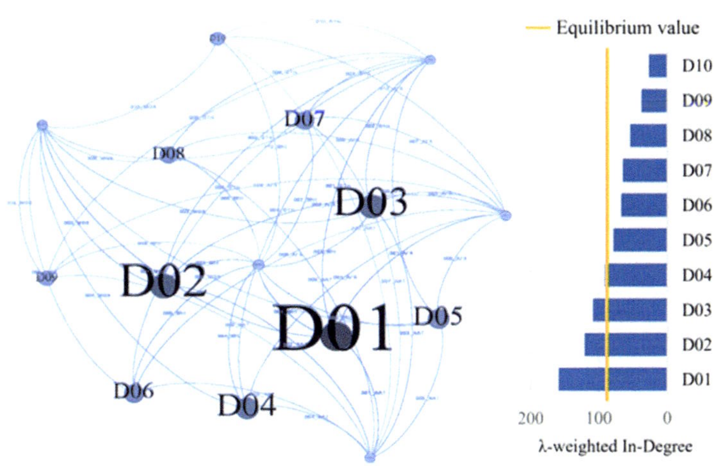

Fig. 2. Network for Weighted Out-degree and its edges' histogram shows socioeconomic status distribution of deprived students (L_0-group).

According with estimates, 21.5% of students are in L_0-class, a prevalence rate of 0.215 corresponding to a $LI = 6.32$ and shows an intensity of deprivation $\lambda = 0.225$, i.e., in average, L_0-student lacks 0.68 standard deviations (SD) of the minimum learning.

To estimate the size of the gap at the macro level, Fig. 2 also shows the percentages of students in each level of achievement for each SES decile. As can be observed, the proportion of students in each decile decreases monotonically as the SES of the group increases, being for the first decile (D01), 39% of students, and only 8% in D10. This difference of 31% points is equivalent to the fact that for each rich family student who does not learn the minimum, there are 5 poor in the same situation.As will be shown later, this situation deepens in rural areas, where the ratio increases to one rich student for every 7 poor students.

3 The Relationship Between Type of School and SES

When studying schools as integrated units, the impact of SES becomes even more evident, in Fig. 3 each school is represented by a circle whose size is proportional to the number of its students enrolled, the source of funding is distinguished by the color: green for private, blue for public. The average SES of students is located on the horizontal axis and the average score of LSA on the vertical axis. In the right side, the two whisker-box plots show dispersion for both indexes.

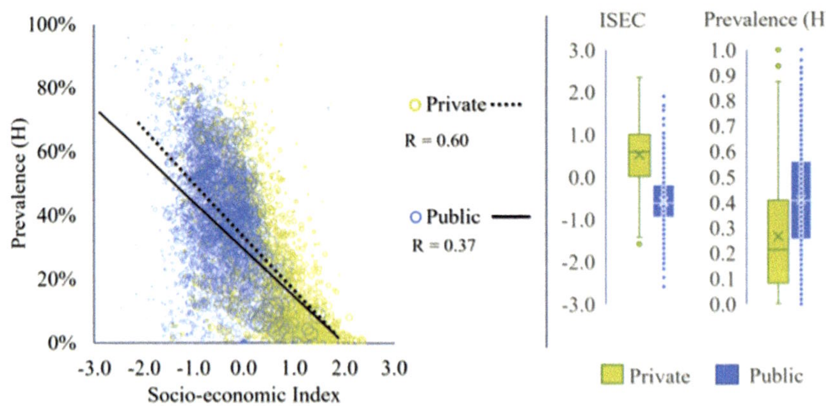

Fig. 3. Relationship between learning prevalence of deprivation and socioeconomic status at school level and box plots disaggregated by type of funding.

The negative correlation between SES and prevalence rate ($R = -0.55$, $p < 0.001$) shows the separation of SES classes in groups of students who have different learning opportunities inside and outside schools, which helps to understand how inequality is gestated in a structural way in the country: schools with high SES predominate in the private sector and it is also there that the lowest levels of deprivation are presented, in this sector the correlation coefficient between the SES and H index is ($R = -0.60$, $p < 0.001$). On the contrary, public schools that serve the poorest students have higher prevalence rates and a lower correlation ($R = -0.39$, $p < 0.001$), which could indicate that, a weight of which the deprivation of learning is higher for the whole group, there is less inequality motivated by the socioeconomic origin of the student.

The socioeconomic gap between private (0.64) and public (−0.18) schools accumulates 0.82 standard deviations (SD), in addition, the prevalence rate in the public sector ($H = 0.219$) is 1.7 times that of the private sector ($H = 0.374$). So, this confirms that public schools not only serve the poorest students in the country, but as they do so in the most depressed and most difficult areas, attendance is an extra challenge reflected in prevalence rates, while private schools concentrate on students in the top quintiles and the prevalence in most cases does not exceed 30.

4 Inequality Gaps and Marginalized Population Groups

For having a more detailed and in-depth analysis, a selection of the two opposite SES population groups were selected as attractor nodes in the network - deciles D01 (Green) and D10 (Pink). In Fig. 4, the network integrates the different ethnic groups of the country, disaggregated by rural and urban areas. Both, the nodes and the labels, are proportional to the Weighted Out-Degree, and $\lambda(LI_j)$ is weighting the edges.

The representation is based on the Eigen Centrality to measure the influence of a factor in terms of the number of edges with the population groups and that appear as other nodes within the network [18,20]. This measure is very valuable because, by counting how well connected a node is, and how many links have its connections through the network, the preponderance of the factors in the population groups becomes very clear in identifying the effect of the three educational deprivators: (1) rural areas, (2) types of funding of schools, and (3) ethnicity.

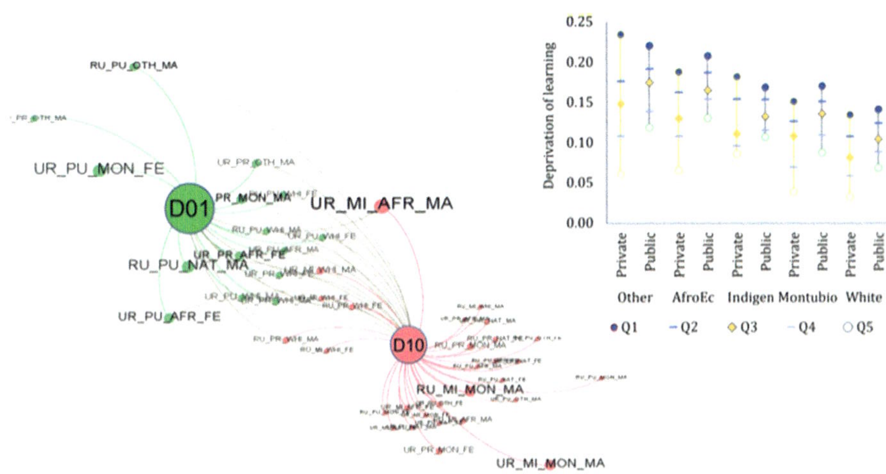

Fig. 4. Network of subpopulation groups splatted in rich and poor students with ? weighted learning scores and its gaps synthesis.

For complementing information, the plot in the right side of Fig. 4 shows that gaps between public and private sectors are quite pronounced, especially among the poorest students in both systems. In addition, although in the public sector there is less variability among quintiles, in all cases, Q1 and Q2 suffer the highest levels of deprivation. As can be seen, the lack of learning is found in the Afro-Ecuadorian population, Montubio's people, indigenous and other groups, before than people identified as White, the gaps are greater than 40. Of special interest is the case when the modularity is represented, in this network, the parameter estimated was 0.073 and resolution −0.254, which produces two communities: the richest and the poorest. This result is particularly important for public policies because it might allow policymakers to work directly with families in a group-oriented strategy to avoid presenting same actions for completely different problems. It is also remarkable that private schools serving indigenous and other minority groups coming from Q1, show higher deprivation rate than graduates of the public system in the same Q1 level. This result points out a tremendous social deception and suggests an urgent migration of those students to the public system in order to review the operation of these schools, given their low performance in a group with so many disadvantages. A racial dramatic case is also found: Afro-Ecuadorians with the lowest level of deprivation are those located in Q5, however, the poorest white students attending public schools show a level of deprivation equivalent. This approach allows to measure the magnitude with which the lower deciles dominate in the interactions with the population groups through the edges [19]. Furthermore, the network has directed edges and its average weighted Out-Degree might be seen as a covariation-measure of the deciles with population of non-learning students, so, it is possible to compare the values thrown by the network with the value that could be expected on this parameter in conditions of equity where the factors would not produce differences.

5 Discussion

With this new kind of analysis, we have developed a model for finding answers to classical questions in educational research using free available data for a Latin American country, providing a direct method to recognize the structure of inequality, as well as the relationship between social determinants for educational deprivation and the conditional distribution of learning outcomes. Given that equity is a major focus of government policies around the world and that it is promoted by international agencies with the aim of transforming educational systems, attending the wide diversity of students in each country and the whole region is a big challenge and in this paper we have presented an analysis that offers a lot of valuable evidence showing the deep lack in this dimension, highlighting that is a structural problem that goes beyond educational policy. Using the network concept of modularity in defining groups with the same kind of challenges, might help to policymakers in selecting those factors that might be so much relevant for one group than to others, overall in some areas where the intra-class variability is very low using complementarian topological and statistical analysis. Given the relationship between the DLI and SES, and that deprivation is almost eight times higher for poorest than for richest students, it is confirmed that the exercise of educational rights is a function of SES and that the gap is wider when considering the types of financing and that this phenomenon gets even worst for minority ethnic groups.

References

1. OECD: Equity in Education: Breaking Down Barriers to Social Mobility, PISA. OECD Publishing, Paris (2018)
2. UN DESA: The Sustainable Development Goals Report 2018. UN, New York (2018)
3. McGrath, S., Nolan, A.: SDG 4 and the child's right to education. NORRAG NEWS **54**, 122 (2017)
4. Sanchez-Restrepo, H.: Equity: the focal point of educational quality. National Educational Evaluation Policy Gazette in Mexico, Year 4, no. 10, pp. 42–44 (2018)
5. United Nations Children's Fund (UNICEF). The investment case for education and equity. Washington, DC (2015)
6. Keeley, B., Little, C.: The State of the World's Children 2017: Children in a Digital World. UNICEF, vol. 3. United Nations Plaza, New York (2017)
7. Samman, E.: SDG progress: fragility, crisis and leaving no one behind: report (2018)
8. Gray, J., Kruse, S., Tarter, C.J.: Enabling school structures, collegial trust and academic emphasis: antecedents of professional learning communities. Educ. Manag. Adm. Leadersh. **44**(6), 875–891 (2016)
9. Laboratorio Latinoamericano de Evaluacion de la Calidad de la Educacion - LLECE (2018). Agenda 2018. http://www.unesco.org/new/en/santiago/press-room/newsletters/newsletter-laboratory-for-assesSSent-of-the-quality-of-education-llece/
10. Hopfenbeck, T.N., et al.: Lessons learned from PISA: a systematic review of peer-reviewed articles on the programme for international student assessment. Scand. J. Educ. Res. **62**(3), 333–353 (2018)
11. Barabasi, A.-L., Pasfai, M.: Network Science. Cambridge University Press, Cambridge (2016)
12. Borsboom, D., Molenaar, D.: Psychometrics. In: Wright, J. (ed.) International Encyclopedia of the Social & Behavioral Sciences, vol. 19, pp. 418–422. Elsevier, Amsterdam (2015)
13. Li, F., et al.: Model selection methods for mixture dichotomous IRT models. Appl. Psychol. Meas. **33**(5), 353–373 (2009)
14. Bracho, T.: Índice de dficit en competencias Avanzamos hacia la garanta del derecho a la educacin? en Reformas y Polticas Educativas, no. 4, septiembre–diciembre 2017, FCE, Mexico (2017)
15. Marsman, M., Borsboom, D., Kruis, J., Epskamp, S., van Bork, R., Waldorp, L.J., Maas, H.V.D., Maris, G.: An introduction to network psychometrics: relating ising network models to item response theory models. Multivar. Behav. Res. **53**(1), 15–35 (2018)
16. Costantini, G., Epskamp, S., Borsboom, D., Perugini, M., Mottus, R., Waldorp, L.J., Cramer, A.O.J.: State of the art personality research: a tutorial on network analysis of personality data in R. J. Res. Pers. **54**, 13–29 (2015)
17. Kossakowski, J.J., Epskamp, S., Kieffer, J.M., van Borkulo, C.D., Rhemtulla, M., Borsboom, D.: The application of a network approach to health-related quality of life (HRQoL): introducing a new method for assessing HRQoL in healthy adults and cancer patient. Qual. Life Res. **25**, 781–792 (2016)
18. Forbush, K., Siew, C., Vitevitch, M.: Application of network analysis to identify interactive systems of eating disorder psychopathology. Psychol. Med. **46**(12), 2667–2677 (2016)

19. Gibson, L., Koenig, A.: Neighboring groups and habitat edges modulate range use in Phayre's leaf monkeys (Trachypithecus phayrei crepusculus). Behav. Ecol. Sociobiol. **66**(4), 633–643 (2012)
20. van Borkulo, C.D., Borsboom, D., Epskamp, S., Blanken, T.F., Boschloo, L., Schoevers, R.A., Waldorp, L.J.: A new method for constructing networks from binary data. Sci. Rep. **4**(5918), 1–10 (2014)

Network-Based Delineation of Health Service Areas: A Comparative Analysis of Community Detection Algorithms

Diego Pinheiro[1(✉)], Ryan Hartman[3], Erick Romero[1], Ronaldo Menezes[2], and Martin Cadeiras[1]

[1] Department of Internal Medicine, University of California, Davis, USA
pinsilva@ucdavis.edu

[2] Department of Computer Science, University of Exeter, Exeter, UK

[3] Independent Researcher, Washington D.C., USA

Abstract. A Health Service Area (HSA) is a group of geographic regions served by similar health care facilities. The delineation of HSAs plays a pivotal role in the characterization of health care services available in an area, enabling better planning and regulation of health care services. Though Dartmouth HSAs have been the standard delineation for decades, previous work has recently shown an improved HSA delineation using a network-based approach, in which HSAs are the communities extracted by the Louvain algorithm in hospital-patient discharge networks. Given the known heterogeneity of communities extracted by different community detection algorithms, a comparative analysis of community detection algorithms for optimal HSA delineation is lacking. In this work, we compared HSA delineations produced by community detection algorithms using a large-scale dataset containing different types of hospital-patient discharges spanning a 7-year period in the USA. Our results replicated the heterogeneity among community detection algorithms found in previous works, the improved HSA delineation obtained by a network-based, and suggested that Infomap may be a more suitable community detection for HSA delineation since it finds a high number of HSAs with high localization index and a low network conductance.

Keywords: Hospital-Patient Discharge Networks · Community detection algorithms · Health Service Area · HSA delineation

1 Introduction

A Health Service Area (HSA) is as a group of geographic regions in which residing patients most often receive healthcare services from similar health care facilities. It was first introduced in 1973 by Wennberg and Gittelsohn as a more meaningful unit of analysis for healthcare data than administrative geographic divisions

H. Barbosa et al. (Eds.): *Complex Networks XI*, SPCOM, pp. 359–370, 2020.
https://doi.org/10.1007/978-3-030-40943-2_30

such as counties [13]. By examining variations in expenditures among HSAs delineated in Vermont, the authors showed, for instance, that the expenditure per capita among HSAs varied from $54 to $162 and such variation, however, had no correlation with age-adjusted mortality.

In 1996, Wenneberg then proposed the *Dartmouth Atlas of Health Care in the United States* [12] which is the current standard HSA delineation in the US. Effectively, each Dartmouth-HSAs are delineated in three steps. First, each health care facility is assigned to its respective city/town. Then, each ZIP Code is assigned to the city/town of the health care facility from which residing patients receive most of their healthcare services. As a result, each Dartmouth-HSA is the group of ZIP Codes associated with the same city/town. Finally, enclave ZIP Codes, if any, are assigned the city/town of its adjacent ZIP Codes to unsure geographic contiguity.

Recently, Hu et al. [7] has proposed a network-based approach to HSA delineation in which a Hospital-Patient Discharge Network (HPDN) was built and the Louvain community detection algorithm was subsequently applied to find communities (i.e., HSAs) with the highest network modularity. In their HPDN, nodes represent distinct ZIP Codes and links represent the total number of discharges between the ZIP Codes of health care facilities and patient residences. Using claims-based hospital discharges in Florida, the authors demonstrated that Louvain-HSAs presented, for instance, a higher localization index than Dartmouth-HSAs, which is a measure of internal validity that quantifies the proportion of patients receiving services from health care facilities located within the same HSA in which they reside.

Yet, a comparative analysis that looks at the effectiveness of various community detection algorithms is still needed in order to attain optimal network-based HSA delineation. Such comparative analysis is needed given owing to the heterogeneity of communities extracted by different community detection algorithms [6]. Through a comparative analysis was previously provided for grouping hospitals [3], the underlying networks differ from HPDNs in two fundamental aspects: nodes were hospitals instead of geographical regions, and links were patients sharing between hospitals instead of the total number of hospital discharges.

In this paper, we conducted a comparative analysis of network-based community detection algorithms for HSAs delineated; we focus on four of the commonly used community detection algorithms, namely, Block Model [8], Infomap [10], Louvain [1], and Speaker-Listener Label Propagation Algorithm (SLPA) [14]. A claims-based patient-hospital discharge data was used; it contains a total of 124,970,471 discharges over a 7-year period in California, USA. Our results replicated the existent heterogeneity of communities extracted by different algorithms of community detection and reinforced the use of a network-based approach to HSA delineation. The results demonstrated, for instance, that Infomap was the most suitable algorithm because it was capable of delineating a high number of HSAs, while still presenting a high localization index and a low network conductance.

2 Methodology

A network-based delineation of Health Service Areas (Fig. 1) consists of (A) modeling claims-based patient-hospital discharge data as networks of ZIP Code Tabulation Areas (ZCTAs), (B) applying a diverse set of community detection algorithms, and (C) performing a comparative analysis of the extracted communities (i.e., HSAs) according to multiple quality metrics of HSA delineation.

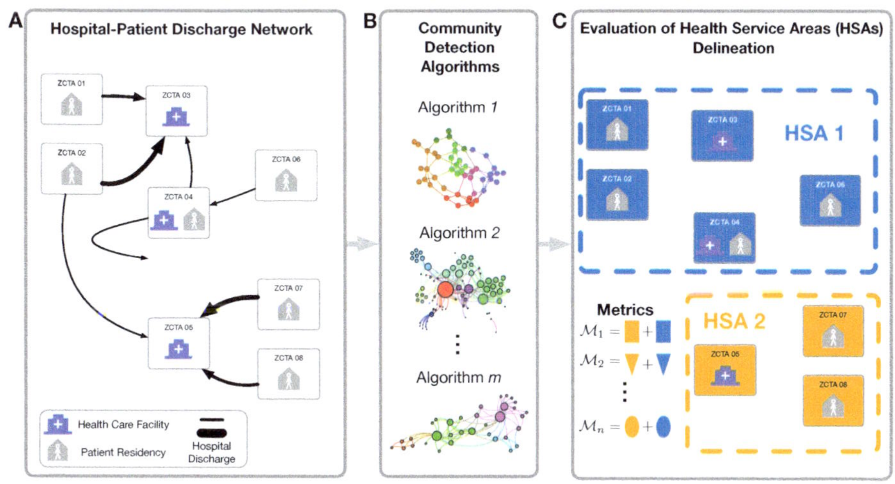

Fig. 1. Network-based delineation of Health Service Areas (HSAs). (**A**) Hospital discharge data is used to build Hospital-Patient Discharge Networks (HPDN) in which nodes represent ZIP Code Tabulation Areas (ZCTAs), and links represent the total number of discharges between the ZCTA locations of health care facilities and patient residencies. (**B**) Community detection algorithms are applied over HDPNs to delineate HSAs. (**C**) Delineated HSAs are evaluated according to multiple quality metrics of HSA delineation.

2.1 Hospital-Patient Discharge Networks

Hospital-Patient Discharge Networks (HPDNs) were built using claims-based hospital discharge data obtained from the California Health and Human Services Agency (CHHS) [2]. This dataset is publicly available and contains a total of 124, 970, 471 hospital-patient discharges of different types spanning a 7-year period from 2012 through 2018. The four discharge types are *Inpatient from ED*, *Inpatient*, *Ambulatory Surgery*, and *ED Only*. Each data point contains the type of discharge, the year, the name of the facility (e.g., Alameda Hospital, University of California Davis Medical Center). Also, it contains the 5-digit ZIP Codes (e.g., 94501, 95831) of both facility location and patient residency as well as the respective number of discharges between them. A discharge is the process by which patients undergo after the provision of the healthcare service when they leave the hospital. The healthcare services may require admission to the hospital for a overnight stay, inpatient, or may not require hospitalization, outpatient. A visit to the Emergent Department (ED), which is considered outpatient, may require further hospitalizations and thus become inpatient.

Hospital discharges to patient residency ZIP Codes other than those with 5-digits were excluded (2.6%); these were mainly discharges of patients from states other than California (e.g., ARIZONA, NEVADA (state), Other USA), from locations outside the US (e.g., OUTSIDE USA), from unknown locations (e.g., UNKNOWN), and from homeless population (e.g., HOMELESS). ZIP Codes are a collection of delivery routes maintained USA Postal Service and ZIP Code Tabulation Area (ZCTAs) are actual generalized areal representations maintained by the USA Census Bureau. Therefore, for each ZIP Code of both health care facility and patient residency, the corresponding ZCTA was obtained using the ZIP Code to ZCTA Crosswalk provided from the Uniform Data System Mapper [11].

A separate weighted and undirected HPDN network was built for each type of hospital discharge and for each year according to the methodology proposed by Hu et al. [7]. In each network, nodes are ZCTAs, links are the total number of discharges between the ZCTAs of health care facilities and patient residencies. The HPDN is an undirected network because a link w_{ij} encodes the total number of discharges *between* ZCTAs i and j without arbitrarily distinguishing whether the direction is due to a health care facility at i discharging patients residing at j or patients residing at i going to a facility at j for health care services. Overall, 28 Hospital-Patient Discharge Network (HPDN) were built, one for each combination of 4 discharge types and 7 years.

2.2 Community Detection

In a network-based HSA delineation, each HSA correspond to a distinct community extracted by a community detection algorithm from a hospital-patient discharge network (HPDN) [7]. While the existence of communities in real-world networks is agreed upon, there is no generally accepted definition of what a community is, or what the most appropriate way to find them is [5]. Some algorithms take a stronger approach to community detection by looking for cliques, which are a group of nodes for which there is a link between every pair of nodes [4]; other approaches just look for more densely connected subgraphs within the network such that a community is a subset of nodes within a network that are densely connected to each other when compared to the rest of the network.

The lack of a general definition of what a community should be is also mirrored in the existent heterogeneity of communities extracted by different community detection algorithms [6] and as such requires a comparison among community detection algorithms, particularly when the communities found are being used in important issues such as HSA delineation. Four commonly used algorithms were selected: Louvain Modularity, Infomap, Stochastic Block Model, and Speaker-listener Label Propagation. These algorithms were selected as they provide four very distinct approaches to community detection and have their implementations easily provided by their authors.

The *Louvain* algorithm [1] finds communities that maximizes the network modularity. This quantifies the extent to which the density of links within the found communities excessively surpasses that of what would be expected if links were placed at random. Trying all possible partitions is not computationally feasible, and Louvain modularity takes a heuristic approach by maximizing the local modularity of smaller communities that are only subsequently joined if such aggregation leads to an increased modularity. These smaller communities start as individual nodes and are iteratively joined together into greater communities until a single community containing the whole network is reached.

The *Stochastic Block Model* [8] uses a maximum likelihood estimator to infer the block structure of the network. This algorithm attempts to recover the hierarchical block structure of the network where each block represents a community. The block model used in this study is the degree-corrected variant given the weighted aspect of HPDNs and that previous work has shown that the such variant tends to perform better on empirical networks [9].

The *Infomap* algorithm [10] finds communities that maximizes the map equation instead of modularity. The map equation quantifies the length of the coding scheme necessary to communicate the sequence of movements of random walkers within the network. In essence, if a community structure exists, random walkers will tend to become trapped within these communities because movements within-communities are more likely than between-communities. Therefore, the coding scheme necessary to communicate the sequence of movements can be reduced by taking into account the community structure as every time a walker enters into a community, the community is identified, and a smaller community-specific coding scheme is used to quantify within-community movements.

The *Speaker-Listener Label Propagation Algorithm* (SLPA) [14] is a localized community detection algorithm based on the concept of label propagation. SLPA finds communities by initially assigning each node to a unique label. Nodes then iteratively changes their label to the label most often used by its neighbors. Initially, this label exchange rule promotes the formation of smaller consensus groups which will subsequently compete with one another for node members depending on the balance between their within-group and between-group interactions. In contrast to the other algorithm's, SLPA does not require prior information about the network, nor does it attempt to maximize any metric as a proxy for well-defined communities. Instead, it only relies on the network structure to identify the communities. While SLPA is able to find overlapping communities, the post processing threshold was set to $r = 0.5$ to ensure the extraction of non-overlapping communities and thus provide a better comparison to the other non overlapping algorithms used.

The application of the 4 community detection algorithms to the 28 HPDNs yielded a total of 112 HSA delineations. Though only a subset of HPDNs and HSA delineations are presented, all of the code, datasets, networks, and analysis are available on the Open Science Framework (OSF) repository of this project at https://doi.org/10.17605/OSF.IO/GW73Y.

2.3 Evaluation of Health Service Areas (HSA) Delineation

The quality of a HSA delineation can be evaluated according to multiple and often conflicting metrics [3, 7]. The following four delineation metrics of a HSA c were used: the number of communities N_c, the localization index $LI(c)$, network conductance $C(c)$, and the total number of discharges $D(c)$.

In a network-based HSA delineation, the *total number of delineated HSAs* N_c is determined by the specific community detection algorithm used as it is the number of communities extracted. The N_c is an important metric since it determines the number of distinct meaningful units of analysis ultimately uncovered and, as in any community detection problem, trivial solutions such as 1 (one) and n, the total number of nodes, are generally undesired [6].

The *localization index* $LI(c)$ of a community c quantifies the proportion of patients seeking and receiving health care services from hospitals within the HSA in which they reside. In a community c with a higher $LI(c)$, residents mostly seek healthcare from hospital within the HSA c. Formally, the localization index $LI(c)$ of community c can be defined as

$$LI(c) = \frac{D(c,c)}{D(c)}, \tag{1}$$

in which $D(r,s)$ is the number of discharges of patients residing at ZCTAs within community r that are discharged from hospitals at ZCTAs within community s, and $D_c = \sum_s^{N_c} D(c,s)$ is the total number of discharges from patients living within community c.

The network conductance $C(c)$ of a community c is a network-based measure which quantifies the extent to which c is a well-formed community by comparing the total links running within community c relative to links running from c to other communities. Conductance is based on the degree-based definition of a community [5]. Formally, the conductance $C(c)$ of a community c can be calculated as

$$C(c) = \frac{w^{ext}(c)}{w(c)}, \tag{2}$$

in which, $w(c) = \sum_{i \in C,j} W_{ij}$ is the total strength of links originating within community c, $w^{ext}(c) = \sum_{i \in C, j \notin C} W_{ij}$ is the external strength, and W_{ij} is the strength of the link connecting nodes i and j.

Aside from the aforementioned metrics, the total number of discharges $D(c)$ from patients living in one of the ZCTAs found within community c is also calculated. Ideally, a community detection algorithm would maximize the number of HSAs where each has a high localization index and a low conductance. Yet, as the total number of communities increases from one to n, the typical value of localization index decreases from 1 to 0 and the typical value of network conductance increases from 0 to 1.

To provide a reliable estimate for all communities found by a single community detection algorithm, $B = 1,000$ bootstrap samples with replacement were draw from the distribution of each metric to calculate a mean value for the localization index $\langle LI(c) \rangle$, network conductance $\langle C(c) \rangle$, and total number of discharges $\langle D(c) \rangle$. The standard deviation was also provided for each of the aforementioned estimators.

3 Results and Discussion

The network statistics of each individual HPDN varied among discharge types and over the years (Table 1). Considering the year of 2012, for instance, *Inpatient from ED* and *ED Only* HPDNs had comparable total number of nodes n. Their number of links m were $49,000$ and $127,000$, respectively, suggesting that a *ED Only* HPDN has a higher number of distinct ZCTA pairs for which hospital discharges occurred between hospital locations and patient residencies. Also, their the total network link strength w were 1.6 million and 9.2 million, respectively, and their network density ρ were 0.0345 and 0.0883, respectively, suggesting that the healthcare service underlying a *ED Only* HPDN has a higher demand and is more dense. Over the years, the average shortest path length l and clustering coefficient c of HPDNs slightly decreased.

Overall, HSA delineation results (Fig. 2 and Table 2) have reinforced the superiority of network-based HSA delineation as well as confirmed the heterogeneity among communities extracted by different community detection algorithms

Table 1. Statistics of Hospital-Patient Discharge Networks (HPDNs). For each type of discharge and for each year, the following network measures were quantified: the total number of nodes (n), the total number of links (m), the total weight (w), the network density (ρ), the average shortest path length (l), as well as the clustering coefficient (c). Other types of discharges and network metrics are available on the Open Science Framework (OSF) repository of this project at https://doi.org/10.17605/OSF.IO/GW73Y.

Type of discharge	Year	n	m	w	ρ	l	c
Inpatient from ED	2012	1.69×10^3	4.92×10^4	1.63×10^6	3.45×10^{-2}	2.81	1.19×10^{-3}
	2013	1.69×10^3	4.97×10^4	1.63×10^6	3.49×10^{-2}	2.72	1.12×10^{-3}
	2014	1.69×10^3	5.05×10^4	1.64×10^6	3.56×10^{-2}	2.71	1.02×10^{-3}
	2015	1.68×10^3	5.25×10^4	1.72×10^6	3.70×10^{-2}	2.89	9.74×10^{-4}
	2016	1.75×10^3	5.37×10^4	1.74×10^6	3.50×10^{-2}	2.70	9.53×10^{-4}
	2017	1.75×10^3	5.38×10^4	1.75×10^6	3.52×10^{-2}	2.67	9.84×10^{-4}
	2018	1.75×10^3	5.41×10^4	1.75×10^6	3.54×10^{-2}	2.62	9.31×10^{-4}
ED only	2012	1.69×10^3	1.27×10^5	9.25×10^6	8.83×10^{-2}	2.15	2.14×10^{-4}
	2013	1.69×10^3	1.28×10^5	9.65×10^6	8.95×10^{-2}	2.18	2.07×10^{-4}
	2014	1.69×10^3	1.33×10^5	1.03×10^7	9.25×10^{-2}	2.20	1.90×10^{-4}
	2015	1.69×10^3	1.39×10^5	1.12×10^7	9.66×10^{-2}	2.16	1.78×10^{-4}
	2016	1.76×10^3	1.42×10^5	1.15×10^7	9.15×10^{-2}	2.15	1.81×10^{-4}
	2017	1.76×10^3	1.43×10^5	1.17×10^7	9.25×10^{-2}	2.15	1.88×10^{-4}
	2018	1.76×10^3	1.42×10^5	1.14×10^7	9.17×10^{-2}	2.16	1.96×10^{-4}

Table 2. Comparison of HSAs delineated using the community detection algorithms Block Model, Infomap, Louvain, and SLPA in terms of in terms of the number of communities (n_c) as well as the typical values of the localization index $\langle li \rangle$, network conductance $\langle c \rangle$, and total number of discharges $\langle d \rangle$. The discharge types presented are *Inpatient from ED* and *ED Only* for the years of 2012 and 2018. The other discharge types and years are available on the Open Science Framework (OSF) repository of this project at https://doi.org/10.17605/OSF.IO/GW73Y.

Type of discharge	Year	Community detection	$N(c)$	$\langle LI(c) \rangle$	$\langle C(c) \rangle$	$\langle D(c) \rangle$
Inpatient from ED	2012	Block Model	35	0.47	0.82	51,450
		Infomap	70	0.77	0.18	25,539
		Louvain	20	0.86	0.13	89,235
		SLPA	110	0.69	0.25	16,392
	2018	Block Model	35	0.47	0.74	54,090
		Infomap	62	0.80	0.15	30,397
		Louvain	15	0.87	0.13	125,833
		SLPA	111	0.65	0.28	16,830
ED Only	2012	Block Model	33	0.45	0.84	311,009
		Infomap	90	0.77	0.22	114,117
		Louvain	24	0.92	0.09	430,396
		SLPA	139	0.70	0.28	74,049
	2018	Block Model	34	0.45	0.83	359,908
		Infomap	76	0.84	0.15	160,869
		Louvain	24	0.90	0.09	512,044
		SLPA	126	0.73	0.25	97,810

even in the same dataset. Such results are not only consistent with community detection comparisons from previous works [6], but also advocate for a further comparison among community detection algorithms.

The comparison of HSA delineations (Fig. 2) involves the evaluation of conflicting metrics such as a higher number of communities and localization index. Considering the discharge type *ED Only* and year 2018 (Table 2), for instance, the difference between the number Louvain-HSAs and SLPA-HSAs was 4-fold, with 24 Louvain-HSAs and 126 SLPA-HSAs. This fewer number of Louvain-HSAs using hospitals discharge in California is also consistent with the fewer number of Louvain-HSAs in the previous work from Hu et al. [7] using hospital discharges in Florida. Though the localization index of Louvain-HSAs (.90) was 7% higher than that of Infomap-HSAs (.84), 52 more Infomap-HSAs were delineated, representing a 2-fold increase.

By examining their geographical patterns (Fig. 3), the respective geographical areas of Louvain-HSAs correspond to a wider and more discontinuous geographical areas than those of Infomap-HSAs. Yet, Block Model-HSAs appear to be the poorest HSAs delineated not only because they are the fewest

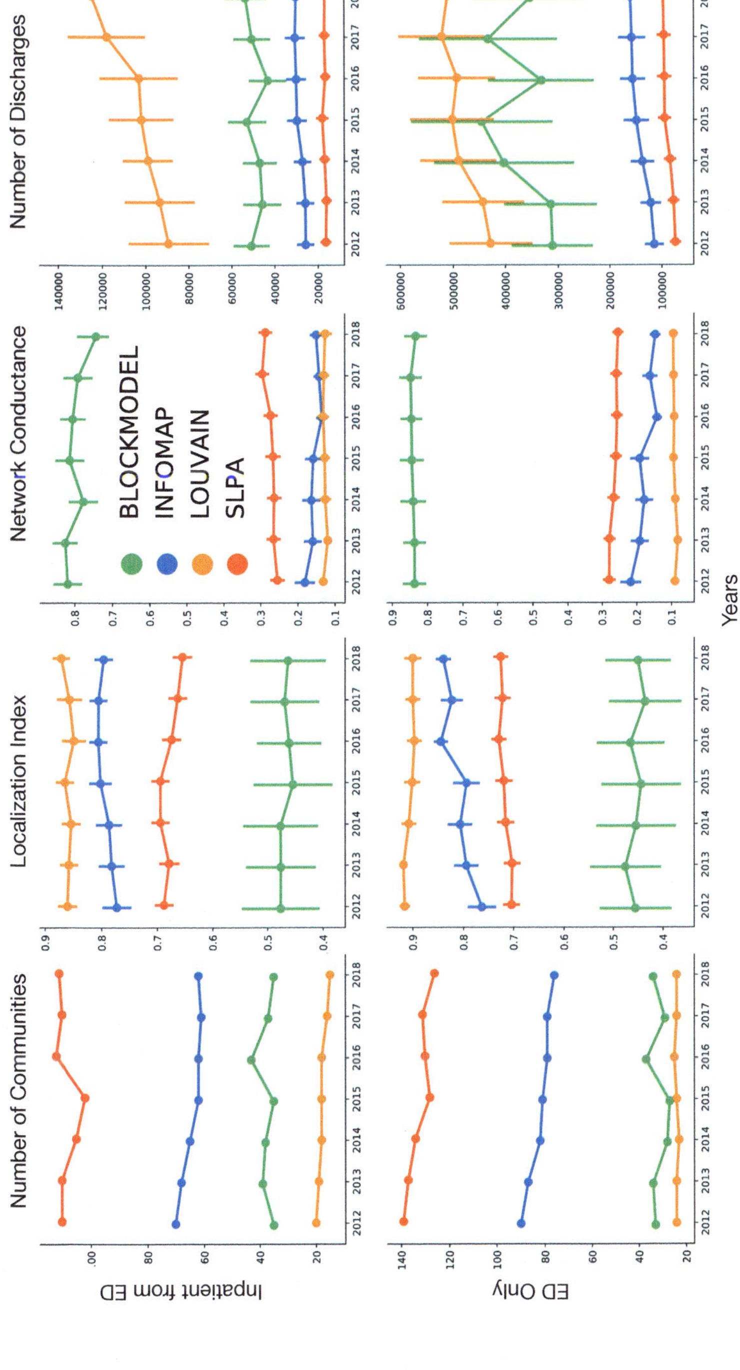

Fig. 2. Comparison of HSAs delineated using Block Model, Infomap, Louvain, and SLPA community detection algorithms. The HSA delineations were compared in terms of the number of communities (N_C), localization index ($LI(c)$), network conductance ($C(c)$), and total number of discharges ($N_D(i,j)$). The discharge types presented are *Inpatient from ED* (top) and *ED Only* (bottom). All results are available on the Open Science Framework (OSF) repository of this project at https://doi.org/10.17605/OSF.IO/GW73Y.

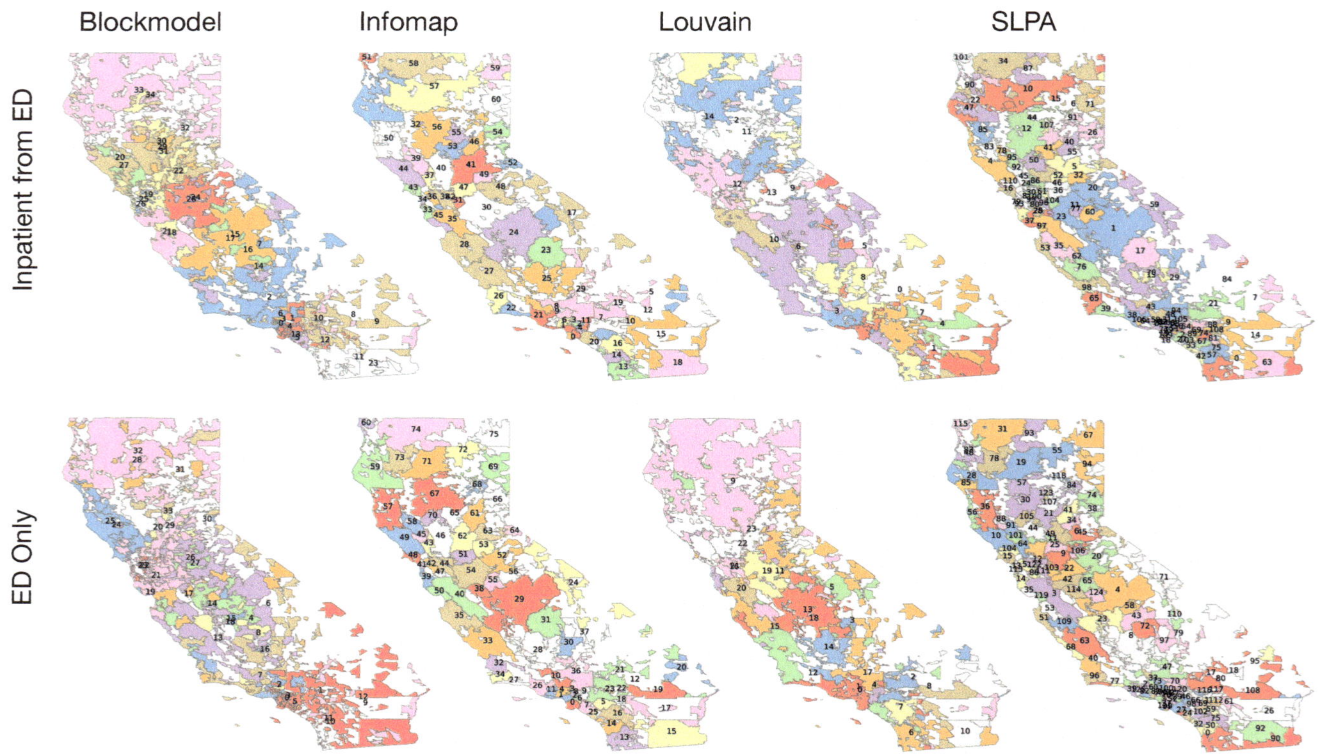

Fig. 3. Maps of HSA delineated using the community detection algorithms Block Model, Infomap, Louvain, and SLPA for discharge types *Inpatient from ED* (top) *ED Only* (bottom) in 2018. Each HSA is displayed as a geographical boundary aggregating one or more ZCTAs along with its respective Id. HSAs were then colored using a color blind palette with 9 distinct colors. All maps are available on the Open Science Framework (OSF) repository of this project at https://doi.org/10.17605/OSF.IO/GW73Y

number of HSAs delineated, which corresponds to a wider and more discontinuous area, but also as their localization index is lower than .5 which raises concerns about the internal validity of the HSAs delineated. Lastly, Block Model-HSAs presented the highest variability regarding their localization index, conductance, and total number of discharges. Using the nested version of Block Model has not changed these results. Conversely, SLPA-HSAs are the highest number of HSAs but their localization index is typically less than .7, which is higher than that of Block Model-HSAs but still raises concerns regarding the internal validity of SLPA-HSAs.

The hospital discharge data and the Hospital-Patient Discharge Networks (HPDNs) inherently represent a flow between hospitals and patients and the apparent superior results achieved by Infomap may be related to the fact that, instead of modularity, the Infomap optimizes the map equation and thus takes into account the local flows emerging from the movements of random walkers trapped within the HPDN communities. Interestingly, Infomap-HSAs presented improved localization index and conductance over time.

4 Conclusions

Health Service Areas (HSAs) are the meaningful units of analysis for improving the scientific basis of both clinical practice and policy decision making in the delivery of health care. The optimal delineation of HSAs is necessary to create not only more meaningful units of analysis, but also to characterize medical practices with greater accountability regarding their respective community needs and shared care practices.

As the delineation approach of HSAs shift from the Dartmouth towards a network-based, further work will be needed to establish a comprehensive methodology for network-based HSA delineation, which should include (i) a broader set of community detection algorithms, (ii) hospital discharge data from states other than California, and integration with other healthcare datasets of utilization, expenditures, and outcomes.

References

1. Blondel, V.D., Guillaume, J.-L., Lambiotte, R., Lefebvre, E.: Fast unfolding of communities in large networks. J. Stat. Mech.-Theory Exp. **2008**(10) (2008). https://sci-hub.tw/https://iopscience.iop.org/article/10.1088/1742-5468/2008/10/P10008/pdf
2. CHHS. California Health and Human Services Agency. https://data.chhs.ca.gov. Accessed 2019 Nov 15
3. Everson, J., Hollingsworth, J.M., Adler-Milstein, J.: Comparing methods of grouping hospitals. Health Serv. Res. **24**(4), 333–339 (2019)
4. Fortunato, S.: Community detection in graphs. Phys. Rep. **486**(3–5), 75–174 (2010)
5. Fortunato, S., Hric, D.: Community detection in networks: a user guide. Phys. Rep. **659**, 1–44 (2016)

6. Hartman, R., Faustino, J., Pinheiro, D., Menezes, R.: Assessing the suitability of network community detection to available meta-data using rank stability. In: WI, pp. 162–169 (2017)
7. Hu, Y., Wang, F., Xierali, I.M.: Automated delineation of hospital service areas and hospital referral regions by modularity optimization. Health Serv. Res. **53**(1), 236–255 (2018)
8. Peixoto, T.P.: Hierarchical block structures and high-resolution model selection in large networks. Phys. Rev. X **4**(1), 011047 (2014). https://journals.aps.org/prx/abstract/10.1103/PhysRevX.4.011047
9. Peixoto, T.P.: Model selection and hypothesis testing for large-scale network models with overlapping groups. Phys. Rev. X **5**(1), 011033 (2015). https://journals.aps.org/prx/abstract/10.1103/PhysRevX.5.011033
10. Rosvall, M., Bergstrom, C.T.: Maps of random walks on complex networks reveal community structure. Proc. Nat. Acad. Sci. **105**(4), 1118–1123 (2008)
11. UDS. Uniform Data System Mapper. Zip code to ZCTA crosswalk. https://www.udsmapper.org/zcta-crosswalk.cfm. Accessed 2019 Nov 15
12. Wennberg, J.: The Dartmouth Atlas of Health Care. American Hospital Association, Chicago (1996)
13. Wennberg, J., Gittelsohn, A.: Small area variations in health care delivery: a population-based health information system can guide planning and regulatory decision-making. Science **182**(4117), 1102–1108 (1973)
14. Xie, J., Szymanski, B.K., Liu, X.: SLPA: uncovering overlapping communities in social networks via a speaker-listener interaction dynamic process. CoRR, abs/1109.5720 (2011)

Diversity Analysis Exposes Unexpected Key Roles in Multiplex Crime Networks

A. S. O. Toledo[1,2](✉), Laura C. Carpi[1], and A. P. F. Atman[1,3]

[1] Programa de Pós-Graduação em Modelagem Matemática e Computacional, Centro Federal de Educação Tecnológica de Minas Gerais, Belo Horizonte, Brazil
asotoledo@hotmail.com

[2] Instituto Brasileiro de Segurança Pública, Sao Paulo, Brazil

[3] Instituto Nacional de Ciência e Tecnologia de Sistemas Complexos, Rio de Janeiro, Brazil

Abstract. The study of criminal networks seeks new approaches and answers to meet the growing demand of society. In this paper, we present an innovative analysis of crime occurrences in the State of Minas Gerais, Brazil, collected from a Public Security Intelligence database, from the point of view of statistical physics and complex networks. We built the network of these individuals by considering the hierarchy, type of crime and relationships reported within criminal organizations. When modeling the crime database as a complex network, it was possible to identify criminal groups of individuals, and better understand the structure of criminal organizations. We apply multiplex and node diversity analysis to map the criminal structure in layers according to the type of crime. Surprisingly, some key elements pointed out by this analysis had not yet been identified previously, as major actors. This work represents a significant improvement in the methodology and data mining of the criminal database of the state of Minas Gerais and can be applied to any similar database.

Keywords: Crime network · Diversity · Multiplex network · Computational modeling

1 Introduction

Organized crime, such as child pornography, drug trafficking, identity theft and cybercrime, is one of the biggest threats to society around the world. Government agencies, scholars, and law enforcement, seek for efficient ways to break or control these illegal structures, requiring proactive interventions to ride away the organizations supporting them [1].

Modeling and simulating human societies is a challenging task as individuals behave differently when acting together, following different and often unexpected behaviors [2]. The application of the network approach to study criminal phenomena

The authors thank Brazilian funding agencies CAPES and CNPq. APFA thanks CNPq grant number 308792/2018-1.

H. Barbosa et al. (Eds.): *Complex Networks XI*, SPCOM, pp. 371–382, 2020.
https://doi.org/10.1007/978-3-030-40943-2_31

has been increasing rapidly, expanding the definitions of organized crime [3–9]. Techniques of network theory have been successfully used to explain criminal phenomena, mainly by investigating interactions between actors [10, 11]. However, the study in this area still remains in more descriptive analysis.

Traditional structural measures, such as those that quantifies centrality, have been successfully used to identify central actors in criminal structures, measuring the connectivity extent of an individual in a network, and thus reflecting its influence in the entire criminal organization [12]. Centrality analysis of criminal networks has described several interesting features. Some works revealed, for example, that criminal leaders possess low degree centrality as an strategy to reduce visibility within the organization and, consequently, reducing vulnerability. Individuals with high betweenness centrality values are essential for the preservation of the network connectivity as they are part of a great number of shortest paths connecting nodes [12–14]. Low betweenness and high degree centrality values correspond to more visible and therefore more vulnerable individuals; marginal individuals are characterized by both, low degree and betweenness centrality values; and remarkably central and highly visible individuals possess high values of both measures [12].

During the last decade, the vision that networks do not act in isolation but in an extensive interacting network of networks, has modified the way we conceive these systems. Interconnected networks are represented by multilayer structures, in which, each layer represents a single network [15, 16]. Biological, social, economic, climatic and transport networks are some examples where their nodes, simultaneously participate in different dynamics, interacting at different levels [17–19].

Multiplex networks correspond to a particular case of multilayer networks in which the structure is composed by the same set of nodes in all layers, and where interactions the between them only occur with their replica in the different layers [20–23], providing a more realistic representation of the phenomena.

Examples of these systems include, among many others, multiplex social networks, where the same group of individuals participate in various social networks (Facebook©, Twitter©, Instagram©), and air transportation networks, in which the same set of airports are connected by different airline companies [24, 25].

Organized crime can be considered a multiplex phenomenon in different aspects [12]. On the one hand, the individuals are embedded in a structure composed by multiple social relations, such as activities, friendships, etc. [26]. On the other hand, they are also embedded in a criminal structure where the actors participate simultaneously in different crimes.

Most works in criminal networks aggregate the relational information into a single structure, loosing valuable details that may be crucial for the identification of individuals that execute multiple important roles, masked for those that possess high centrality but maybe less power within the organization.

The objective of this work is to study a criminal network as a multiplex structure in which nodes correspond to criminal individuals, and layers correspond to different crimes. We use the concept of diversity in multiplex networks proposed in [27], to identify individuals that are crucial for preserving the network functioning, independently of their centrality values. The node diversity analysis is able to reveal interesting new features, providing information to identify individuals with important roles in the network, not identified by traditional measures.

2 Criminal Network

In this study, we use the *Links of Occurrences* database, collected through observations of intelligence agents, by the Military Police of Minas Gerais, Brazil. This database contains 152,973 registrations of individuals related to different crimes and it also contains information regarding situations in which individual have been connected (individuals are seen together, caught together, are part of the same gang, phone calls, among others). As an example, Fig. 1 show a crime network containing 34,505 nodes, and 46,239 links that were created if any of these connections exists in the database.

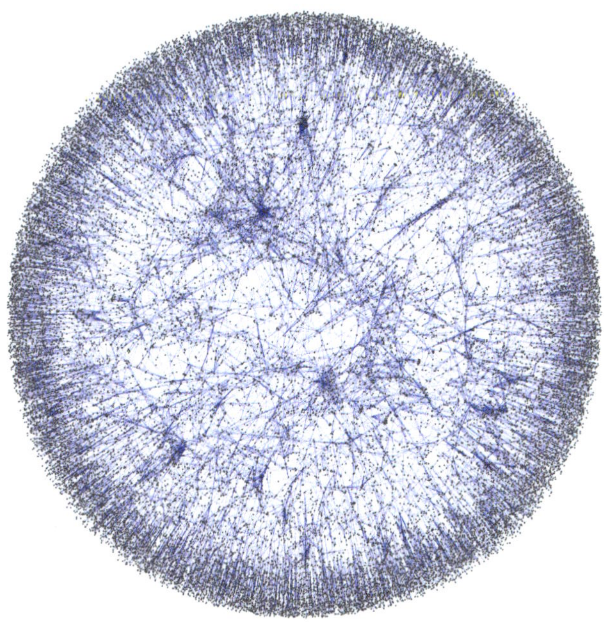

Fig. 1. Crime network containing 34,505 nodes (individuals) and 46,239 links (different types of connections present in the database).

From the original data, we selected a group composed by 92 individuals that were the target of an intelligence operation that occurred in 2013, and that concluded in the subsequent arrest of several of them. During this intelligence operation, elements 51 and 80 were identified as the leaders of the criminal group composed by offenders (individuals who participated in crimes) and co-offenders (co-workers). Data extraction and processing were possible through the Python algorithm developed for this work. This new network containing 92 individuals is called, from now on, Intelligence Network (IN).

2.1 Centrality Measures of the Crime Network

Degree centrality, betweenness centrality, and PageRank, have been used in this study to identify the central individuals of the crime network. Degree centrality is the number of direct contacts the individual has [28], betweenness centrality quantifies the number of times an individual appears as a bridge along the shortest path between two other individuals [28], and pageRank is a variation of eigenvector centrality, and is an effective way to catalog the relevance of an individual within a network [28]. To have a robust measure, the three centrality values have been normalized, and by calculating the mean value, it was possible to identify individuals 80, 51, 48, 11, 47, 81, 19, 2, 44 and 90, as the more central ones, in decreasing order. These results are showed in Fig. 2, in which, centrality values are represented by the size of the nodes.

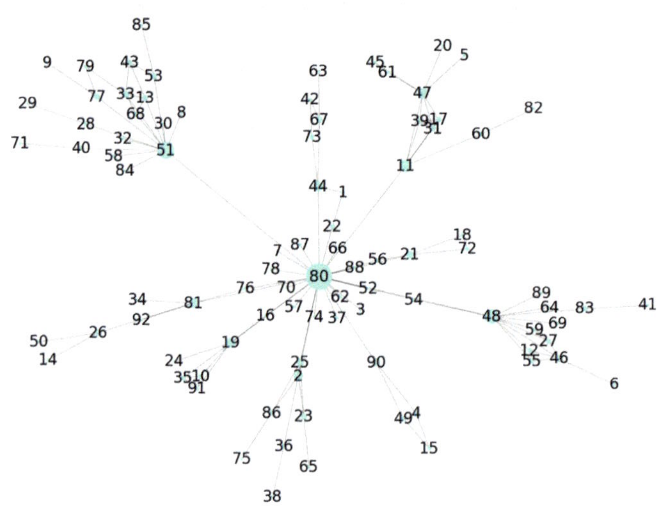

Fig. 2. Intelligence Network in which the size of the nodes is proportional to the mean value considering three centrality measures (Degree centrality + Betweenness centrality + PageRank).

Subgroups or communities are natural occurrences in crime networks that are associated with their internal organization strategies. The presence of subgroups can be easily identified in this crime network. Subgroups in criminal networks limit the information share, preventing leakage and detection. The more extensive the crime network, the more relevant the presence of subgroups. These considerations suggest that the analysis of subgroups in crime networks can provide useful insights regarding its internal structure, helping in the planning of efficient preventive and repressive strategies [29]. From a community analysis, as it is shown in Fig. 3, it can be seen that IN is divided into nine subgroups, and that the central individuals identified by centrality measures, are located in different subgroups, inducing the conclusion that they have similar roles in different sub-organizations.

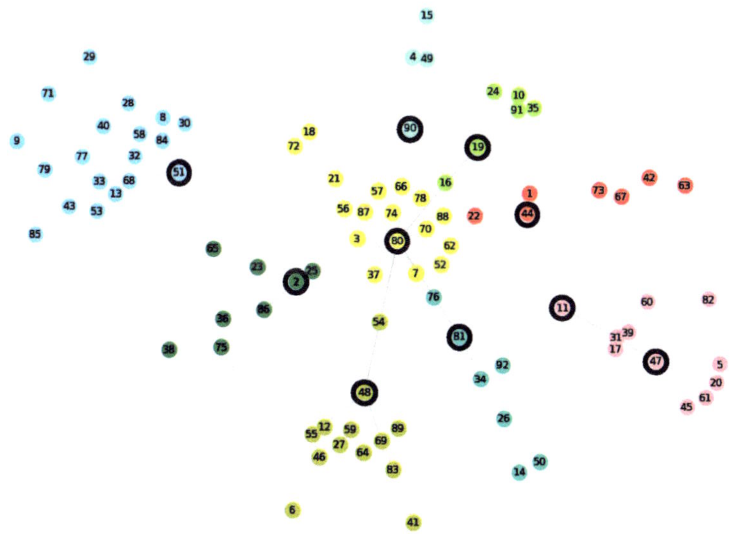

Fig. 3. Community structure of the Intelligence Network, in which the different colors represent different communities. Central individuals, identified by centrality measures are highlighted in bold.

2.2 Diversity Analysis of the Multiplex Crime Network

Several measures have been proposed in the literature to study the concept of diversity [30–34], usually used to represent the variety of entities forming a system. In the specific case of multiplex networks, diversity refers to the variety of connectivity configurations, the elements that constitute the network possess. A method to quantify diversity in multiplex networks was proposed in [27], which is based on the computation of dissimilarities of layers and nodes, and in concepts defined in [30, 31]. The importance of diversity measures in networks is to identify the elements that are critical to maintain the functionality of the system.

A set of relational data was organized in a multiplex structure, containing information about the nine crimes practiced by individuals of the IN (Fig. 4).

To conduct a more detailed analysis of individuals participation in different crimes, and to identify strong relationships, the crimes of Ideological Falsehood, Arms Trafficking, Kidnapping, Weapon Carrying, Homicide, Theft and Drug Trafficking were considered as networks in the different layers. In Fig. 5 one can visualize the organization of the multiplex crime network.

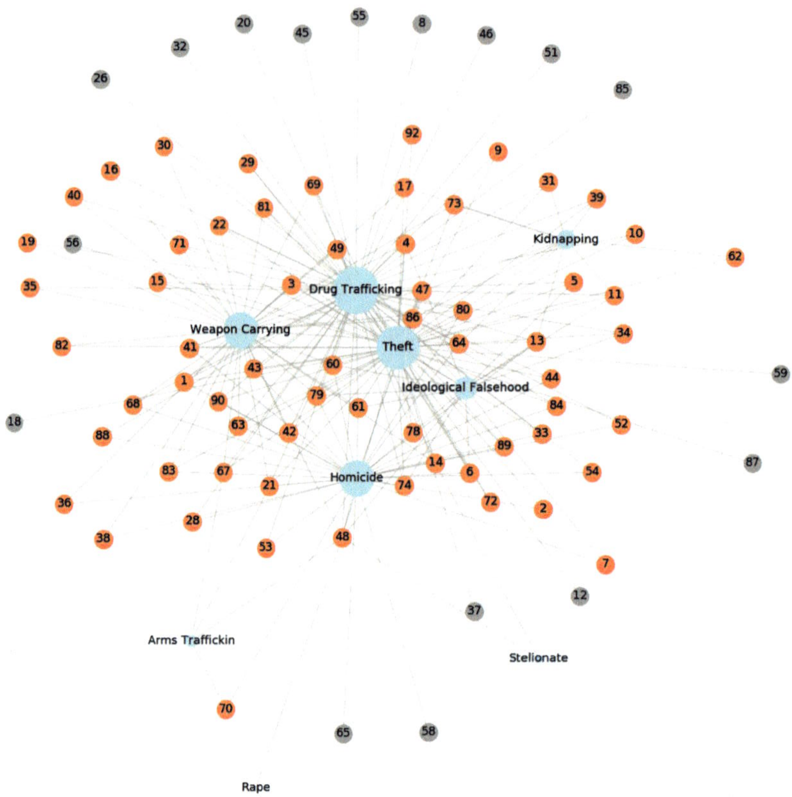

Fig. 4. Identification of the nine crime modalities with their respective participants. Individuals in orange correspond to those who participate in more than one crime simultaneously.

Following the methodology presented in [27], we first computed the distance of each node in the different layers, where $\mathscr{D}_i(\overline{p},\overline{q})$ is the distance of node i in layers $\overline{p}$ e $\overline{q}$, called as Node Dissimilarity (ND), and between layers $\mathscr{D}(\overline{p},\overline{q})$, called the Layer Dissimilarity (LD). ND quantifies the differences of the connectivity patterns of the node i in the layers $\overline{p}$ and $\overline{q}$ (Eq. 1).

$$\mathscr{D}_i(\overline{p},\overline{q}) = \frac{\sqrt{\mathscr{J}(\mathscr{N}_i^{\overline{p}},\mathscr{N}_i^{\overline{q}})} + \sqrt{\mathscr{J}(T_i^{\overline{p}},T_i^{\overline{q}})}}{2\sqrt{\log(2)}}, \tag{1}$$

where $\mathscr{J}$ is the Jensen-Shannon divergence that measures the dissimilarity between probability distribution functions ($PDFs$) [35]. $\mathscr{N}_i^{\overline{p}}$ is the Distance Distribution (NDD) of the node i in the layer $\overline{p}$: $\mathscr{N}_i^{\overline{p}}(d)$ is the fraction of nodes that are the distance d (minimum path) of node i in layer $\overline{p}$. $\mathscr{T}^{\overline{p}}$ corresponds to the Transition Matrix of the $\overline{p}$. $\mathscr{T}^{\overline{p}}$ is the Adjacency Matrix of the $\overline{p}$ layer, re-scaled by the degree of each node.

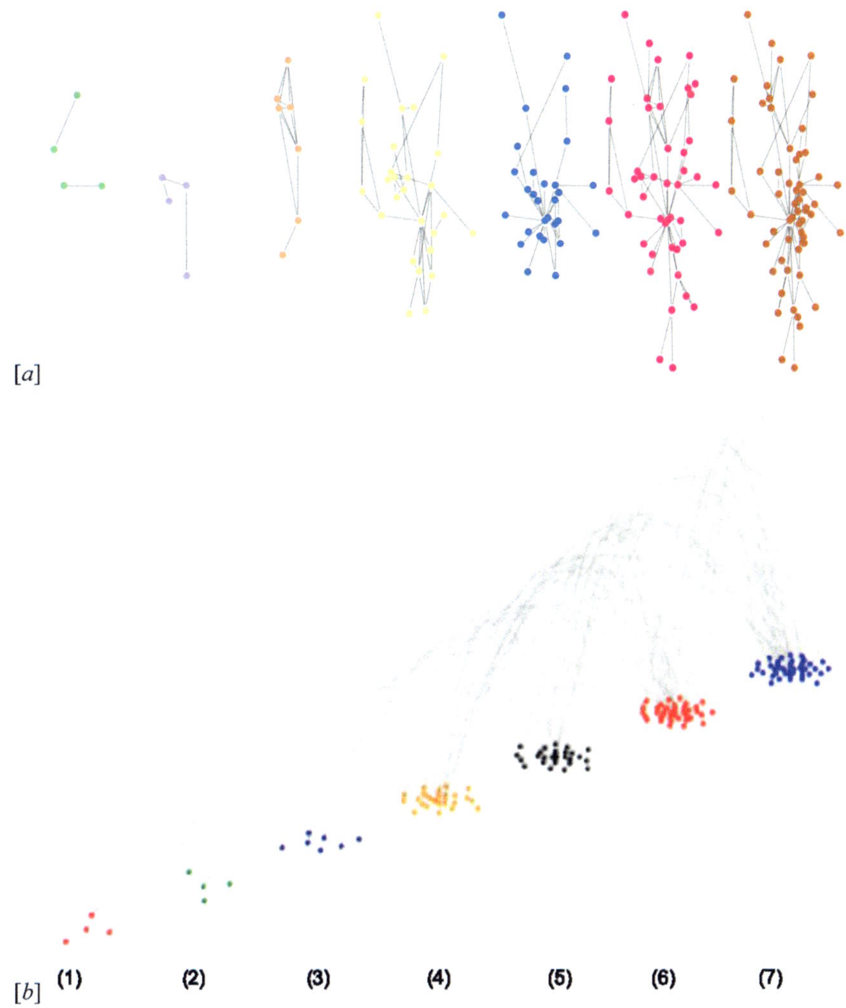

Fig. 5. Multiplex crime network represented in the structure are: (1) Ideological Falsehood, (2) Arms Trafficking, (3) Kidnapping, (4) Weapon Carrying, (5) Homicide, (6) Theft and (7) Drug Trafficking, containing 77 individuals: **[a]** with 232 edges in layers multiplex; **[b]** with 159 edges between layers multiplex

When $\mathscr{D}_i(\overline{p},\overline{q}) = 0$ the *PDFs* are identical, that is, the i node has equal connection patterns in the layers p and q. When $\mathscr{D}_i(\overline{p},\overline{q}) = 1$ the dissimilarities of i in the layers p and q are extreme, where the i node is disconnected in a layer (not active) and connected in some way to all the nodes of the other.

Layer dissimilarity quantifies the differences between the connectivity patterns of the $\overline{p}$ and $\overline{q}$ layers. Dissimilarity of the layers is simply $\mathscr{D}_i(\overline{p},\overline{q})$ considering all nodes (Eq. 2).

$$\mathscr{D}(\overline{p},\overline{q}) = \langle \mathscr{D}_i(\overline{p},\overline{q}) \rangle_i \tag{2}$$

$\mathscr{D}(\overline{p},\overline{q}) = 0$ indicates that the layers p and q are identical and $\mathscr{D}(\overline{p},\overline{q}) = 1$ indicates that one of the layers is fully connected, while the other is completely disconnected. $\mathscr{D}$ has metric properties.

The model proposed in [30] and reformulated in [31] characterizes the measurement of diversity as a function of dissimilarity, which can be defined as a distance [27, 35].

Let $\tilde{S}$ be the set of all entities, which in the context of multiplex networks can be the set of nodes or the set of layers. Assuming we have the set $S \subset \tilde{S}$ and we can compute the distances between all of them, we define the distance between the element $\overline{g} \notin S$ and the set S, $\mathscr{D}(\overline{g}, S)$, as the shortest distance between $\overline{g}$ and any element of the set S:

$$\mathscr{D}(\overline{g}, S) = \min_{\overline{s_i} \in S} \mathscr{D}(\overline{g}, \overline{s_i}) \tag{3}$$

The distance of an element that does not belong to the set is defined as the minimum distance from that element to the set. In the Eq. 3 when we consider the nodes, the distance $\mathscr{D}(\overline{g}, \overline{s_i})$ is the ND, presented in Eq. 1; when the entities are the layers, $\mathscr{D}(\overline{g}, \overline{s_i})$ is the LD, presented in Eq. 2.

The diversity function $U : \tilde{S} \rightarrow \mathbb{R}_+$, is defined recursively as $U(S) = max_{\overline{s_i} \in S}\{U(S \setminus \overline{s_i}) + \mathscr{D}(\overline{s_i}, S \setminus \overline{s_i})\}$ for all $S \in \tilde{S}$ com $|S| \geq 2$ with $|S|$, represents the cardinality of the set e $U(S) = 0$, for all $S \in \tilde{S}$ such that $|S| = 1$.

We use U_i to refer to the diversity of nodes i in the different layers and U to refer to the diversity of layers. When an element, node or layer, is removed, the diversity of the system decreases. If $\overline{g}$ is removed from $S \cup \overline{g}$, the loss of diversity is at least equal to $D(\overline{g}, S)$, $U(S \cup \overline{g}) \geq U(S) + D(\overline{g}, S)$.

By a method based on a distance lexicographic order [31], the set $\mathscr{O}(S) = \{\overline{s_1}, \overline{s_2}, \ldots, \overline{s_{|S|}}\}$ is obtained, which indicates the elements in order of their contribution to the diversity of the system. $U_i(S)$ is used to refer to the diversity of the connection patterns that the node i has in the different layers and $U(S)$.

The diversity values of the nodes corresponding to the crime network are presented in Fig. 6.

Diversity was computed at two different levels, local diversity (U_i), which refers to the diversity of the connections of a node in the various layers; and global diversity (U), which refers to the diversity of relationships in the system.

Surprisingly the individual 48 stands out in the network since it has the highest value of ND. Individual 48 is the one who committed the highest number of crimes (Ideological Falsehood, Arms Trafficking, Weapon Carrying, Homicide, Theft and Drug Trafficking), and has high redundancy of connections with the individual 80, which has higher values of centrality in the network, consequently generating a high degree of reliability to the other individuals in the network. Another fact that highlights the individual 48 is the practice of the Crime of Arms Trafficking, indispensable for the organization of other crimes and self-protection in the inhospitable environment of criminality.

The redundancy of links between two individuals in different crimes in the multiplex network may imply in trust; the higher the redundancy, the greater the level of trust

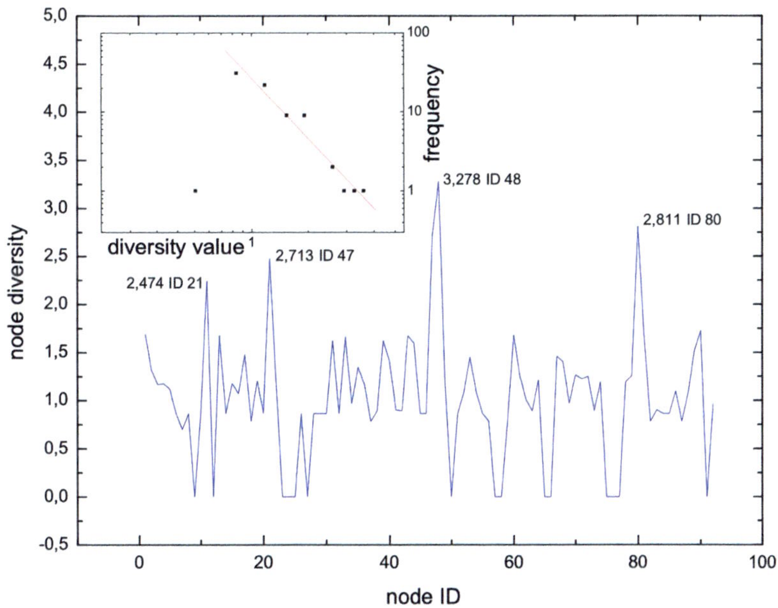

Fig. 6. The node diversity of individuals in the multiplex crime network. The individuals with higher diversity values are in decreasing order: 48, 80, 47, 21. The insert graph shows Power Law followed by the diversity values.

between them, representing stronger bonds. The ND values, identifies individuals 48, 80, 47 and 21 as the most diverse ones.

In the multiplex study, individual 51 loses importance as a hub (highlighted by the measures of centrality), for having committed only the crime of Drug Trafficking, therefore participating in just one layer. Individual 21 assumes an important role, as it is present in 4 layers of the multiplex structure (arms trafficking, arms transport, homicide, theft, and drug trafficking). Node 21 was not identified by the centrality measures.

The order of contribution to the global diversity of the multiplex system was: (1) Drug Trafficking, (2) Homicide, (3) Theft, (4) Weapon Carrying, (5) Kidnapping, (6) Arms Trafficking, (7) Ideological Falsehood. This order shows the ranking of the crimes that most diversity brings to the network. While centrality measures identify central individuals from a one-dimensional point of view, multiplex analysis addresses multi-dimensional features allowing the identification of key individuals by considering characteristics as participation in different crimes, connectivity diversity and link redundancy. The multiplex study highlights the individuals who most influence the network, that is, the individuals whose actions make the network more effective in the practice of crimes, whose role rests with the individual 48, considering its ND value.

As high diverse nodes mainly overlaps with central nodes, a centrality/diversity correlation study has been performed, finding $r = 0.46$, where r corresponds to the Pearson correlation coefficient, indicating a weak correlation (see Fig. 7).

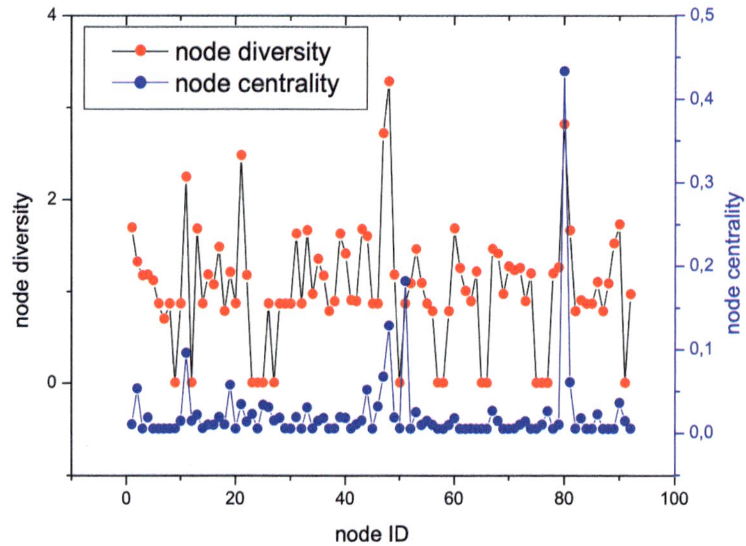

Fig. 7. Red dots represent the diversity values of each node in the multiplex structure, and blue dots represent the node centrality value considering the aggregated network.

3 Conclusion

We present in this article an innovative approach to analyse criminal statistics from the Public Security database via complex network analysis.

The multiplex analysis allows the identification of nodes with main roles in the criminal network, by using the concept of diversity. We found that, nodes with not so high centrality values, but high node diversity execute crucial roles in the network for the achievement of an effective functioning. We show with these results that it is possible to study criminal networks from a real databases, to create models and simulations to characterize the structures and the relationships between participants, and to identify key elements that are hidden by traditional network measures.

With the proposed method, we believe that police have gained an important tool to identify, visualize and qualify criminal activity in near real time, increasing the chances of a successful interruption of crime.

We hope to apply the method to other databases and eventually integrate the current methodology with social data to evaluate hidden connections between individuals, and also, hidden key actors.

References

1. Duijn, P.A.C, Kashirin, V., Sloot, P.M.: The relative ineffectiveness of criminal network disruption. Sci. Rep. **4**, 4238 (2014)
2. Morselli, C.: Inside Criminal Networks. Springer (2009)

3. Varese, F.: The structure and the content of criminal connections: the Russian Mafia in Italy. Eur. Sociol. Rev. **29**, 899–909 (2012)
4. Oliveira, M., Bastos-Filho, C., Menezes, R.: The scaling of crime concentration in cities. Public Libr. Sci. **12**, e0183110 (2017)
5. Ribeiro, H.V., Alves, L.G., Martins, A.F., Lenzi, E.K., Perc, M.: The dynamical structure of political corruption networks. J. Complex Netw. **6**, e0183110 (2018)
6. Ren, X.-L., Gleinig, N., Helbing, D., Antulov-Fantulin, N.: TGeneralized network dismantling. Natl. Acad. Sci. **116**, 6554–6559 (2019)
7. da Cunha, B.R., Gonçalves, S.: Topology, robustness, and structural controllability of the Brazilian federal police criminal intelligence network. SpringerOpen **3**, 36 (2018)
8. D'Orsogna, M.R., Perc, M.: Statistical physics of crime: a review. Phys. Life Rev. **12**, 1-21 (2015)
9. Alves, L.G.A., Ribeiro, H.V., Lenzi, E.K., Mendes, R.S.: Distance to the scaling law: a useful approach for unveiling relationships between crime and urban metrics. Public Libr. Sci. **8**, e69580 (2013)
10. Papachristos, A.V.: The coming of a networked criminology. Adv. Criminol. Theory **17**, 101-140 (2011)
11. Bouchard, M., Amirault, J.: Advances in research on illicit networks. Glob. Crime Theory **14**, 119–122 (2013)
12. Diviák, T., Dijkstra, J.K., Snijders, T.A.B.: Structure, multiplexity, and centrality in a corruption network: the Czech Rath affair. Trends Organ. Crime **22**, 274–297 (2019)
13. Morselli, C., Roy, J.: Brokerage qualifications in ringing operations. Criminology **46**, 71–98 (2008)
14. Morselli, C.: Assessing vulnerable and strategic positions in a criminal network. J. Contemp. Crim. Justice **26**, 382–392 (2010)
15. Kurant, M., Thiran, P.: Layered complex networks. Phys. Rev. Lett. **96**, 045104 (2006)
16. Gao, J., Buldyrev, S.V., Havlin, S., Stanley, H.E.: Robustness of a network formed by n interdependent networks with a one-to-one correspondence of dependent nodes. Phys. Rev. E **85**, 066134 (2012)
17. Bargigli, L., Di Iasio, G., Infante, L., Lillo, F., Pierobon, F.: The multiplex structure of interbank networks. Quant. Finance **15**, 673-691 (2015)
18. De Domenico, M., Sasai, S., Arenas, A.: Mapping multiplex hubs in human functional brain networks. Front. Neurosci. **10**, 326 (2016)
19. Cantini, L., Medico, E., Fortunato, S., Caselle, M.: Detection of gene communities in multinetworks reveals cancer drivers. Sci. Rep. **15**, 17386 (2015)
20. De Domenico, M., Solé-Ribalta, A., Cozzo, E., Kivelä, M., Moreno, Y., Porter, M.A., Gómez, S., Arenas, A.: Mathematical formulation of multilayer networks. Phys. Rev. X **3**, 041022 (2013)
21. Battiston, F., Nicosia, V., Latora, V.: Structural measures for multiplex networks. Phys. Rev. E **89**, 032804 (2014)
22. Kivelä, M., Arenas, A., Barthelemy, M., Gleeson, J.P., Moreno, Y., Porter, M.A.: Multilayer networks. J. Complex Netw. **2**, 203–271 (2014)
23. Battiston, F., Nicosia, V., Latora, V., San Miguel, M.: Layered social influence promotes multiculturality in the Axelrod model. Sci. Rep. **7**, 1809 (2017)
24. Guo, Q., Cozzo, E., Zheng, Z., Moreno, Y.: Levy random walks on multiplex networks. Sci. Rep. **6**, 37641 (2016)
25. Oliveira, I.M., Carpi, L.C., Atman, A.P.F.: The multiplex efficiency index: unveiling the Brazilian air transportation multiplex network–BATMN. arXiv preprint arXiv:1910.11974 (2019)
26. Papachristos, A.V., Wildeman, C.: Network exposure and homicide victimization in an African American community. Am. J. Public Health **104**, 143-150 (2014)

27. Carpi, L.C., Schieber, T.A., Pardalos, P.M., Marfany, G., Masoller, C., Díaz-Guilera, A., Ravetti, M.G.: Assessing diversity in multiplex networks. Sci. Rep. **9**, 4511 (2019)
28. Scott, J.: Social Network Analysis. Sage (2015)
29. Morselli, C., Giguère, C., Petit, K.: The efficiency/security trade-off in criminal networks. Soc. Netw. **29**, 143–153 (2007)
30. Weitzman, M.L.: The quarterly journal of economics. On Divers. **107**, 363-405 (1992)
31. Bossert, W., Pattanaik, P.K., Xu, Y.: The measurement of diversity. Centre de recherche et développement en économique, Université de Montréal (2001)
32. Fu, Y.-H., Huang, C.-Y., Sun, C.-T.: Using global diversity and local topology features to identify influential network spreaders. Phys. A **433**, 344–355 (2015)
33. Gao, J., Barzel, B., Barabási, A.-L.: Universal resilience patterns in complex networks. Nature **530**, 307 (2016)
34. Raducha, T., Gubiec, T.: Predicting language diversity with complex networks. PloS One **13**, e0196593 (2018)
35. Schieber, T.A., Carpi, L., Díaz-Guilera, A., Pardalos, P.M., Masoller, C., Ravetti, M.G.: Quantification of network structural dissimilarities. Nat. Commun. **8**, 4511 (2017)

Science of Science

Policy-Relevant Science: The Depth and Breadth of Support Networks

Bruce A. Desmarais[1](✉) and John A. Hird[2]

[1] The Pennsylvania State University, University Park, State College, PA 16801, USA
bdesmarais@psu.edu

[2] University of Massachusetts Amherst, Amherst, MA, USA
jhird@umass.edu
http://brucedesmarais.com/

Abstract. Proponents of basic science argue that objective scientific understanding can inform improvement in public policy. We gather data on scientific research cited in official benefit-cost analyses produced by US federal regulatory agencies to justify policy decisions between 2008 and 2012. We construct a science-policy network in which benefit-cost analyses and the studies they cite are the nodes, and citations represent the edges. We assess two features of each scientific publication in the network; how frequently is it used; and how broadly it spans across the network, as measured by betweenness centrality. We ask which author affiliations and funders are associated with the best-cited and farthest spanning publications. Elite universities and major government funders support publications that are most heavily cited, but the farthest spanning articles are written by scientists with non-academic affiliations and sponsored by non-governmental funders. These results suggest that bias towards academically affiliated investigators should be scrutinized by major funding organizations if a major objective is to support science that is used by policymakers.

Keywords: Scientometrics · Policy networks · Regulation

1 Introduction

Science improves knowledge of the physical, biological, and social worlds and contributes to society's betterment through industrial innovations and improved public policy. Policymakers use science in various ways to reach policy decisions [27], yet the mechanisms by which science is engaged and its contributions to policy outcomes are poorly understood. Allegations of manipulation and misuse are rampant [3,8]. Furthermore, while the quality of science is the premier

This work was supported in part by NSF grants 1558661, 1637089, 1619644, and 1360104. The data and code used in this study can be downloaded at https://doi.org/10.7910/DVN/IY9B1T.

H. Barbosa et al. (Eds.): *Complex Networks XI*, SPCOM, pp. 385–392, 2020.
https://doi.org/10.1007/978-3-030-40943-2_32

concern of the scientific community, research on the applicability of scientific findings to public policymaking has been underway for decades, with a central challenge being the ability to trace the direct links between policy and science [18]. Recently-developed data on scientific citations in the benefit-cost analyses produced in the process of making significant federal regulations in the U.S. [9–11] offers a promising approach to directly connecting policy and science. This approach to measuring science use in public policy has been also been followed in the study of sub-national government in the U.S. [19]. In this article, we examine the properties of policymakers' invocation of science in justifying a broad range of regulatory policy decisions over time through the lens of a new policy-science network dataset. The scholarly literature on policymaking recognizes the importance of the network conceptualization of the policy process [5,15,23]. As Sandström and Carlsson (p. 505) [24] aptly summarize, "the web of interactions within policy producing structures is an important aspect to consider when explaining policy outcomes".

We construct a network using citations to the scientific literature by US federal regulatory agencies' public justifications of regulations. All major regulations since 1981 are required, through a series of Executive Orders, to be accompanied by Regulatory Impact Analyses (RIAs), which must stipulate the problem to be addressed, the proposed rule's anticipated benefits and costs, the relevant alternative approaches, and other factors. Among other criteria, EO 12866 [1] states, "Each agency shall base its decisions on the best reasonably obtainable scientific, technical, economic, and other information concerning the need for, and consequences of, the intended regulation". Data on agency citations to the scientific literature in RIAs allow us to track the use of science across agencies and across policy areas, including the environment, health, transportation, defense, labor, and more. In the current article, we investigate where policy-relevant research finds support. We ask two related questions. First, what are the sources of support for research that is cited most heavily by RIAs? Second, what are the sources of support for the research that is used most broadly in RIAs (i.e., across policy domains and agencies)?

The data include citations to scientific literature from 104 Regulatory Impact Analyses completed between 2008–2012.[1] For the purpose of this study, "scientific literature" is defined to be an article found in the Web of Science citation index supported by Thompson Reuters, a definition that is commonly employed in the bibliometric literature [7,11,13,25]. These include regulations from a wide range of agencies and, therefore, disparate substantive policy areas. In addition to collecting and coding scientific citations for each RIA, we use the Web of Science [14] to collect citation attributes of the cited research, which include the authors' affiliation(s), and the sources of financial support acknowledged in the articles.

[1] This timeline is admittedly limited, and presents an opportunity for future research to extend the data to assess changes in the patterns we observe

2 Network Description and Exploration

The complete policy-science network we construct is visualized in Fig. 1. The network includes 104 RIAs and 823 scientific articles. As can be seen, EPA is the most regular user of scientific research. There is little cross-agency and even cross-RIA overlap in the science that is used, outside of EPA. In what follows, we study the support of the research represented in this network, identifying the funders and affiliations behind the bulk of the research, and pay particular attention to the supporters of research that span multiple RIAs. In Fig. 2 we present the list of the most prominent funders and author affiliations associated with articles cited. In terms of affiliations, we see a list of several elite universities including Harvard, Yale, Cornell, Columbia and Duke, as well as the most research-active government agency, the Environmental Protection Agency, and Canada's large government agency responsible for public health policy: Health Canada. In terms of funders, we see a list of the prominent governmental research sponsors in the US, Canada, Europe and California, including the National Science Foundation, the National Institutes of Health, EPA, the National Science and Engineering Research Council of Canada, and the European Commission. These results are largely unsurprising: the supporters of research that are associated with the large volume of scientific publications cited in RIAs include elite research universities and the largest funders of research on the planet. However, analyzing just the volume of citations leaves out an important component of the impact story: the diversity of policy that is informed by scientific research. Seen from the perspective of the supporter of the research, the return on investment in terms of policy impact depends heavily on the breadth of policy areas influenced by individual articles.

In order to assess the diversity of policy in which the research supported by a given affiliation or sponsor is used, we use methods that have been heavily adopted in bibliometric network analysis to measure the interdisciplinarity of research, with RIAs forming the disciplinary landscape. If a scientific publication has interdisciplinary impact, we expect to see it cited in RIAs that do not otherwise cite similar bodies of literature. This constitutes an analysis of the breadth of the impact of the research supported by the funders and affiliations in our data. Breadth is particularly important in this network, as most of the network is composed of largely separate components that are not connected to each other via citations. Indeed, 671 of the 823 scientific papers are cited by just one RIA. The network analytic measure, betweenness centrality, is used to assess the degree to which an article spans diverse RIAs. The betweenness centrality of a node in a network is the number of shortest paths between other nodes on which the node sits, adjusted for alternative shortest paths. Betweenness centrality measurement represents the state-of-the-art approach to assessing interdisciplinarity in scientific citation networks [4,20–22,26]. In the current application, we assess the betweenness centrality of an article by the number of shortest paths between RIAs on which it sits.

In Fig. 3 we present the top affiliations and funders in terms of the average number of shortest paths between RIAs on which supported articles sit.

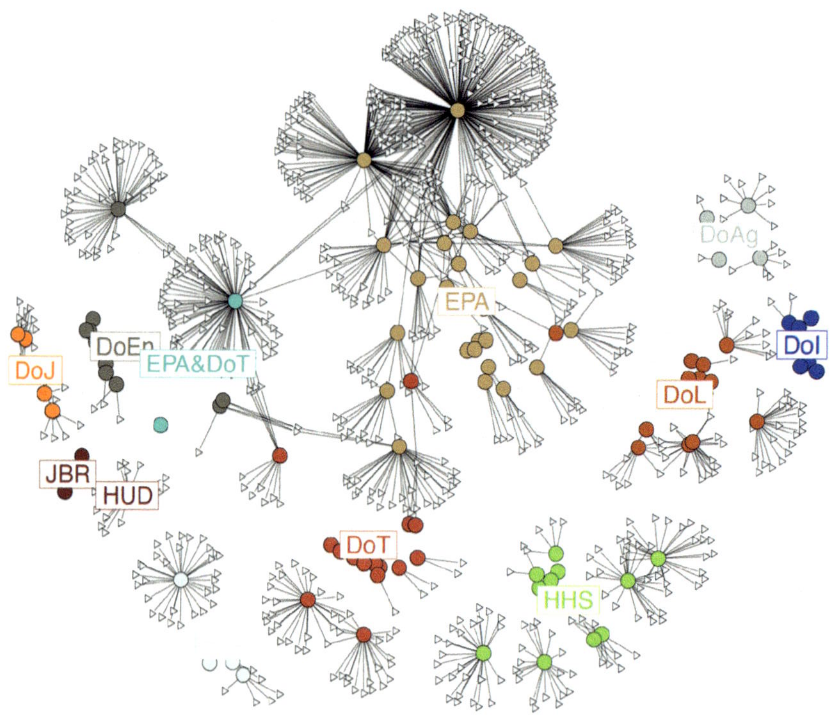

Fig. 1. Network of Regulatory Impact Analyses connected by agencies. Triangles are scientific publications. RIAs produced by the same agency are color-coordinated. Lines indicate the article is cited in the RIA.

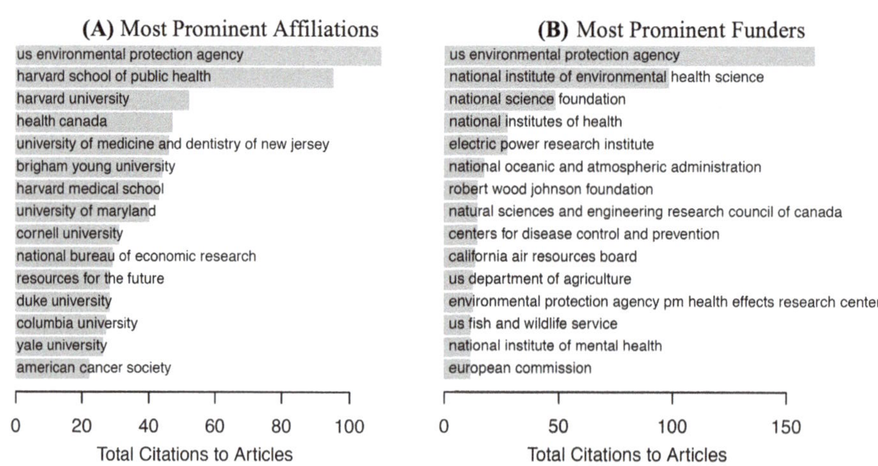

Fig. 2. Affiliations of authors of cited articles and sponsors acknowledged in cited articles are depicted. x-axis gives the number of articles in which the affiliation or sponsor appears.

These are the supporters of research whose cited articles exhibit the broadest impact in the network of RIAs and scientific publications. These lists are substantially different from the lists of affiliations and funders that support the most heavily cited articles. Notable among affiliations is the dearth of universities. Among funders, there are many fewer US federal government sources of support. These results suggest that the studies with the broadest impact on regulatory policy are conducted by researchers who are not affiliated with universities and are funded by non-US-government sponsors.

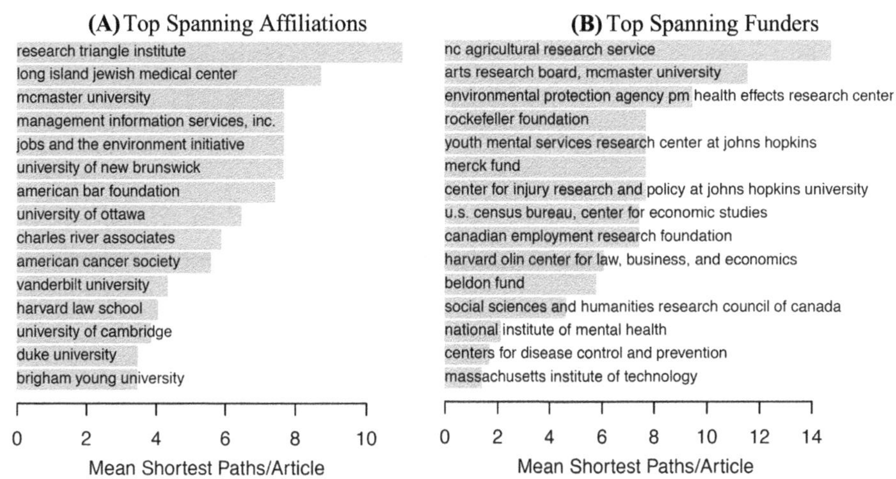

Fig. 3. Affiliations of authors of cited articles and sponsors acknowledged in cited articles are depicted. x-axis gives the weighted average number of shortest pathways between RIAs of which supported articles are part.

3 Formal Tests for Reach in the Science-Policy Network

We conduct formal tests of the hypotheses that are suggested by our analysis of the top supporters of broadly used research. We do this through the use of a hypothesis testing method referred to as a conditional uniform graph test (CUG test) [2]. In a CUG test, networks are simulated that control for structural features of the observed network, in order to test whether a feature of interest is statistically unusual given the other features that are being controlled. It is common in the analysis of betweenness to control for the number of connections to and/or from a node, as more connected nodes will tend to sit on more shortest paths due simply to their prominence. The concept of a potential boundary spanner has been used to identify nodes that have high betweenness centrality relative to degree centrality [12,16,17]. As Bigrigg et al. [6] (p. 5) note, boundary spanning potential assesses, "how likely is it if the node is removed that the network has a greater chance of being partitioned into major subnetworks."

Using this concept and measure, we test whether the articles that hold together the different parts of the regulatory policy network are less likely to be supported by US government funders and authored by scientists affiliated with universities. In the CUG tests presented in Fig. 4, we simulate comparison sets of networks that are randomly re-wired, but we hold fixed the number of articles cited by each RIA and the number of RIAs citing each article.

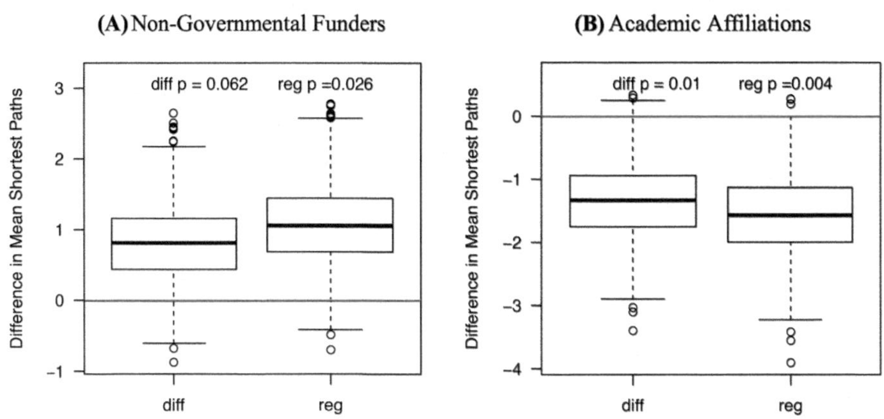

Fig. 4. Boxplots depict the distributions of differences between the average betweenness centrality of articles supported by the respective category—(A) non-governmental funders, (B) Academically affiliated authors—in the observed network and the average betweenness centrality in the respective group in the simulated networks. 1,000 simulated differences are depicted. The p-values reported are 2 times the lesser of the proportion of differences greater than zero and the proportion of differences less than zero.

The results in Fig. 4 give the differences in the average shortest paths on which articles sit, for articles supported by non-governmental and US federal government funders (A) and articles authored by those with academic affiliations and those without (B). Each plot contains two boxes. The first box "diff" reflects univariate results, calculating the difference between the articles with and without the respective feature (i.e., academic affiliation, government funding). The second box "reg" reflects multiple regression results in which the effect of the focal feature is calculated, controlling for the other feature, in a linear regression model. The hypothesis testing results confirm what we found in the descriptive analysis. Research supported by non-governmental funders sits, on average, on one more shortest path between RIAs than research that is supported by a US Federal government funder. Research that is authored by those with academic affiliations sits on 1–2 fewer shortest paths than research authored by those with non-academic affiliations.

4 Conclusion

Our results offer several important implications regarding the sources of support for policy-relevant science. First, our results demonstrate the importance to policymaking of non-university organizations that pursue basic research, such as the Research Triangle Institute. Future science of science research should investigate what it is about the selection of projects and/or the communication of results, that leads to research conducted by non-university affiliates to be more broadly used in policymaking. The second major result is that research supported by the top US government funders is less broadly influential than research supported by other funders. Future research should consider whether features of the funding process at large sponsors such as NSF or NIH steer the focus of the awards away from basic research with relevance to public policymaking.

References

1. Executive Order No. 12866: Regulatory Planning and Review. 58 FR 51735 (1993)
2. Anderson, B.S., Butts, C., Carley, K.: The interaction of size and density with graph-level indices. Soc. Netw. **21**(3), 239–267 (1999)
3. Baker, B.: Politicizing science: what is the role of biologists in a hyperpartisan world? BioScience **64**(3), 171–177 (2014)
4. Barnett, G.A., Huh, C., Kim, Y., Park, H.W.: Citations among communication journals and other disciplines: a network analysis. Scientometrics **88**(2), 449–469 (2011)
5. Berardo, R., Scholz, J.T.: Self-organizing policy networks: Risk, partner selection, and cooperation in estuaries. Am. J. Polit. Sci. **54**(3), 632–649 (2010). https://doi.org/10.1111/j.1540-5907.2010.00451.x. http://onlinelibrary.wiley.com/doi/10.1111/j.1540-5907.2010.00451.x/abstract
6. Bigrigg, M.W., Carley, K.M., Manousakis, K., McAuley, A.: Routing through an integrated communication and social network. In: MILCOM 2009-2009 IEEE Military Communications Conference, pp. 1–7. IEEE (2009)
7. Colebunders, R., Kenyon, C., Rousseau, R.: Increase in numbers and proportions of review articles in tropical medicine, infectious diseases, and oncology. J. Assoc. Inf. Sci. Technol. **65**(1), 201–205 (2014)
8. Scientific Integrity in Policy Making. Union of Concerned Scientists, Cambridge, MA (2004). https://www.ucsusa.org/sites/default/files/2019-09/scientific_integrity_in_policy_making_july_2004_1.pdf
9. Costa, M., Desmarais, B.A., Hird, J.A.: Science use in regulatory impact analysis: the effects of political attention and controversy. Rev. Policy Res. **33**(3), 251–269 (2016)
10. Costa, M., Desmarais, B.A., Hird, J.A.: Public comments' influence on science use in US rulemaking: the case of epa's national emission standards. Am. Rev. Public Adm. **49**(1), 36–50 (2019)
11. Desmarais, B.A., Hird, J.A.: Public policy's bibliography: the use of research in us regulatory impact analyses. Regul. Gov. **8**(4), 497–510 (2014)
12. Diesner, J., Carley, K.M.: A methodology for integrating network theory and topic modeling and its application to innovation diffusion. In: 2010 IEEE Second International Conference on Social Computing, pp. 687–692. IEEE (2010)

13. Eysenbach, G.: Citation advantage of open access articles. PLoS Biol. **4**(5), e157 (2006)
14. Harzing, A.W., Alakangas, S.: Google scholar, scopus and the web of science: a longitudinal and cross-disciplinary comparison. Scientometrics **106**(2), 787–804 (2016)
15. Heclo, H.: Issue networks and the executive establishment. In: King, A. (ed.) The New American Political System. American Enterprise Institute, Washington (1978)
16. Hutchins, C.E., Benham-Hutchins, M.: Hiding in plain sight: criminal network analysis. Comput. Math. Organ. Theor. **16**(1), 89–111 (2010)
17. Jin, J.H., Park, S.C., Pyon, C.U.: Finding research trend of convergence technology based on Korean R&D network. Expert Syst. Appl. **38**(12), 15159–15171 (2011)
18. Kenneth, P., Schwandt, T.A., Straf, M.L.: Using Science as Evidence in Public Policy. The National Academies Press, Washington (2012). National Research Council
19. Koontz, T.M.: The science-policy nexus in collaborative governance: use of science in ecosystem recovery planning. Rev. Policy Res. **36**(6), 708–735 (2019)
20. Leydesdorff, L.: Betweenness centrality as an indicator of the interdisciplinarity of scientific journals. J. Am. Soc. Inf. Sci. Technol. **58**(9), 1303–1319 (2007)
21. Leydesdorff, L., Rafols, I.: Indicators of the interdisciplinarity of journals: diversity, centrality, and citations. J. Informetr. **5**(1), 87–100 (2011)
22. Orosz, K., Farkas, I.J., Pollner, P.: Quantifying the changing role of past publications. Scientometrics **108**(2), 829–853 (2016)
23. Provan, K.G., Veazie, M.A., Staten, L.K., Teufel-Shone, N.I.: The use of network analysis to strengthen community partnerships. Public Adm. Rev. **65**(5), 603–613 (2005)
24. Sandstrom, A., Carlsson, L.: The performance of policy networks: the relation between network structure and network performance. Policy Stud. J. **36**(4), 497–524 (2008). https://doi.org/10.1111/j.1541-0072.2008.00281.x. http://onlinelibrary.wiley.com/doi/10.1111/j.1541-0072.2008.00281.x/abstract
25. Sugimoto, C.R., Thelwall, M.: Scholars on soap boxes: science communication and dissemination in TED videos. J. Am. Soc. Inf. Sci. Technol. **64**(4), 663–674 (2013)
26. Tonta, Y., Darvish, H.R.: Diffusion of latent semantic analysis as a research tool: a social network analysis approach. J. Informetr. **4**(2), 166–174 (2010)
27. Weiss, C.H.: The many meanings of research utilization. Public Adm. Rev. **39**(5), 426–431 (1979)

Characterizing the Dynamics of Academic Affiliations: A Network Science Approach

Josemar Faustino[1](✉), Nandini Iyer[2], Juan Mendonza[3], and Ronaldo Menezes[4]

[1] Florida Institute of Technology, Melbourne, FL, USA
jcruz@biocomplexlab.org
[2] University of Illinois at Urbana–Champaign, Champaign, IL, USA
nandini2@illinois.edu
[3] Central Washington University, Ellensburg, WA, USA
mendozajua@cwu.edu
[4] University of Exeter, Exeter, UK
r.menezes@exeter.ac.uk

Abstract. Affiliation exchanges are inherent phenomena of academia that underlies some of the structural characteristics of academic institutions. Important questions involving the dynamics of institutions' prestige, the concentration of funding, the spread of scientific ideas, and interdisciplinarity are all related to which institution academics will move upon completion of their training, or when looking for a better environment to pursue their research endeavors. In this work, we investigated the phenomena of affiliation switches in academia and its relationship with research topics in Computer Science. Drawing from publication data spanning over thirty years, we mapped the connections among academic institutions as networks. We report on the stability of network properties over the years in spite of the exponential growth of Computer Science in recent decades. Additionally, we contextualize the structural properties of networks and topic modeling results with the current academic landscape. Altogether, our results help to characterize the phenomena of academic affiliation exchanges through the macro perspective of network science and natural language processing.

Keywords: Academic affiliations · Group dynamics · Network structure

1 Introduction

The human nature is one of association to social groups coupled with a desire to move from group to group in search of better social contexts. This phenomenon is true in many institutions of modern society. In academia, these institutions are organizations, from which people switch affiliations to and from, throughout their academic careers. The exchange of academics among educational and

H. Barbosa et al. (Eds.): *Complex Networks XI*, SPCOM, pp. 393–404, 2020.
https://doi.org/10.1007/978-3-030-40943-2_33

research institutions is fundamentally related to the formal process of knowledge generation and also affects academics' career decisions, education directions, and scientific discoveries in general.

Changing affiliations is an integral part of academic life. Academic mobility has been found to promote a healthy exchange of ideas [26], disseminating knowledge and practices that would otherwise be retained in the places in which they emerged [22]. When one academic moves between institutions, such movement affects both institutions involved. The vacuum of knowledge left, as well as the mixing of knowledge in the new institution, gives rise to complex interactions. Affiliation switching in academia is an inherently social phenomenon as there are many reasons behind the decision of an academic to choose institution A over institution B. These include perceived institution's prestige, tradition in a particular topic of research, funding opportunities, salary negotiations, and many others [6]. All of these factors combined, affect academic mobility, which itself can affect the factors that determined it in the first place. The circular nature of these influences is very characteristic of complex networked systems [20].

The complexity arising from affiliation switches in academia, to the best of our knowledge, has not yet been investigated. Research works that address academic mobility, have approached it from a top-down perspective [22,30]. That is, they identify a structural problem in the academia and scrutinize existing datasets that have the potential to shed light on the problem at hand. This approach has been used to characterize the systemic inequality in faculty hiring [6]. They used web-crawled data from departments of history, business, and computer science to build "faculty-hiring networks". Their main findings are that prestigious institutions dominate the job-market of faculty hiring and that even in the digital age, scientific ideas spread mainly by hiring.

Here, we investigate the structural properties of a network of institutions linked by academic affiliation switches in the field of Computer Science. Our dataset extends for over 30 years, covering some of the fundamental developments of the Computer Science field. To do so, we cleaned the data, applying a wide variety of data curation techniques, including the collection of additional data. By leveraging the power of large datasets combined with techniques from data mining, natural language processing, and network science, we put together a set of results that further help the understanding of the complex social phenomena embedded in the network of switching academic affiliations.

2 Methods

In order to model the dynamics of academic affiliations, we consider a framework of group to group interaction previously used to model interactions among political parties [8]. That is, whenever an individual from one group, in this case, an academic institution, joins another, we determine that there is a relationship between these groups, assuming that upon the affiliation switch they have a latent level of commonality in their constituency.

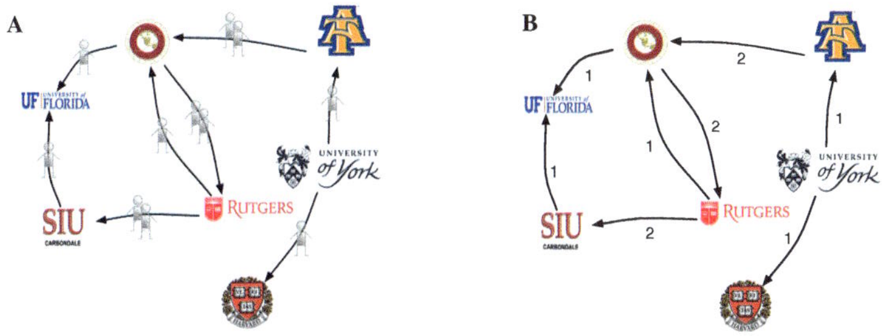

Fig. 1. The affiliation switches of academics between institutions **(A)** yields a network of institutions **(B)**. Each affiliation change is a connection between two institutions. The result is a network where nodes represent institutions and edges represent the connections among them. The weight w_{ij}^{y} of an edge represents the ij total number of academics switching from i to j in year y.

2.1 Network Model of Academic Affiliation Switching

The academic networks are composed of institutions representing nodes and links representing the relationship among the institutions whenever academics switch affiliation between any pair of institutions, as shown in Fig. 1. We formally describe the process as follows. Let $V = \{1, \ldots, n\}$ be the set of academics institutions and $G^y = (V, E)$ be the affiliation exchange graph. A directed weighted edge $(i, j)^y \in E$ connects institution i to institution j if there exists an academic who switched affiliation from institution i to institution j in the year y. The weight w_{ij}^{y} is the total number of academics switching from i to j in year y.

This process generates one network per year for every year in the data. To aide the understanding of the academic switching phenomena, we computed a set of structural network metrics such as WCGC, weighted degree distribution, betweenness centrality, clustering coefficient, average path length, and community structure.

2.2 Dataset

Acquisition. The original data was acquired through web-crawling [7] to study collaboration networks. The publications' metadata was collected from the Association for Computing Machinery (ACM) Digital Library, which included publications from 1951 to 2010. Table 1 provides a summary of the dataset. Notice that [7] reports 62,758 authors since their aim was in publications with more than one author. In this case, for authors with multiple affiliations, we only considered the first affiliation as it should be the author's most relevant appointment.

Data Curation. A common problem with the publication's metadata is the variety of ways in which authors report affiliation. For example, an author affiliated with the University of California Los Angeles might publish under affiliation *UCLA*, *UC Los Angeles*, *UC at Los Angeles*, to name just a few.

Table 1. Summary of raw and curated datasets

Description	Raw data	Curated data	Curation reduction
Publications	554,706	481,403	13%
Authors	160,693	147,995	8%
Institution affiliations	107,302	5,531	95%

Further to the above, we have affiliations reported in different languages. This is fairly common for institutions where English is not the primary language. For instance, an author can publish a paper under affiliation *University of Montreal*, and the same author can publish under *Université de Montréal* or even *University of Montréal.* The punctuation and combination of English and other languages for the affiliation adds another layer of complexity in the standardization procedure.

To curate the affiliation names, we applied the following techniques:

(a) **Removal of records with short strings:** many records had very short strings, often times containing only the country code, or a few meaningless characters due to encoding issues.
(b) **Additional data collection and translation:** to address the issue with institutions with names in different languages, we collected additional data from www.4icu.org. This dataset contained 13,600 institutions with their English and primary language names (when other than English). We parsed that data and matched with the ACM dataset. This allowed us to standardize all the institutions with an English name.
(c) **Aggregation and Self-cleaning:** using aggregation, we could identify the most common occurrence of affiliation names, and then we combined substringing with Levenshtein distance to find and merge duplicated entries.
(d) **Manual verification:** in many cases, a manual renaming of aggregated fields was necessary. For instance, in 1970, the University of Paris split into thirteen institutions. Many authors published papers interchangeably using the university number name, e.g., *University of Paris 5* or its other name *Paris Descartes University.* Furthermore, in recent years many of these schools are merging back again. For the purpose of this work, we renamed the affiliations to the most up-to-date nomenclature. Additionally, research institutes such as Italy's *Consiglio Nazionale delle Ricerche* (CNR), or France's *Centre National de la Recherche Scientifique* (CNRS), were all standardized as one institution, regardless of geographical location. The same process was adopted for private research laboratories such as IBM Research, Microsoft Research, and similar.

The process itemized above was applied at once for *a* and *b* and was applied iteratively for *c* and *d* until we reduced the data to numbers described on Table 1. We do not claim that the dataset is complete, the intention was to curate it to a point that allows an attempt to study the underlying phenomena.

2.3 Publication's Topic Modeling via Community Analysis

Aiming to investigate the structural properties of the academic affiliation networks with respect to research topics, we applied a topic modeling technique to categorize the publications. Using the publication's title and/or DOI (Document Object Identifier) we matched the publication records in the ACM dataset with Semantic Scholar data [1]. We were able to match 62% of publications in the ACM dataset with the Semantic Scholar dataset. For the remained unmatched documents we used the publication title instead of abstract.

The most common method for topic modeling is Latent Dirichlet Allocation (LDA) [3]. However, recent work showed fundamental issues with the assumptions required in LDA, which assumes that latent topics follow a Dirichlet distribution. They also propose a robust approach to address the problem [11]. The new method is based on network science and uses the Stochastic Block Model method for community detection to find topics [13,25].

We connected words and documents (abstracts) generating a bipartite network of 35,958 unique words connected to 452,766 documents, which yields 15,667,313 weighted edges. Then, we applied the SBM community detection method to retrieve the research topics. Since each publication has a topic distribution, we have the means to infer topics per institution and per author over time.

3 Results

3.1 Network Structural Analysis

The dataset from which the networks are built spans through a crucial period allowing the networks to capture snapshots of Computer Science as it developed and established itself as a field. During this time, scientific research in computing experimented unprecedented growth. Within a 30-year period, the number of institutions that published research of any form in Computer Science grew by 15-fold. Table 2 shows such growth in terms of scientific publications as an increasing number of academic institutions are listed as affiliations. The growth is represented with the higher number of nodes and edges in networks of subsequent years. Table 2 also contains the results for the average clustering coefficient and the average path length. The former measures the extent of connectivity among the neighbors of a particular node, and the latter is the mean number of hops from any two pairs of nodes given by a shortest path algorithm [24]. The combined low values of these metrics indicate properties of small-world networks, which are pervasive in many complex networked systems [15,29]. Furthermore, the low variability of these results for an extended period of time indicates the existence of a stable process underlying the small-world properties which persist over time without global perturbations [17].

As the networks become larger over the years, the weighted degree captures the scale of in and out flow of academics in a particular institution while betweenness measures the centrality of the institutions in the graph as shown in Fig. 2.

Table 2. Summary of network metrics for full network and giant component (GC). Blocks represent the number of communities as found by the SBM method, Avg. CF is the average clustering coefficient and Avg. PL is the average path length.

Year	Nodes	Edges	% GC size	Blocks	Avg. CF $\pm\sigma$	Avg. PL $\pm\sigma$
1980	173	179	0.53	1	0.10 ± 0.02	4.45 ± 1.70
1990	508	823	0.80	1	0.10 ± 0.01	4.59 ± 1.50
2000	1071	2474	0.87	2	0.12 ± 0.01	4.29 ± 1.24
2010	2636	9971	0.93	13	0.11 ± 0.01	3.91 ± 0.98

Unlike the average path length and clustering coefficient, the probabilities of weighted degree distribution and betweenness centrality change considerably over time, that is because these metrics are sensitive to changes with respect to size (number of nodes and edges). Furthermore, the skewed pattern of the weighted degree distribution is indicative of a "rich get richer" phenomena [2], which is clear from the stretch in the distributions in the networks of recent years.

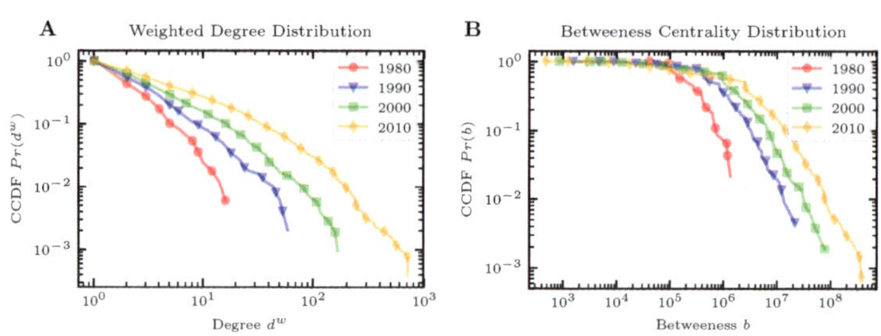

Fig. 2. Weighted degree distribution **(A)** and betweenness centrality **(B)**. Betweenness centrality is calculated as an undirect and unweighted graph on connected nodes. Complement Cumulative Distributions ($1-CDF$) for the networks of 1980, 1990, 2000, and 2010. As the field of computer science grows, the probability of finding a node with a higher weighted degree or betweenness centrality increases.

In general, the pattern of the distributions is preserved across different networks, the difference over time being the curve shift due to the larger number of nodes and edges. Although recently, there have been discussions about the generality of the specific distributions of these phenomena, "rich get richer" or preferential attachment in the context of networks is known to occur in many different complex systems [14, 23]. In this context, it can be argued that productive academics are attracted to prestigious institutions, these academics tend to publish more and therefore attract additional funding, thus creating more research opportunities. This leads to a concentration of academic prestige in a positive feedback loop akin to circular causality [28]. Nevertheless, recent research has shown that perceived prestige of traditional academic institutions overwhelmingly favors its members in academia [27] and also hinders the spread of good scientific ideas. That is, highly prestigious institutions receive more praise and credit than what is deserved, in detriment of smaller ones [22].

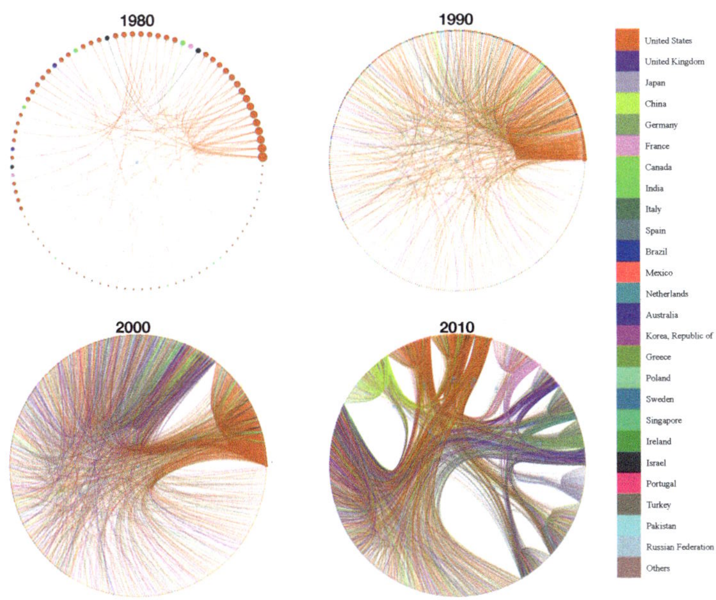

Fig. 3. Stochastic Block Model community detection results for 1980, 1990, 2000 and 2010. An aggregated organization of the network appears after in the year of 2000 and only becomes apparent in 2010.

The structure of communities in academic networks was investigated using hierarchical SBM. Figure 3 shows the evolution of group structure over time. For the 1980s and 1990s, a well-defined group structure is absent, such structure only appears in the 2000s and is clearly delineated in 2010. Likewise, well-defined country patterns emerge within the communities, with the US, Asian and European countries easily identified as the majority of institutions within the community. American institutions appear to connect to institutions from all over the world as they are present in practically all of the identified groups. This can be attributed to the tradition of American academic institutions in attracting investigators seeking better academic environments [16]. The overall organization of academic affiliation switches has a group of well-connected institutions in the US that is detached as a community in the year of 2000. In 2010, that group seems to split into smaller communities as the network becomes better defined.

3.2 Topic Modeling

The progress of scientific knowledge is inherently attached to the academics' decision of which problem to tackle. However, the factors that influence the decisions of why certain research topics take precedence over others are rather diverse [9]. Computer science is traditionally sensitive to investigate topics relevant to the industry, given the intertwined relationship of industry and academia concerning the spread of computing technologies in modern society. However, in the last couple of decades, computer science percolated into other areas giving rise to highly interdisciplinary fields such as bioinformatics [19], computational chemistry [5], cognitive science [12], and computational social science [18]. We investigated the properties of the research topics in Computer Science and its relationship with the structural characteristics of academic switching networks.

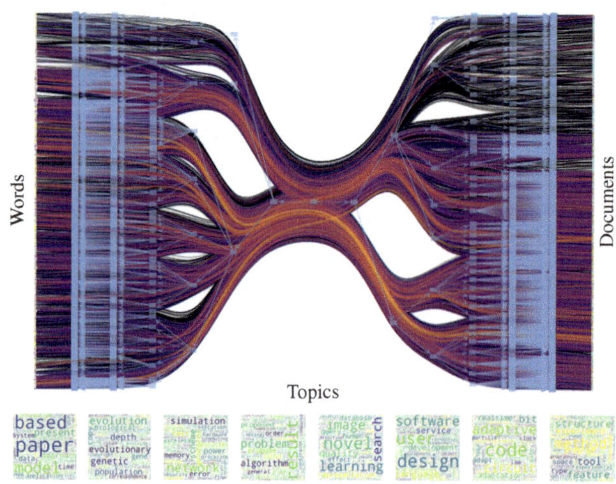

Fig. 4. Hierarchical structure of topics as given the inference of communities in the bipartite network of words and documents (papers' abstracts) using the Stochastic Block Model algorithm with node overlap. Words are clustered on the left, and documents are clustered on the right. The blue rectangles are topics, the centermost rectangle represents the entirety of computer science areas, as we move from the center to the periphery, the branches represent topics in more specific areas of knowledge. The word clouds in the bottom are the third level from the center (8 topics) sized by frequency.

Topic modeling allows for the organization of research topics from academic publications into a hierarchical and quantified manner. Figure 4 is a sample from the word-document graph depicting such structure as inferred by the SBM algorithm (see Methods in Sect. 2 for details). The topics are represented by the tree of rectangles, which spans from the center square node into the periphery leaf nodes (right are documents, left are words), going from general to specific fields. The bottom of the figure depicts the third level (8 blocks) of topics with the most

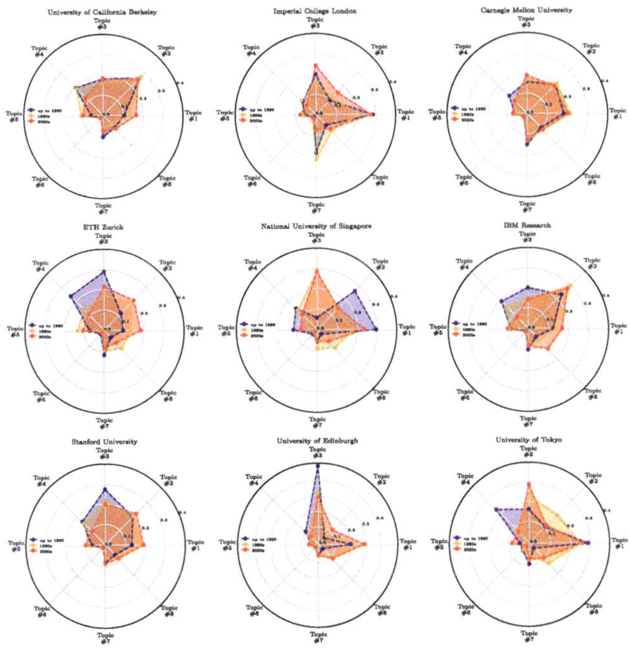

Fig. 5. Radial plots with the distribution of papers among eight publication topics extracted from the level three of the Stochastic Block Model for community detection (Fig. 4) The patterns of publication from each institution can be captured by its distribution among different topics.

frequent words in bigger size in the word cloud. The link between publication and topic allows us to extrapolate the attribution topic → publication, into topic → author, and also topic → institution. Figure 5 shows individual spider plots containing the distribution of topics in academic institutions with high output in Computer Science research. Each axis in the radial plot are one topic and the colored area represents the distribution of all the papers published by the institution within the decade. Notice that differently from Fig. 2 in which the curves are snapshots of each year, here we aggregated all the publications in the referenced time period.

In Fig. 6, we selected academics changing affiliations and named as authors in more than ten publications. The minimum number of publications is an arbitrary choice assumed to provide a reasonable baseline that allows for capturing the distribution of topics per incoming author over time. Also, the numbers reported in Fig. 6 are computed out of the fourth level of topics in the hierarchy of topics in Fig. 4 (55 topics). The curves are CCDF distributions of number of topics per institution every first year of the decade, not all institutions meet the requirement of having incoming authors with more than ten publications in the year, in which cases we plotted the curves only for more recent years (e.g.,National University of Singapore). Overall, we report a trend in increasing

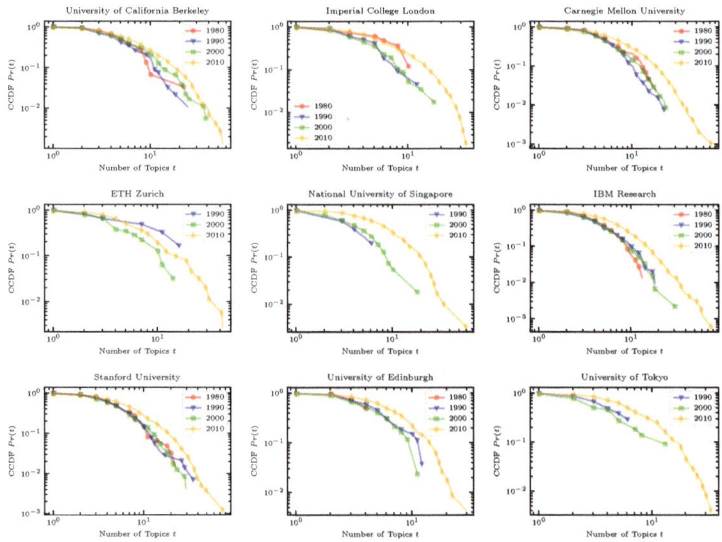

Fig. 6. Distribution of topics from incoming authors with more than ten academic publications. A trend in increasing the number of topics emerges as a pattern of hiring for most institutions in recent years.

the number of topics for incoming authors in the majority of institutions. This is likely attributed to the fact that collaborations are becoming more common in academia [21,31]. Also, our choices of a minimum number of publications are likely to select authors occupying faculty/permanent positions. In an increasingly competitive job market in academia, with a record-breaking number of PhDs awarded every year [10], young investigators seeking to settle in permanent positions are faced with stricter requirements [4]. The capacity to publish steadily and proficiently demonstrate scholarship and accolade which are perceived as necessary to propel the prestige of the institution they aim to join. Such an environment in combination with an increase in collaboration and the percolation of computer science techniques into other areas is a possible explanation for the higher number of topics explored by academics switching affiliations.

4 Conclusion

In this work, we explored the phenomena of affiliation switches through the lens of group to group interaction using a large longitudinal dataset of Computer Science publications. We found that despite the sheer scale process of the field in recent decades, the network of institutions based of affiliation switches maintained its structural properties over time, suggesting the existence of a stable underlying social mechanism. We also found that authors switching their affiliations in more recent years tend to tackle a wider variety of problems than authors in previous decades, which is possibly related to higher interdisciplinarity due to the general increase in collaboration in academia. But also to the increasing competition in the academic job-market leading higher demands in publication requirements to acquire a stable position in academic institutions.

Acknowledgments. All authors acknowledge partial support from NSF grant No. 1560345. JF acknowledges support from CAPES Foundation grant No. 99999.001043/2014-05. RM acknowledges support from the Army Research Office grant No. W911NF-17-1-0127-P00001. Any opinions, findings, and conclusions or recommendations expressed in this material are those of the authors and do not necessarily reflect the views of the funding agencies.

References

1. Ammar, W., et al.: Construction of the literature graph in semantic scholar. In: Conference of the Association for Computational Linguistics (2018). https://doi.org/10.18653/v1/n18-3011
2. Barabási, A.L., Albert, R.: Emergence of scaling in random networks. Science **286**(5439), 509–512 (1999). https://doi.org/10.1126/science.286.5439.509
3. Blei, D.M., Ng, A.Y., Jordan, M.I.: Latent Dirichlet allocation. J. Mach. Learn. Res. **3**(Jan), 993–1022 (2003)
4. Brechelmacher, A., Park, E., Ates, G., Campbell, D.F.J.: The Rocky Road to Tenure - Career Paths in Academia, pp. 13–40. The Changing Academy/Springer, Heidelberg (2015). https://doi.org/10.1007/978-3-319-10720-2_2
5. Cech, T.R., Rubin, G.M.: Nurturing interdisciplinary research. Nat. Struct. Mol. Biol. **11**(12), 1166–1169 (2004). https://doi.org/10.1038/nsmb1204-1166
6. Clauset, A., Arbesman, S., Larremore, D.B.: Systematic inequality and hierarchy in faculty hiring networks. Sci. Adv. **1**(1), e1400005 (2015). https://doi.org/10.1126/sciadv.1400005
7. Divakarmurthy, P., Menezes, R.: The Effect of Citations to Collaboration Networks, pp. 177–185. Springer, Heidelberg (2013). https://doi.org/10.1007/978-3-642-30287-9_19
8. Faustino, J., Barbosa, H., Ribeiro, E., Menezes, R.: A data-driven network approach for characterization of political parties' ideology dynamics. Appl. Netw. Sci. **4**(1), 48 (2019). https://doi.org/10.1007/s41109-019-0161-0
9. Fortunato, S., et al.: Science of science. Science **359**(6379), eaao0185 (2018). https://doi.org/10.1126/science.aao0185
10. Foundation, N.S.: Doctorate Recipients from U.S. Universities: 2017. No. NSF 19-301 in Survey of Earned Doctorates, National Science Foundation, December 2018. https://ncses.nsf.gov/pubs/nsf19301/report
11. Gerlach, M., Peixoto, T.P., Altmann, E.G.: A network approach to topic models. Sci. Adv. **4**(7), eaaq1360 (2018). https://doi.org/10.1126/sciadv.aaq1360
12. Griffiths, T.L.: Manifesto for a new (computational) cognitive revolution. Cognition **135**, 21–23 (2015). https://doi.org/10.1016/j.cognition.2014.11.026
13. Hartman, R., Faustino, J., Pinheiro, D., Menezes, R.: Assessing the suitability of network community detection to available metadata using rank stability. In: Proceedings of the ICWI 2017, pp. 162–169. ACM Press, New York (2017). https://doi.org/10.1145/3106426.3106493
14. Jeong, H., Néda, Z., Barabási, A.L.: Measuring preferential attachment in evolving networks. EPL (Europhys. Lett.) **61**(4), 567 (2003). https://doi.org/10.1209/epl/i2003-00166-9

15. Kleinberg, J.: The small-world phenomenon: an algorithmic perspective. In: Proceedings of the 32nd Annual ACM Symposium on Theory of Computing, STOC 2000, pp. 163–170. ACM, New York (2000). https://doi.org/10.1145/335305.335325
16. Knight, J., Wit, H.D.: Internationalization of higher education: past and future. Int. High. Educ. **1**(95), 2–4 (2018). https://doi.org/10.6017/ihe.2018.95.10715
17. Kossinets, G., Watts, D.J.: Empirical analysis of an evolving social network. Science **311**(5757), 88–90 (2006). https://doi.org/10.1126/science.1116869
18. Lazer, D., et al.: Computational social science. Science **323**(5915), 721–723 (2009). https://doi.org/10.1126/science.1167742
19. Lewis, J., Bartlett, A., Atkinson, P.: Hidden in the middle: culture, value and reward in bioinformatics. Minerva **54**(4), 471–490 (2016). https://doi.org/10.1007/s11024-016-9304-y
20. Lloret-Climent, M., Nescolarde-Selva, J.: Data analysis using circular causality in networks. Complexity **19**(4), 15–19 (2014). https://doi.org/10.1002/cplx.21480
21. Milojević, S.: Principles of scientific research team formation and evolution. PNAS **111**(11), 3984–3989 (2014). https://doi.org/10.1073/pnas.1309723111
22. Morgan, A.C., Economou, D.J., Way, S.F., Clauset, A.: Prestige drives epistemic inequality in the diffusion of scientific ideas. EPJ Data Sci. **7**(1), 40 (2018). https://doi.org/10.1140/epjds/s13688-018-0166-4
23. Newman, M.E.J.: Clustering and preferential attachment in growing networks. Phys. Rev. E **64**(2), 025102 (2001). https://doi.org/10.1103/PhysRevE.64.025102
24. Newman, M.E.J.: Networks: An introduction. Oxford University Press, Oxford, March 2010. https://doi.org/10.1093/acprof:oso/9780199206650.001.0001
25. Peixoto, T.: Hierarchical block structures and high-resolution model selection in large networks. Phys. Rev. X **4**(1), 1–18 (2014). https://doi.org/10.1103/PhysRevX.4.011047
26. Rotabi, R., Danescu-Niculescu-Mizil, C., Kleinberg, J.: Competition and selection among conventions. In: Proceedings of the 26th ICWWW, IWWWC 2017, pp. 1361–1370 (2017). https://doi.org/10.1145/3038912.3052652
27. Skinner, B.: First-order transition in a model of prestige bias. arXiv:1910.05813 [cond-mat, physics:physics], October 2019
28. Thomas, R.: Circular causality. IEE Proc. - Syst. Biol. **153**(4), 140–153 (2006). https://doi.org/10.1049/ip-syb:20050101
29. Watts, D.J., Strogatz, S.H.: Collective dynamics of 'small-world' networks. Nature **393**(6684), 440–442 (1998). https://doi.org/10.1038/30918
30. Way, S.F., Morgan, A.C., Larremore, D.B., Clauset, A.: Productivity, prominence, and the effects of academic environment. PNAS **116**(22), 10729–10733 (2019). https://doi.org/10.1073/pnas.1817431116
31. Wuchty, S., Jones, B.F., Uzzi, B.: The increasing dominance of teams in production of knowledge. Science **316**(5827), 1036–1039 (2007). https://doi.org/10.1126/science.1136099